Reasoning Robots

APPLIED LOGIC SERIES

VOLUME 33

SCOPE OF THE SERIES
Logic is applied in an increasingly wide variety of disciplines, from the traditional subjects of philosophy and mathematics to the more recent disciplines of cognitive science, computer science, artificial intelligence, and linguistics, leading to new vigor in this ancient subject. Kluwer, through its Applied Logic Series, seeks to provide a home for outstanding books and research monographs in applied logic, and in doing so demonstrates the underlying unity and applicability of logic.

The titles published in this series are listed at the end of this volume.

Reasoning Robots

The Art and Science of Programming Robotic Agents

by

MICHAEL THIELSCHER
Technische Universität Dresden, Germany

A C.I.P. Catalogue record for this book is available from the Library of Congress.

ISBN 10 1-4020-3068-1 (HB)
ISBN 10 1-4020-3069-X (e-book)
ISBN 13 978-1-4020-3068-1 (HB)
ISBN 13 978-1-4020-3069-X (e-book)

Published by Springer,
P.O. Box 17, 3300 AA Dordrecht, The Netherlands.

www.springeronline.com

Printed on acid-free paper

Printed in the Netherlands.

Contents

Preface

The creation of intelligent robots is surely one of the most exciting and challenging goals of Artificial Intelligence. A robot is, first of all, nothing but an inanimate machine with motors and sensors. In order to bring life to it, the machine needs to be programmed so as to make active use of its hardware components. This turns a machine into an autonomous robot. Since about the mid nineties of the past century, robot programming has made impressive progress. State-of-the-art robots are able to orient themselves and move around freely in indoor environments or negotiate difficult outdoor terrains, they can use stereo vision to recognize objects, and they are capable of simple object manipulation with the help of artificial extremities.

At a time where robots perform these tasks more and more reliably, we are ready to pursue the next big step, which is to turn autonomous machines into **reasoning robots**. A reasoning robot exhibits higher cognitive capabilities like following complex and long-term strategies, making rational decisions on a high level, drawing logical conclusions from sensor information acquired over time, devising suitable plans, and reacting sensibly in unexpected situations. All of these capabilities are characteristics of human-like intelligence and ultimately distinguish truly intelligent robots from mere autonomous machines.

What are Robotic Agents?

A fundamental paradigm of Artificial Intelligence says that higher intelligence is grounded in a mental representation of the world and that intelligent behavior is the result of correct reasoning with this representation. A **robotic agent** is a high-level control program for a robot—or, for that matter, for a proactive software agent—in which such mental models are employed to draw logical conclusions about the world. Intelligent robots need this technique for a variety of purposes:

- Reasoning about the current state.

  What follows from the current sensor input in the context of the world model?

- Reasoning about action preconditions.

  Which actions are currently possible?

- Reasoning about effects.

  What holds after an action has been taken?

- Planning.

  What needs to be done in order to achieve a given goal?

- Intelligent troubleshooting.

  What went wrong and why, and what could be done to recover?

Research on how to design an automatic system for reasoning about actions has a long history in Artificial Intelligence. The earliest formal model for the ability of humans to solve problems by reasoning has been the so-called situation calculus, whose roots go back to the early sixties. In the late sixties this model has been used to build an automatic problem solver. However, this first implementation did not scale up beyond domains with a small state space and just a few actions because it suffered from what soon became a classic in Artificial Intelligence, the so-called frame problem. In a nutshell, the challenge is to describe knowledge of effects of actions in a succinct way so that an automatic system can efficiently update an internal world model upon the performance of an action. The frame problem has haunted researchers for many years, and only in the early nineties the first satisfactory solutions have emerged. These formal models for reasoning about actions are now being developed into actual programming languages and systems for the design of robotic agents.

One successful approach to the frame problem is provided by a formalism known as **fluent calculus**. This book is concerned with this model of rational thought as a way to sepcify mental models of dynamic worlds and to reason about actions on the basis of these models. Very recently the calculus has evolved into the programming method and system **FLUX**, which supports the problem-driven, top-down design of robotic agents with the cognitive capabilities of reasoning, planning, and intelligent troubleshooting.

Why Fluent Calculus?

Fluent calculus originates in the classical situation calculus. It provides the formal underpinnings for an effective and computationally efficient solution to the fundamental frame problem. To this end, fluent calculus extends situation calculus by the basic notion of a state, which allows to define effects very naturally in terms of how an action changes the state of the world. Based on classical predicate logic, fluent calculus is a very versatile formalism, which captures a variety of phenomena that are crucial for robotic agents, such as incomplete knowledge, nondeterministic actions, imprecise sensors and effectors, and indirect effects of actions as well as unexpected failures when an action is performed in the real, unpredicatble world.

Why FLUX?

FLUX is a Prolog-based method for programming robotic agents based on the expressive action theory of fluent calculus. It comprises a way of encoding incomplete world models along with a technique for updating these models according to a declarative specification of the acting and sensing capabilities of a robot. Using a powerful constraint solver, a generic base system provides general reasoning facilities, so that the agent programmer can focus on specifying the application domain and designing the intended high-level behavior. Allowing for concise programs and supporting modularity, FLUX is eminently suitable for programming complex strategies for robotic agents. Thanks to a restricted expressiveness and a sound but incomplete inference engine, the system exhibits excellent computational behavior. In particular, it scales up well to long-term control thanks to the underlying principle of "progression," which means to continually update a (possibly incomplete) world model upon the performance of an action. Appealing to a declarative programming style, FLUX programs are easy to write, understand, and maintain. The book includes the details of the base system written in Prolog, so that it can be easily adapted and extended according to one's own needs.

Further Aspects

This introductory motivation would not be complete without the admission that the current state-of-the-art in research on reasoning robots still poses fundamentally unsolved problems. Maybe the most crucial issue is the interaction between cognitive and low-level control of a robot, in particular the question about the origin of the symbols, that is, the names for individuals and categories that are being used by a robotic agent. In this book we take a pragmatic, top-down approach, which requires the designer of a system to predefine the grounding of the symbols in the perceptual data coming from the sensors of a robot. Ultimately, a truly intelligent robot must be able to handle this symbol grounding problem by itself in a more flexible and adaptive manner. Another important issue is the lack of self-awareness and true autonomy. The reasoning robots considered in this book are able to follow complex strategies and they are capable of devising and executing their own plans. Still both the intended behavior and the boundaries for the plans are determined by the programmer. Ultimately, a truly self-governing robot must be able to reflect and adapt its behavior when facing new and unforeseen situations.

What's in this Book?

This book provides an in-depth and uniform treatment of fluent calculus and FLUX as a mathematical model and programming method for robotic agents. As theory and system unfold, the agents will become capable of dealing with incomplete world models, which require them to act cautiously under uncertainty; they will be able to explore unknown environments by logically reasoning about sensor inputs; they will plan ahead some of their actions and react sensibly to action failure.

The book starts in Chapter 1 with an introduction to the axiomatic formalism of fluent calculus as a method both for specifying internal models of dynamic environments and for updating these models upon the performance of actions. Based on this theory, the logic programming system FLUX is introduced in Chapter 2 as a method for writing simple robotic agents. The first two chapters are concerned with the special case of agents having complete knowledge of all relevant properties of their environment. In Chapters 3 and 4, theory and system are generalized to the design of intelligent agents with **incomplete knowledge** and which therefore have to act under uncertainty. Programming these agents relies on a formal account of knowledge given in Chapter 5, by which conditions in programs are evaluated on the basis of what an agent knows rather than what actually holds. Chapter 6 is devoted to **planning** as a cognitive capability that greatly enhances flexibility and autonomy of agents by allowing them to mentally entertain the effect of different action sequences before choosing one that is appropriate under the current circumstances. Chapters 7 and 8 are both concerned with the problem of **uncertainty** due to the fact that effects of actions are sometimes unpredictable, that state properties in dynamic environments may undergo changes unnoticed by an agent, and that the seonsors and effectors of robots are always imprecise to a certain degree. Chapter 9 deals with a challenge known as the **ramification problem**. It arises in complex environments where actions may cause chains of indirect effects, which the agent needs to take into account when maintaining its internal world model. Chapter 10 is devoted to an equally important challenge known as the **qualification problem**. It arises in real-world applications where the executability of an action can never be predicted with absolute certainty since unexpected circumstances, albeit unlikely, may at any time prevent the successful performance of an action. The solution to this problem enables agents to react to unexpected failures by generating possible explanations and revising their intended course of actions accordingly. Finally, Chapter 11 describes a **system architecture** in which robotic agents are connected to a low-level, reactive control layer of a robot. The appendix contains a short user manual for the FLUX system.

The only prerequisite for understanding the material in this book is basic knowledge of standard first-order logic. Some experience with programming in Prolog might also be helpful. Sections marked with * contain technical details that may be skipped at first reading or if the reader's main interest is in programming. The book has its own webpage at

www.fluxagent.org

where the FLUX system and all example programs in this book are available for download.

Acknowledgments

This book would not have been possible without the help of many people who contributed in various ways to its success. The author wants to especially thank Wolfgang Bibel, Matthias Fichtner, Norman Foo, Michael Gelfond, Axel Großmann, Sandra Großmann, Birgit Gruber, Yi Jin, Matthias Knorr, Markus Krötzsch, Yves Martin, Maurice Pagnucco, Martin Pitt, Ray Reiter, Erik Sandewall, and Stephan Schiffel.

Chapter 1

Special Fluent Calculus

Imagine a robot sitting in a hallway with a number of offices in a row. The robot is an automatic post boy, whose task is to pick up and deliver in-house mail exchanged among the offices. To make life easier for the robot, the documents that are being sent from one office to another are all together put into one standardized delivery package. The robot is equipped with several slots, a kind of mail bag, each of which can be filled with one such package. Initially, however, there may be many more delivery requests than the robot can possibly carry out in one go, given its limited capacity. Figure 1.1 depicts a sample scenario in an environment with six offices and a robot with three mail bags. A total of 21 packages need to be delivered. The question is how to write a control program which sends the robot up and down the hallway and tells it where to collect and drop packages so that in the end all requests have been satisfied.

Actually, it is not too difficult to come up with a simple, albeit not necessarily most efficient, strategy for the "mailbot:" Whenever it finds itself at some office for which it carries one or more packages, then these are delivered. Conversely, if the robot happens to be at some place where items are still waiting to be collected, then it arbitrarily picks up as many as possible, that is, until all bags are filled. If no more packages can be dropped nor collected at its current location, the robot makes an arbitrary decision to move either up or down the hallway towards some office for which it has mail or where mail is still waiting to be picked up.

Following this strategy, our robot in Figure 1.1, for example, chooses to pick up all three packages in the first room and then to move up to room number 2, where it can deliver one of them. Thereafter, it may select the package which is addressed from room 2 to 3, and to move up again. Arriving at room number 3, the robot can drop both the package coming from room 1 and the one from room 2. This leaves the robot with two empty bags, which it fills again, and so on. At the end of the day, there will be no unanswered requests left and the robot will have emptied again all its mail bags.

The simple strategy for the mailbot is schematically depicted in Figure 1.2. Using a pseudo-language, the algorithm is implemented by the following pro-

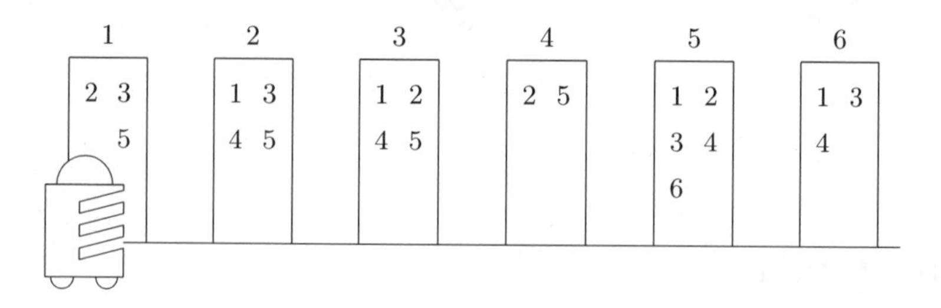

Figure 1.1: The initial state of a sample mail delivery problem, with a total of 21 requests.

---

gram:

loop
  if possible to drop a package
    then do it
  else if possible to pick up a package
    then do it
  else if can pick up or drop a package up (resp. down) the hallway
    then go up (resp. down)
  else stop
end loop

Obviously, this program requires our mailbot to evaluate conditions which depend on the current state of affairs. For in order to decide on its next action, it always needs to know the current contents of its mail bags, the requests that are still open, and its current location. Since these properties constantly change as the program proceeds and not all of them can be directly sensed, the robot has to keep track of what it does as it moves along. For this purpose, our mailbot is going to maintain an **internal model** of the environment, which throughout the execution of the program conveys the necessary information about the current location of all packages that have not yet been delivered. The model needs to be updated after each action in accordance with the effects of the action. With regard to the scenario in Figure 1.1, for instance, if the robot starts with putting a particular package into one of its mail bags, this will be recorded as a modification of the model, namely, by removing the corresponding request and adding the information into which mail bag the package in question has been put. Likewise, whenever a package is actually taken out of a mail bag, this has to be done mentally as well, in order for the robot to know that this bag can be filled again.

Fluent calculus lays the theoretical foundations for programming robotic agents like our mailbot that base their actions on internal representations of the state of their environment. The theory provides means to encode internal models for agents and to describe actions and their effects. As the name suggests, the

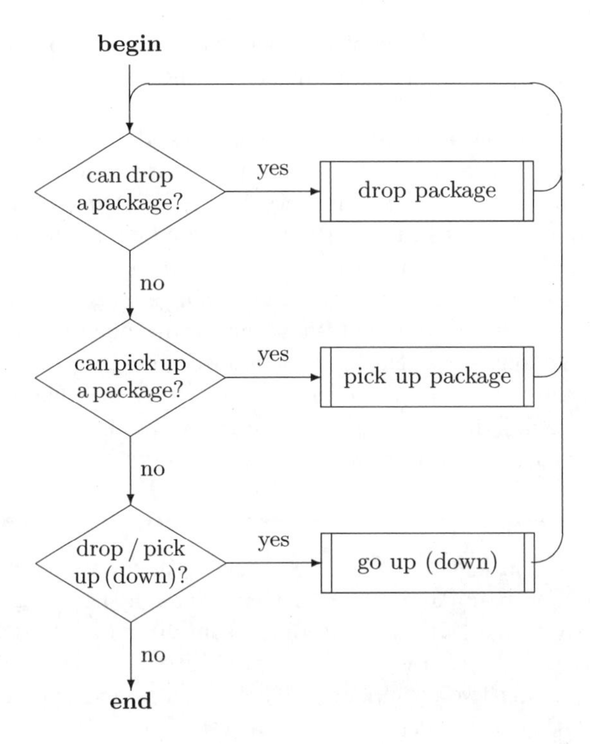

Figure 1.2: A high-level algorithm for mail delivery.

calculus also tells us how to **calculate the update** of a model whenever an action is performed. In this first chapter, we will focus on a special variant of fluent calculus which is suitable for agents that always know all relevant facts of their environment. The mail delivery robot falls into this category if we assume that all requests are given initially and do not change during program execution.

The chapter is divided into three sections, which introduce the three basic components of a problem-oriented description of a robot domain:

1.1. the states of the environment;

1.2. the actions of the robot;

1.3. the effects of the actions on the environment.

1.1 Fluents and States

Typically, the state space tends to be quite large even in simple robot environments. For our mail delivery scenario with just six offices and three mail bags, for example, we can distinguish more than $2 \cdot 10^{12}$ states, considering all possible combinations of requests, fillings of the bags, and locations of the robot. This number increases exponentially with the size of the office floor or the capacity of the robot. In many applications there is even no upper bound at all for the state space. It is therefore advisable for agents to adopt a principle known as "logical atomism," which means to break down states into atomic properties. The state components of the mail delivery world, for example, are the current requests, the contents of the mail bags, and the location of the robot.

By convention, atomic properties of states are called **fluents**, thereby hinting at their fleeting nature as time goes by. Maintaining a model of the environment during program execution is all about how the various fluents change due to the performance of actions. Fluent calculus is named after these state components and provides means to calculate the changes they undergo when the agent acts.

Generally speaking, fluent calculus is a language that uses standard predicate logic with a few customary conventions regarding sorts and equality.

Sorts can be used in logic to define the range of the arguments of predicates and functions. For example, a function *Pickup* might be specified as taking two arguments of some sort called $\mathbb{N}$ (for: natural numbers), yielding a term of another sort called ACTION; similarly, predicate *Poss* (for: possible) may be specified as taking two arguments of sort ACTION and another sort called STATE. Sorts need not be disjoint; e.g., natural numbers $\mathbb{N}$ are a sub-sort of integers $\mathbb{Z}$, which in turn are contained in the real numbers $\mathbb{R}$. Variables in formulas are sorted, too, and may be substituted by terms of the right sort only. For example, in the formula $(\exists x, z)\, Poss(Pickup(2, x), z)$, the variable x must be of sort $\mathbb{N}$ and z of sort STATE. For conventional sorts like the natural numbers we use the standard arithemtic operations with their usual interpretation. We also use the **equality predicate** "=" (defined for all pairs of elements of the same sort), which is assumed to be interpreted as the identity relation.

For modeling states of an environment, fluent calculus uses two basic sorts, namely, for fluents and for states. Intuitively, any collection of fluents makes a state in which exactly these fluents hold. Formally, each fluent itself constitutes a (singleton) state, and there is a binary connection function that allows to form a new state by taking the union of two states. Furthermore, a special constant denotes the empty state.

Definition 1.1 A triple $\langle \mathcal{F}, \circ, \emptyset \rangle$ is a **fluent calculus state signature** if

- $\mathcal{F}$ finite, non-empty set of function symbols into sort FLUENT
- $\circ$: STATE × STATE ↦ STATE
- $\emptyset$: STATE.

A **fluent** is a term of sort FLUENT. A **state** is a term of sort STATE, with FLUENT being a sub-sort of STATE. □

For the sake of better readability, the binary connection function will be written in infix notation, as in $z_1 \circ z_2$. Throughout the book, fluent variables will be written by the letters f or g and state variables by the letter z, all possibly with sub- or superscripts. Conventions like this will allow us to refrain, in most cases, from explicitly stating the sorts of the variables in formulas.

Example 1 Numbering both the offices and the robot's mail bags, a state in the mail delivery world will be described with the help of the following four fluents:

$$\begin{array}{rll} At: & \mathbb{N} \mapsto \text{FLUENT} & At(r) \mathrel{\hat{=}} \text{robot is at room } r \\ Empty: & \mathbb{N} \mapsto \text{FLUENT} & Empty(b) \mathrel{\hat{=}} \text{mail bag } b \text{ is empty} \\ Carries: & \mathbb{N} \times \mathbb{N} \mapsto \text{FLUENT} & Carries(b, r) \mathrel{\hat{=}} \text{bag } b \text{ has a package for } r \\ Request: & \mathbb{N} \times \mathbb{N} \mapsto \text{FLUENT} & Request(r_1, r_2) \mathrel{\hat{=}} \text{request from } r_1 \text{ to } r_2 \end{array}$$

Using the connection function, the initial state depicted in Figure 1.1 can then be formalized by this term:

$$\begin{array}{l} At(1) \circ (Empty(1) \circ (Empty(2) \circ (Empty(3) \circ \\ \quad (Request(1,2) \circ (Request(1,3) \circ \ldots \circ Request(6,4) \ldots))))) \end{array} \tag{1.1}$$

□

Defining fluents is a matter of design. The fluent *Empty*, for example, is somewhat redundant because emptiness of a mail bag could also be indirectly expressed as not carrying anything in this bag, using the fluent *Carries*. On the other hand, the additional fluent both makes specifications more readable and allows to check conditions more efficiently.

In agent programs, conditions which refer to the state of the outside world are based on the notion of fluents to *hold* in states. For example, it is impossible for the mailbot to put some package into mail bag b if fluent $Empty(b)$ does not hold in the current state. Likewise, the contents of some bag b can be delivered just in case $Carries(b, r)$ and $At(r)$ hold for the same r, that is, the robot needs to be at the very room to which the package is addressed. The property of a fluent to hold in a state is formally expressed by an equational formula which says that the state needs to be decomposable into the fluent plus some arbitrary other sub-state. Since this notion is almost ubiquitous in fluent calculus formulas, it is abbreviated by a **macro**:

$$Holds(f : \text{FLUENT}, z : \text{STATE}) \stackrel{\text{def}}{=} (\exists z')\, z = f \circ z' \tag{1.2}$$

Macros constitute a kind of surface language which helps making logical axioms both more readable and handier. Whenever a macro occurs in a formula, it has to be taken as a mere abbreviation for the actual sub-formula it stands for. As we unfold the theory of fluent calculus, further macros will be introduced for the benefit of the reader.

Some words on the logic notation used in this book. **Formulas** are built up from atoms and the standard logical connectives, stated in order of decreasing priority: $\forall$ (**universal quantification**), $\exists$ (**existential quantification**), $\neg$ (**negation**), $\wedge$ (**conjunction**), $\vee$ (**disjunction**), $\supset$ (**implication**), $\equiv$ (**equivalence**), and $\top$ and $\bot$ (**true** and **false**, respectively). Throughout the book, both predicate and function symbols start with a capital letter whereas variables are in lower case, sometimes with sub- or superscripts. Sequences of terms like $x_1, \ldots, x_n$ are often abbreviated as $\vec{x}$. Variables outside the range of the quantifiers in a formula are implicitly assumed **universally quantified**, unless stated otherwise.

In order to capture the intuition of identifying states with the fluents that hold, the special connection function of fluent calculus obeys certain properties which closely resemble those of the union operation for sets.

Definition 1.2 The **foundational axioms** Σ_{sstate} **of special fluent calculus** are,[1]

1. **Associativity** and **commutativity**,

$$\begin{aligned}(z_1 \circ z_2) \circ z_3 &= z_1 \circ (z_2 \circ z_3) \\ z_1 \circ z_2 &= z_2 \circ z_1\end{aligned}$$

2. **Empty state axiom**,

$$\neg Holds(f, \emptyset)$$

3. **Irreducibility**,

$$Holds(f, g) \supset f = g$$

4. **Decomposition**,

$$Holds(f, z_1 \circ z_2) \supset Holds(f, z_1) \vee Holds(f, z_2)$$

5. **State equality**,

$$(\forall f)\,(Holds(f, z_1) \equiv Holds(f, z_2)) \supset z_1 = z_2$$

□

Associativity and commutativity imply that it does not matter in which order the fluents occur in a state term. The axiom of associativity permits us to omit parenthesis on the level of "$\circ$" from now on. The empty state axiom says that no fluent is contained in the special state $\emptyset$. The axioms of decomposition

[1] The first "s" in Σ_{sstate} indicates that these axioms characterize the *special* version of fluent calculus.

and irreducibility imply that compound states can be decomposed into single fluents and not further. The very last foundational axiom says that states are indistinguishable if they contain the same fluents.

The foundational axioms of special fluent calculus entail, for example,

$$(\exists b)\, Holds(Empty(b), Z)$$

(where Z shall be the state in (1.1)) since $\Sigma_{sstate} \models (\exists b, z')\, Z = Empty(b) \circ z'$. On the other hand, Σ_{sstate} does not entail

$$(\exists b, x)\,(Holds(At(x), Z) \wedge Holds(Carries(b, x), Z))$$

Hence, in the initial state the robot can put something into one of its mail bags while there is nothing that can be delivered.

The foundational axioms together are consistent, as can be easily shown by constructing a standard model which interprets states as sets of fluents.

Proposition 1.3 Σ_{sstate} *is consistent.*

Proof: Let ι be an interpretation whose domain includes all sets of ground fluents and where the mapping is defined by

1. $\emptyset^\iota = \{\}$,
2. $f^\iota = \{f\}$ for each ground fluent f,
3. $(z_1 \circ z_2)^\iota = z_1^\iota \cup z_2^\iota$.

Recall that macro $Holds(f, z)$ stands for $(\exists z')\, z = f \circ z'$, which is true under ι just in case there exists a set Z' such that $Z = \{F\} \cup Z'$, where $Z = z^\iota$ and $F = f^\iota$. Interpretation ι is then easily verified to be a model for Σ_{sstate}:

1. Set union is an associative-commutative operation.
2. There are no sets $\{F\}$ and Z such that $\{\} = \{F\} \cup Z$.
3. $\{G\} = \{F\} \cup Z$ implies $F = G$.
4. If $\{F\} \cup Z = Z_1 \cup Z_2$, then $F \in Z_1$ or $F \in Z_2$; hence, $\{F\} \cup Z' = Z_1$ or $\{F\} \cup Z' = Z_2$ for some set Z'.
5. For any F, suppose that there exists some Z with $\{F\} \cup Z = Z_1$ if and only if there also exists some Z' such that $\{F\} \cup Z' = Z_2$. Then $F \in Z_1$ iff $F \in Z_2$; hence, $Z_1 = Z_2$. ■

The foundational axioms are also mutually independent, that is, there is no redundancy in Σ_{sstate} (see Exercise 1.2).

The standard model raises the question why we cannot use sets in the first place instead of introducing and axiomatically characterizing a special connection function. The reason is that the approach taken in fluent calculus constitutes an **extensional** definition of sets and set operations, as will be shown soon.

The extensional view is extremely useful when it comes to logical reasoning with incomplete specifications of states,[2] in which case the foundational axioms give rise to short logical derivations. This is in contrast to the standard characterization of sets, which is **intensional** and, for example, provides no means for quickly inferring the result of removing an element from an incompletely specified set. This issue will be raised again in Chapter 3 when introducing general fluent calculus, which deals with incomplete state information.

In special fluent calculus, agents are assumed to always know exactly which fluents hold in their environment. Speaking formally, we are particularly interested in states consisting of an explicit enumeration of fluents.

Definition 1.4 A **finite state** τ is a term $f_1 \circ \ldots \circ f_n$ such that each f_i $(1 \leq i \leq n)$ is a fluent $(n \geq 0)$. If $n = 0$, then τ is $\emptyset$. A **ground state** is a finite state without variables. □

State (1.1), for example, is ground while the term mentioned in footnote 2 is, by definition, not a finite state due to the occurrence of state variable z (in which infinitely many fluents may hold).

It is easy to see that finite states tell us precisely which fluents hold in them.

Proposition 1.5 *Let $\tau = f_1 \circ \ldots \circ f_n$ be a finite state $(n \geq 0)$, then*

$$\Sigma_{sstate} \models Holds(f, \tau) \equiv \bigvee_{i=1}^{n} (f_i = f)$$

Proof: Let f be a fluent. If $\bigvee_{i=1}^{n}(f_i = f)$, then $n \geq 1$ and $(\exists z)\, \tau = f \circ z$ according to associativity and commutativity; hence, $Holds(f, \tau)$ according to macro definition (1.2).

The converse is proved by induction on n. If $n = 0$, then the empty state axiom implies $Holds(f, \tau) \equiv \bot$. Suppose the claim holds for n, and consider $Holds(f, \tau \circ f_{n+1})$. Decomposition implies $Holds(f, \tau) \vee Holds(f, f_{n+1})$. The induction hypothesis and irreducibility imply $\bigvee_{i=1}^{n}(f_i = f) \vee f = f_{n+1}$. ■

Agents maintain a state description as their internal world model, against which conditional program statements are verified. Since their actions affect the outside world, agents need to constantly update their model as they move along. In general, actions cause some fluents to become false and others to become true. Whenever, say, our mailbot in room r_1 grabs and puts into bag b a package for room r_2, then the corresponding fluent $Carries(b, r_2)$ is true afterwards. The fluent $Empty(b)$, on the other hand, ceases to hold, and so should the fluent $Request(r_1, r_2)$ in order that the robot is not tempted to pick up the package again. Addition and removal of fluents from states are therefore two important operations on states. The addition of a fluent f to a state z is easily specified by $z \circ f$. Axiomatizing the subtraction of fluents, on the other hand, is slightly more complicated.

[2] An arbitrary example of an incomplete state is $At(x) \circ Carries(2, x) \circ z$, which says that the robot carries in its second bag a package for the office of its current location. It is not said which office we are talking about, nor does the term indicate what else holds.

Below, we introduce the macro equation $z_1 - (f_1 \circ \ldots \circ f_n) = z_2$ with the intended meaning that state z_2 is state z_1 **minus** the fluents $f_1, \ldots, f_n$. The crucial item in the inductive definition is the second one, by which removal of a single fluent is axiomatized as a case distinction, depending on whether the fluent happens to be true or not. The first and third item define, respectively, the special case of removing the empty state and the generalization to removal of several fluents:

- $z_1 - \emptyset = z_2 \stackrel{\text{def}}{=} z_2 = z_1$;
- $z_1 - f = z_2 \stackrel{\text{def}}{=} (z_2 = z_1 \vee z_2 \circ f = z_1) \wedge \neg \mathit{Holds}(f, z_2)$;
- $z_1 - (f_1 \circ f_2 \circ \ldots \circ f_n) = z_2 \stackrel{\text{def}}{=} (\exists z)\,(z_1 - f_1 = z \wedge z - (f_2 \circ \ldots \circ f_n) = z_2)$ where $n \geq 2$.

The intuition behind the second item is easy to grasp with the equivalent, but less compact, formulation of the right hand side as the conjunction of $\neg \mathit{Holds}(f, z_1) \supset z_2 = z_1$ and $\mathit{Holds}(f, z_1) \supset z_2 \circ f = z_1 \wedge \neg \mathit{Holds}(f, z_2)$.

The following proposition shows that this extensional definition of removal of finitely many elements is correct wrt. the standard intensional definition of set difference.

Proposition 1.6 *Let $\tau = f_1 \circ \ldots \circ f_n$ be a finite state ($n \geq 0$). Then*

$$\Sigma_{sstate} \models z_1 - \tau = z_2 \supset [\mathit{Holds}(f, z_2) \equiv \mathit{Holds}(f, z_1) \wedge \neg \mathit{Holds}(f, \tau)]$$

Proof: Suppose $z_1 - f_1 \circ \ldots \circ f_n = z_2$. The proof is by induction on n.

If $n = 0$, then $z_1 = z_2$; hence, $\mathit{Holds}(f, z_2) \equiv \mathit{Holds}(f, z_1)$ according to the definition of *Holds*. With the empty state axiom it follows that $\mathit{Holds}(f, z_2) \equiv \mathit{Holds}(f, z_1) \wedge \neg \mathit{Holds}(f, \emptyset)$.

Suppose the claim holds for n, and consider $z_1 - (f_{n+1} \circ \tau) = z_2$. Macro expansion implies that there exists some z such that $z_1 - f_{n+1} = z$ and $z - \tau = z_2$, that is,

$$z = z_1 \vee z \circ f_{n+1} = z_1 \tag{1.3}$$

$$\neg \mathit{Holds}(f_{n+1}, z) \tag{1.4}$$

$$z - \tau = z_2 \tag{1.5}$$

Suppose $\mathit{Holds}(f, z_2)$, then equation (1.5) and the induction hypothesis imply $\mathit{Holds}(f, z) \wedge \neg \mathit{Holds}(f, \tau)$. From (1.3) it follows that $\mathit{Holds}(f, z_1)$. Moreover, (1.4) and $\mathit{Holds}(f, z)$ imply $f \neq f_{n+1}$; hence, $\neg \mathit{Holds}(f, f_{n+1} \circ \tau)$ according to the axioms of decomposition and irreducibility.

Conversely, suppose $\mathit{Holds}(f, z_1) \wedge \neg \mathit{Holds}(f, f_{n+1} \circ \tau)$, then decomposition implies $\neg \mathit{Holds}(f, f_{n+1})$ and $\neg \mathit{Holds}(f, \tau)$. From (1.3), decomposition and $\neg \mathit{Holds}(f, f_{n+1})$ it follows that $\mathit{Holds}(f, z)$; hence, (1.5) and $\neg \mathit{Holds}(f, \tau)$ along with the induction hypothesis imply $\mathit{Holds}(f, z_2)$. ■

While the intensional, "piece-wise" definition of set difference is both more common and more general (as it is not restricted to removal of finite sets), the extensional definition is better suited for actually calculating the state term which results from the removal of fluents, in particular if the original state is incompletely specified.

On top of the macro for specifying negative effects, the macro equation $z_2 = (z_1 - \vartheta^-) + \vartheta^+$ abbreviates an axiom which stipulates that state z_2 is state z_1 **minus** the fluents in ϑ^- **plus** the fluents in ϑ^+. Both ϑ^- and ϑ^+ are finite states:

$$z_2 = z_1 - f_1 \circ \ldots \circ f_m + g_1 \circ \ldots \circ g_n \stackrel{\text{def}}{=} (\exists z)\,(z_1 - f_1 \circ \ldots \circ f_m = z \\ \wedge\, z_2 = z \circ g_1 \circ \ldots \circ g_n)$$

State update is thus defined by an *update equation*, which an agent needs to solve in order to know what holds after it has performed an action. The following computation rule is very useful to this end. It says that equal fluents on both sides of an equation can be cancelled out under certain conditions.

Proposition 1.7 (Cancellation Law)

$$\Sigma_{sstate} \models \neg Holds(f, z_1) \wedge \neg Holds(f, z_2) \supset [z_1 \circ f = z_2 \circ f \supset z_1 = z_2]$$

Proof: Suppose that $\neg Holds(f, z_1)$, $\neg Holds(f, z_2)$ and $z_1 \circ f = z_2 \circ f$. We show that $Holds(f', z_1) \equiv Holds(f', z_2)$, which proves the claim according to the axiom of state equality. If $Holds(f', z_1)$, then $Holds(f', z_1 \circ f)$; hence, $Holds(f', z_2 \circ f)$ since $z_1 \circ f = z_2 \circ f$. From $Holds(f', z_1)$ and $\neg Holds(f, z_1)$ it follows that $f' \neq f$. This and $Holds(f', z_2 \circ f)$ implies $Holds(f', z_2)$ due to decomposition and irreducibility. The converse is similar. ■

In case of complete states, solving the state equations is of course fairly trivial as one can just apply the corresponding set operations. As an example, let Z_1 be the initial state (1.1) of our mail delivery problem, and consider the following equation:

$$Z_2 = Z_1 - Empty(3) \circ Request(1,2) + Carries(3,2)$$

This update from state Z_1 to Z_2 formalizes the effect of putting the package from room 1 to 2 into mail bag 3. Macro expansion implies the existence of z', z'' such that

$$\begin{aligned} z' &= Z_1 - Empty(3) \\ z'' &= z' - Request(1,2) \\ Z_2 &= z'' \circ Carries(3,2) \end{aligned}$$

Since $Empty(3)$ holds in Z_1, the first equation implies $z' \circ Empty(3) = Z_1$ and $\neg Holds(Empty(3), z')$. The latter allows to apply the cancellation law after inserting (1.1) for Z_1, which yields

$$z' = At(1) \circ Empty(1) \circ Empty(2) \circ \\ Request(1,2) \circ Request(1,3) \circ \ldots \circ Request(6,4)$$

In the same fashion we conclude

$$z'' = At(1) \circ Empty(1) \circ Empty(2) \circ Request(1,3) \circ \ldots \circ Request(6,4)$$

Finally,

$$Z_2 = At(1) \circ Empty(1) \circ Empty(2) \circ \\ Request(1,3) \circ \ldots \circ Request(6,4) \circ Carries(3,2)$$

where mail bag 3 is no longer empty but contains a package for office 2, and where the request for a delivery from 1 to 2 is no longer present.

Actually, the cancellation law applies in the example just given only if the fluents $Empty(1)$ and $Request(1,2)$ do not equal any other fluent in Z_1. Axioms defining inequality of terms which use different function symbols or arguments are very common in fluent calculus. Informally speaking, they say that different names refer to different objects.

Definition 1.8 A **unique-name axiom** is a formula of the form

$$\bigwedge_{i=1}^{n-1} \bigwedge_{j=i+1}^{n} h_i(\vec{x}) \neq h_j(\vec{y}) \;\wedge\; \bigwedge_{i=1}^{n} [\, h_i(\vec{x}) = h_i(\vec{y}) \supset \vec{x} = \vec{y}\,]$$

abbreviated $UNA[h_1, \ldots, h_n]$. □

The first part of a unique-name axiom stipulates that terms with different leading function symbol are unequal; e.g., $At(r_1) \neq Carries(b, r_2)$. The second part implicitly says that terms are unequal which start with the same function symbol but whose arguments differ; e.g., $Empty(1) \neq Empty(3)$ since $1 \neq 3$ under the standard interpretation of natural numbers and equality. Given uniqueness of names for all fluent functions in the mail delivery world, i.e., $UNA[At, Empty, Carries, Request]$, the conditions of the cancellation law are guaranteed to hold in the example given before.

1.2 Actions and Situations

In agent programming, actions mean the interactions of the agent with its environment. Actions may change the outside world, e.g., when a robot picks up a package or a software agent orders a product over the Internet. Other actions only change the status of the physical agent itself, e.g., when a robot moves to a new position. Finally, actions may just provide the agent with information about the environment, e.g., when a robot senses whether a door is open or a software agent compares prices at different online stores. While a single action can be a very complex behavior on the level of the physical agent, actions are taken as elementary entities on the level of agent programs and in fluent calculus.

Agent programming is all about triggering the agent to do the right action at the right time. Every agent program should produce a sequence of actions to be

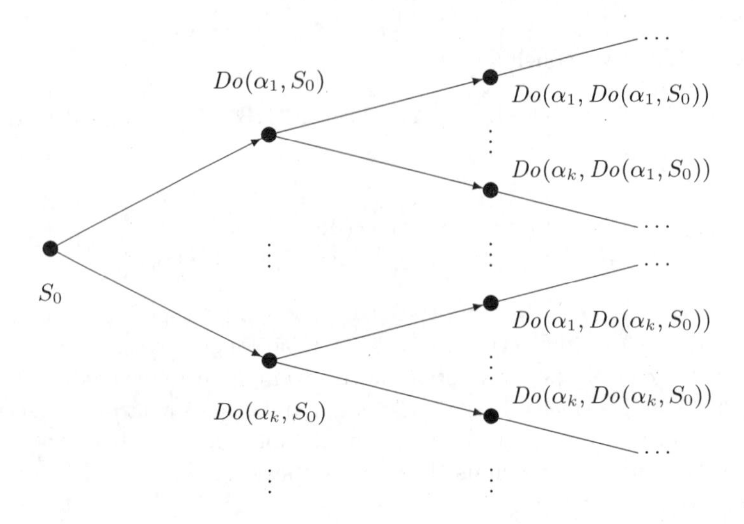

Figure 1.3: A tree of situations.

executed by the agent. When talking about the different stages of an actual run of a program, we refer to the sequence of actions that have been performed up to a point as a *situation*. The special constant S_0 denotes the initial situation, in which the agent has not yet done any interaction with the outside world. The constructor $Do(a, s)$ then maps an action a and a situation s to the situation after the performance of the action. Hence, action sequences are nested terms of the form $Do(a_n, \ldots, Do(a_1, S_0) \ldots)$. The situations can be visualized as the nodes of a tree rooted in S_0; see Figure 1.3. Each branch in this tree is a potential run of a program for the agent.

Most conditions in agent programs are evaluated against the state which the environment is in at the time the program statement is executed. The standard function $State(s)$ is used to denote these states at the different stages of a program run, i.e., at the different situations s. The initial situation in our mail delivery scenario, for instance, is characterized by the equational axiom $State(S_0) = Z$ where Z is the term of (1.1). A successful program for the mail-bot should then generate a sequence of actions $S_n = Do(a_n, \ldots, Do(a_1, S_0) \ldots)$ such that, for some r,

$$State(S_n) = At(r) \circ Empty(1) \circ Empty(2) \circ Empty(3)$$

On the formal side, there are special sorts for both actions and situations. Completing the signature, the standard predicate $Poss(a, z)$ is used to specify the conditions on state z for action a to be executable by the agent.

Definition 1.9 A tuple $\mathcal{S} \cup \langle \mathcal{A}, S_0, Do, State, Poss \rangle$ is a **fluent calculus signature** if

- $\mathcal{S}$ state signature

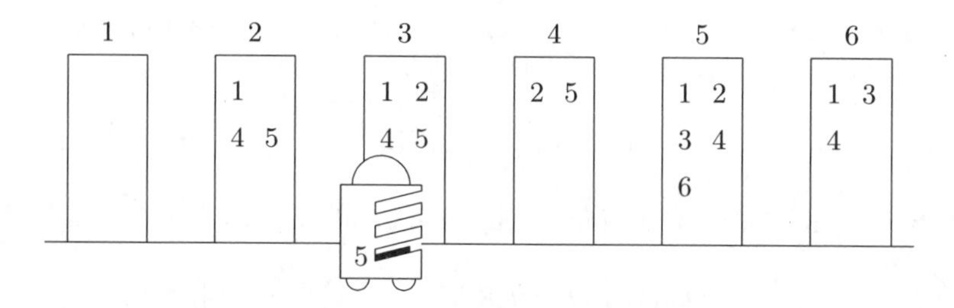

Figure 1.4: The mail delivery robot of Figure 1.1 after having performed the action sequence of situation term (1.6).

- $\mathcal{A}$ finite, non-empty set of function symbols into sort ACTION
- S_0 : SIT
- Do : ACTION × SIT ↦ SIT
- $State$: SIT ↦ STATE
- $Poss$: ACTION × STATE.

An **action** is a term of sort ACTION. A **situation** is a term of sort SIT. □

Throughout the book, action variables will be written by the letter a and situation variables by the letter s, all possibly with sub- or superscripts.

Example 1 (continued) The actions of the mailbot shall be denoted by the following three function symbols:

$$\begin{array}{rll} Pickup: & \mathbb{N} \times \mathbb{N} \mapsto \text{ACTION} & Pickup(b,r) \mathrel{\hat{=}} \text{put in } b \text{ package for } r \\ Deliver: & \mathbb{N} \mapsto \text{ACTION} & Deliver(b) \mathrel{\hat{=}} \text{deliver contents of bag } b \\ Go: & \{Up, Down\} \mapsto \text{ACTION} & Go(d) \mathrel{\hat{=}} \text{move one office up or down} \end{array}$$

Applied to the scenario of Figure 1.1, a successful control program may reach the following situation after nine steps, where the robot has already carried out three of the 21 delivery tasks (cf. Figure 1.4):

$$\begin{array}{l} Do(Deliver(2), Do(Deliver(1), Do(Go(Up), \\ \quad Do(Pickup(1,3), Do(Deliver(1), Do(Go(Up), \\ \quad\quad Do(Pickup(3,5), Do(Pickup(2,3), Do(Pickup(1,2), S_0))))))))) \end{array} \tag{1.6}$$

On the other hand, there are many situations which a program should never generate as they describe action sequences which are impossible to execute. Examples are $Do(Go(Down), S_0)$ or $Do(Deliver(3), Do(Pickup(1,2), S_0))$, assuming that the robot starts in state (1.1). □

For notational convenience, we will sometimes write situation terms using lists of actions according to the following definition:

$$Do([a_1, \ldots, a_n], s) \stackrel{\text{def}}{=} Do(a_n, \ldots, Do(a_1, s) \ldots) \quad (n \geq 0) \tag{1.7}$$

The situation in (1.6), for example, can thus be abbreviated as

$$Do([Pickup(1,2), Pickup(2,3), \ldots, Deliver(2)], S_0)$$

The concept of a situation allows us to talk about the state of an environment at different times. In particular, we can define what it means for a fluent to hold at a certain stage of a program run:

$$Holds(f, s) \stackrel{\text{def}}{=} Holds(f, State(s)) \tag{1.8}$$

Based on this definition, one can formulate arbitrary logical assertions about the state of the world in a situation.

Definition 1.10 A **state formula** $\Delta(z)$ is a first-order formula with free state variable z and without any occurrence of states other than in expressions of the form $Holds(f, z)$, and without actions or situations. If $\Delta(z)$ is a state formula and s a situation variable, then $\Delta(z)\{z/State(s)\}$ is a **situation formula**, written $\Delta(s)$. □

For instance, $(\exists r)\,(Holds(At(r), z) \wedge r \geq 4)$ is a state formula, whereas the formula $(\exists b)\,Holds(Empty(b), z \circ Empty(1))$ is not. Likewise, the implication $(\forall b)\,(Holds(Empty(b), s) \supset \neg(\exists r)\,Holds(Carries(b, r), s))$ is a situation formula, whereas $(\exists r)\,Holds(At(r+1), Do(Go(Up), s))$ is not.

Situation formulas provide an elegant way of distinguishing those combinations of fluents which describe a physical impossibility. Usually there are many formal states which cannot possibly occur in the real world. The mailbot, for example, can never be at more than one location at a time, nor can it ever carry a package in a bag which is empty. Hence, the syntactically perfectly acceptable state $At(1) \circ At(4) \circ Empty(2) \circ Carries(2,3)$ represents a state of affairs that will actually never happen. So-called domain constraints are formulas that need to be satisfied in order for a state to possibly arise in reality.

Definition 1.11 A **domain constraint** is a formula $(\forall s)\Delta(s)$ where $\Delta(s)$ is a situation formula. A state z **satisfies** a set of domain constraints Σ_{dc} wrt. some auxiliary axioms Σ_{aux} iff $\Sigma_{dc} \cup \Sigma_{aux} \cup \Sigma_{sstate} \cup \{State(s) = z\}$ is consistent. □

Auxiliary axioms are formulas such as the underlying unique-name axioms which describe static properties of the domain.

Example 1 (continued) The physical reality of the mail delivery world is characterized by the following domain constraints:

$$\begin{array}{l} (\exists r)\,(Holds(At(r), s) \wedge 1 \leq r \leq n) \\ Holds(At(r_1), s) \wedge Holds(At(r_2), s) \supset r_1 = r_2 \\ Holds(Empty(b), s) \supset \neg Holds(Carries(b, r), s) \wedge 1 \leq b \leq k \\ Holds(Carries(b, r), s) \supset 1 \leq b \leq k \wedge 1 \leq r \leq n \\ Holds(Carries(b, r_1), s) \wedge Holds(Carries(b, r_2), s) \supset r_1 = r_2 \\ Holds(Request(r_1, r_2), s) \supset r_1 \neq r_2 \wedge 1 \leq r_1, r_2 \leq n \end{array} \tag{1.9}$$

Here, n is the number of offices and k the number of mail bags. Put in words, in every situation the robot is at some unique office (first and second constraint), mail bags cannot be both empty and carrying a package (third constraint), the contents of a mail bag must lie within the allowed range and be unique (fourth and fifth constraint), and packages need to have different sender and addressee in the allowed range (last constraint). It is easy to verify that, e.g., state (1.1) satisfies these domain constraints wrt. the unique-name axioms given earlier. □

Fluents are sometimes called **functional** if the domain constraints stipulate that they are unique in some of their arguments, as with $At(r)$ in our example. This is because it is quite natural to say that "in situation s, the location of the robot is 1" instead of "in situation s it holds that the robot is at location 1."

1.3 State Update Axioms

One purpose of the internal model is to help the agent decide which action it can perform at a certain stage. Our mailbot, for example, shall drop and pick up packages whenever possible. This requires the robot to know that it can deliver only packages which it carries and only if it is at the recipient's office. Similar conditions apply to the other actions of the mailbot. So-called precondition axioms are used to formally specify the circumstances under which an action is possible in a state.

Definition 1.12 Consider a fluent calculus signature with action functions $\mathcal{A}$, and let $A \in \mathcal{A}$. A **precondition axiom** for A is a formula

$$Poss(A(\vec{x}), z) \equiv \Pi(z)$$

where $\Pi(z)$ is a state formula with free variables among $\vec{x}, z$. □

Example 1 (continued) The preconditions of the three actions of the mailbot can be formalized by these axioms:

$$\begin{array}{l} Poss(Pickup(b,r),z) \equiv \\ \quad Holds(Empty(b),z) \wedge \\ \quad (\exists r_1)\,(Holds(At(r_1),z) \wedge Holds(Request(r_1,r),z)) \\ \\ Poss(Deliver(b),z) \equiv \\ \quad (\exists r)\,(Holds(At(r),z) \wedge Holds(Carries(b,r),z)) \\ \\ Poss(Go(d),z) \equiv \\ \quad (\exists r)\,(Holds(At(r),z) \wedge [\,d = Up \wedge r < n \vee d = Down \wedge r > 1\,]) \end{array} \tag{1.10}$$

With this specification, it is a logical consequence that, say, in the initial state (1.1) these actions are possible: $Pickup(b,r)$ where $b \in \{1,2,3\}$ and $r \in \{2,3,5\}$, and $Go(Up)$. □

The assertion that an action and a sequence of actions, respectively, is possible in a situation reduces to the corresponding assertion about the states:

$$\begin{array}{rcl} Poss(a,s) & \stackrel{\text{def}}{=} & Poss(a, State(s)) \\ Poss([a_1,\ldots,a_n],s) & \stackrel{\text{def}}{=} & Poss(a_1,s) \wedge \ldots \wedge Poss(a_n, Do([a_1,\ldots,a_{n-1}],s)) \end{array}$$

In order to keep their internal model up to date, agents need to know not only the preconditions of their actions but also how these affect the environment. With this knowledge the agent can update its model after each action to reflect the changes it has effected. The mailbot, for example, needs to know that picking up packages in mail bags causes the packages to be carried. It also needs to know that mail bags cease to be empty when being filled, for otherwise the robot would be tempted to fill it again. On the other hand, collecting a package does not alter the contents of the other bags, nor does it change the location of the robot, etc.

The changes that an action causes are specified by axioms which define the update of states. Under the condition that the action $A(\vec{x})$ in question is possible in a situation s, the so-called state update axiom for this action relates the resulting state, $State(Do(A(\vec{x}),s))$, to the current state, $State(s)$. More precisely, the positive and negative effects of the action are cast into an update equation which encodes the difference between these two states. Some actions have conditional effects, like $Go(d)$ in the mail delivery world where the effect depends on whether $d = Up$ or $d = Down$. State update axioms may therefore combine several update equations, each of which applies conditionally.

Definition 1.13 Consider a fluent calculus signature with action functions $\mathcal{A}$, and let $A \in \mathcal{A}$. A **state update axiom** for A is a formula

$$\begin{array}{l} Poss(A(\vec{x}),s) \supset \\ \quad (\exists \vec{y}_1)\,(\Delta_1(s) \wedge State(Do(A(\vec{x}),s)) = State(s) - \vartheta_1^- + \vartheta_1^+) \\ \quad \vee \ldots \vee \\ \quad (\exists \vec{y}_n)\,(\Delta_n(s) \wedge State(Do(A(\vec{x}),s)) = State(s) - \vartheta_n^- + \vartheta_n^+) \end{array} \tag{1.11}$$

where

- $n \geq 1$;
- for $i = 1, \ldots, n$,
  - $\Delta_i(s)$ is a situation formula with free variables among $\vec{x}, \vec{y}_i, s$;
  - $\vartheta_i^+, \vartheta_i^-$ are finite states with variables among $\vec{x}, \vec{y}_i$.

The terms ϑ_i^+ and ϑ_i^- are called, respectively, **positive** and **negative** effects of $A(\vec{x})$ under **condition** $\Delta_i(s)$. An empty positive or negative effect is usually not explicitly stated. □

Example 1 (continued) The effects of the three actions of the mailbot can be formalized by these state update axioms:

$$\begin{array}{l}
Poss(Pickup(b,r),s) \supset \\
\quad (\exists r_1)\,(Holds(At(r_1),s) \wedge \\
\qquad\qquad State(Do(Pickup(b,r),s)) = \\
\qquad\qquad\quad State(s) - Empty(b) \circ Request(r_1,r) + Carries(b,r)\,) \\
\\
Poss(Deliver(b),s) \supset \\
\quad (\exists r)\,(Holds(At(r),s) \wedge \\
\qquad\qquad State(Do(Deliver(b),s)) = \\
\qquad\qquad\quad State(s) - Carries(b,r) + Empty(b)\,) \\
\\
Poss(Go(d),s) \supset \\
\quad (\exists r)\,(d = Up \wedge Holds(At(r),s) \wedge \\
\qquad\qquad State(Do(Go(d),s)) = State(s) - At(r) + At(r+1)\,) \\
\quad \vee \\
\quad (\exists r)\,(d = Down \wedge Holds(At(r),s) \wedge \\
\qquad\qquad State(Do(Go(d),s)) = State(s) - At(r) + At(r-1)\,)
\end{array} \tag{1.12}$$

Take, for instance, $State(S_0) = (1.1)$, then $Poss(Pickup(1,2),S_0)$ according to the precondition axiom in (1.10). Let $S_1 = Do(Pickup(1,2),S_0)$, then the state update axiom for picking up packages implies

$$State(S_1) = State(S_0) - Empty(1) \circ Request(1,2) + Carries(1,2)$$

A solution to this equation (more precisely, to the set of equations obtained by macro expansion) is given by

$$\begin{array}{rcl}
State(S_1) & = & At(1) \circ Empty(2) \circ Empty(3) \circ Carries(1,2) \circ \\
& & \qquad Request(1,3) \circ \ldots \circ Request(6,4)
\end{array}$$

The reader may verify that likewise the precondition and state update axioms entail $Poss(Go(Up),S_1)$ and, setting $S_2 = Do(Go(Up),S_1)$,

$$\begin{array}{rcl}
State(S_2) & = & At(2) \circ Empty(2) \circ Empty(3) \circ Carries(1,2) \circ \\
& & \qquad Request(1,3) \circ \ldots \circ Request(6,4)
\end{array}$$

Finally, the axiomatization implies that $Poss(Deliver(1), S_2)$ and, setting $S_3 = Do(Deliver(1), S_2)$,

$$\begin{array}{rcl} State(S_3) & = & At(2) \circ Empty(1) \circ Empty(2) \circ Empty(3) \circ \\ & & \quad Request(1,3) \circ \ldots \circ Request(6,4) \end{array}$$

We could go on with this derivation to prove that Figure 1.4 (page 13) indeed depicts the correct state for the given situation. □

The concept of state update axioms tackles a classical problem in Artificial Intelligence, known as the **frame problem**. It arises because mere effect formulas do not suffice to draw conclusions about unchanged fluents. Suppose, for example, action *Pickup* were specified solely by the effect axiom

$$\begin{array}{l} Poss(Pickup(b,r),s) \supset \\ \quad (\exists r_1)\,(Holds(At(r_1),s) \wedge \neg Holds(Empty(b), Do(Pickup(b,r),s)) \\ \qquad\qquad\qquad\qquad\qquad \wedge \neg Holds(Request(r_1,r), Do(Pickup(b,r),s)) \\ \qquad\qquad\qquad\qquad\qquad \wedge\, Holds(Carries(b,r), Do(Pickup(b,r),s))\,) \end{array}$$

Substituting $\{b/1, r/2, s/S_0\}$, this formula alone does not entail whether $At(1)$ or $Empty(2)$ or any other initially true fluent still holds in the resulting situation. This is in contrast to the state equations used in state update axioms, which do imply that the "frame," i.e., the unaffected fluents, continue to hold after performing the action. In other words, a fluent holds in the resulting situation just in case it is a positive effect or it holds in the current situation and is not a negative effect. State update axioms thus constitute a solution to the frame problem.

Theorem 1.14 *Let ϑ^+ and ϑ^- be two finite states, then foundational axioms Σ_{sstate} and $State(Do(a,s)) = State(s) - \vartheta^- + \vartheta^+$ entail*

$$\begin{array}{rcl} Holds(f, Do(a,s)) & \equiv & Holds(f, \vartheta^+) \\ & & \vee \\ & & Holds(f,s) \wedge \neg Holds(f, \vartheta^-) \end{array}$$

Proof: Macro expansion implies the existence of a state z such that

$$\begin{array}{rcl} State(s) - \vartheta^- & = & z \\ State(Do(a,s)) & = & z \circ \vartheta^+ \end{array}$$

According to Proposition 1.6, the first equation implies that $Holds(f,z) \equiv Holds(f,s) \wedge \neg Holds(f,\vartheta^-)$. From the second equation and decomposition it follows that $Holds(f, Do(a,s)) \equiv Holds(f,z) \vee Holds(f,\vartheta^+)$. ■

State update axioms complete the formal specification of agent domains. The various components are summarized in the following definition.

Definition 1.15 Consider a fluent calculus signature with action functions $\mathcal{A}$. A **domain axiomatization** in special fluent calculus is a finite set of axioms $\Sigma = \Sigma_{dc} \cup \Sigma_{poss} \cup \Sigma_{sua} \cup \Sigma_{aux} \cup \Sigma_{init} \cup \Sigma_{sstate}$ where

- Σ_{dc} set of domain constraints;
- Σ_{poss} set of precondition axioms, one for each $A \in \mathcal{A}$;
- Σ_{sua} set of state update axioms, one for each $A \in \mathcal{A}$;
- Σ_{aux} set of auxiliary axioms, with no occurrence of states except for fluents, no occurrence of situations, and no occurrence of $Poss$;
- $\Sigma_{init} = \{State(S_0) = \tau_0\}$ where τ_0 is a ground state;
- Σ_{sstate} foundational axioms of special fluent calculus. □

It is useful to have some guidelines of how to arrive at "good" domain axiomatizations. In particular, state update axioms should be designed in such a way that they never leave doubts about the effects of a (possible) action. To this end—at least for the time being—domain axiomatizations should be deterministic in the following sense.[3]

Definition 1.16 A domain axiomatization Σ is **deterministic** iff it is consistent and for any ground state τ and ground action α there exists a ground state τ' such that

$$\Sigma \models [State(s) = \tau \wedge Poss(\alpha, s)] \supset State(Do(\alpha, s)) = \tau'$$

□

By this definition, the state update axioms in a deterministic domain entail unique successor states for any state and any action that is possible in that state.

The following theorem tells us how to design deterministic domains. Basically, it says that the conditions Δ_i in state update axioms should be set up in such a way that there is always a unique instance of one and only one update equation that applies. Furthermore, this equation should always determine a successor state that satisfies the domain constraints.

Theorem 1.17 *Consider a fluent calculus signature with action functions $\mathcal{A}$, then a domain axiomatization Σ is deterministic if the following holds:*

1. $\Sigma_{sstate} \cup \Sigma_{dc} \cup \Sigma_{aux} \cup \Sigma_{init}$ *is consistent;*
2. $\Sigma_{aux} \models A_1(\vec{x}) \neq A_2(\vec{y})$ *for all* $A_1, A_2 \in \mathcal{A}$ *such that* $A_1 \neq A_2$*;*
3. *for any ground state* τ *and ground action* $A(\vec{r})$ *with state update axiom* (1.11) $\in \Sigma_{sua}$, *if*
   (a) τ *satisfies* Σ_{dc} *wrt.* Σ_{aux} *and*
   (b) $\Sigma_{sstate} \cup \Sigma_{poss} \cup \Sigma_{aux} \models Poss(A(\vec{r}), \tau)$,

[3] Nondeterminism is going to be the topic of Chapter 7.

then there is some $i = 1, \ldots, n$ *and ground terms* $\vec{t}$ *such that*

(c) $\Sigma_{aux} \models \Delta_i(\tau)\{\vec{x}/\vec{r}, \vec{y}_i/\vec{t}\} \wedge [\Delta_i(\tau)\{\vec{x}/\vec{r}\} \supset \vec{y}_i = \vec{t}\,]$,

(d) $\Sigma_{aux} \models \neg(\exists \vec{y}_j)\, \Delta_j(\tau)\{\vec{x}/\vec{r}\}$ *for all* $j \neq i$, *and*

(e) *any state* τ' *with* $\Sigma_{sstate} \cup \Sigma_{aux} \models \tau' = \tau - \vartheta_i^- + \vartheta_i^+\{\vec{x}/\vec{r}, \vec{y}_i/\vec{t}\}$ *satisfies* Σ_{dc} *wrt.* Σ_{aux}.

Proof: Let s be a situation, τ a ground state, and $A(\vec{r})$ a ground action. Suppose $State(s) = \tau$ and $Poss(A(\vec{r}), s)$, then we have to show that there exists a ground state τ' such that Σ entails $State(Do(A(\vec{r}), s)) = \tau'$:

From $Poss(A(\vec{r}), s)$ and $State(s) = \tau$ it follows that $Poss(A(\vec{r}), \tau)$. According to conditions (c) and (d), there is a unique $i = 1, \ldots, n$ and a unique instance $\vec{t}$ for $\vec{y}_i$ such that Σ_{aux} entails $\Delta_i(\tau)\{\vec{x}/\vec{r}, \vec{y}_i/\vec{t}\}$. Hence, the state update axiom for $A(\vec{x})$ implies

$$State(Do(A(\vec{x}), s)) = \tau - \vartheta_i^- + \vartheta_i^+\{\vec{x}/\vec{r}, \vec{y}_i/\vec{t}\}$$

This shows that Σ entails $State(Do(A(\vec{r}), s)) = \tau'$ for some ground state τ'.

To prove that Σ is consistent, consider a model ι for the set of axioms $\Sigma' = \Sigma_{sstate} \cup \Sigma_{aux}$. Since neither of these axioms mentions predicate $Poss$, since a precondition axiom alone cannot be inconsistent, and since any two actions starting with different function symbols cannot be equal, there is also a model ι for $\Sigma' \cup \Sigma_{poss}$. Moreover, since neither of these axioms mentions situations, a model for $\Sigma = \Sigma' \cup \Sigma_{poss} \cup \Sigma_{init} \cup \Sigma_{sua} \cup \Sigma_{dc}$ can be constructed from ι as follows. Let $State(S_0)$ be interpreted by a term τ_0 as stipulated by Σ_{init}. For any ground situation σ such that $State(\sigma)$ is interpreted in ι by a ground state τ which satisfies Σ_{dc}, and for any ground action $A(\vec{r})$ such that $Poss(A(\vec{r}), \tau)$ is true in ι, let $State(Do(A(\vec{r}), \sigma))$ be interpreted by a state τ' which satisfies condition (e). So doing determines a model for Σ_{sua} according to conditions (c) and (d), and for Σ_{dc} according to condition (e) and since $\Sigma_{init} \cup \Sigma_{dc}$ is consistent under $\Sigma_{sstate} \cup \Sigma_{aux}$ according to assumption (a). ■

The value of this theorem lies in reducing overall consistency to consistency of the auxiliary axioms and domain constraints, and to a local condition on state update axioms. Both are usually easy to verify.

Example 1 (continued) Let the mailbot domain consist of domain constraints (1.9); precondition axioms (1.10); state update axioms (1.12); auxiliary axioms *UNA*[*At*, *Empty*, *Carries*, *Request*], *UNA*[*Pickup*, *Deliver*, *Go*], and *UNA*[*Up*, *Down*]; and initial state axiom $State(S_0) = (1.1)$. This domain is deterministic:

1. Since state (1.1) satisfies the domain constraints Σ_{dc} wrt. the unique-name axioms Σ_{aux}, the initial state axiom along with the foundational axioms, Σ_{dc}, and Σ_{aux} are consistent.

2. The unique-name axioms entail that different actions are unequal.

3. To verify item 3 of Theorem 1.17, let τ be a state which satisfies the domain constraints of the mail delivery world, and consider the three actions of the mailbot:

   - Suppose $Poss(Pickup(b, r), \tau)$. The domain constraints imply that there is a unique r_1 such that $Holds(At(r_1), \tau)$. Let

     $$\tau' = \tau - Empty(b) \circ Request(r_1, r) + Carries(b, r)$$

     Since τ satisfies the first two domain constraints in (1.9), so does τ'. The third, fourth, and fifth domain constraints are satisfied in τ' wrt. b since $Carries(b, r)$ holds and $Empty(b)$ does not. These three constraints are satisfied in τ' wrt. any other bag since they are satisfied in τ. Finally, since τ satisfies the last domain constraint, so does τ'.
   - The argument for $Deliver(b)$ is similar.
   - Suppose $Poss(Go(d), \tau)$. The domain constraints imply that there is a unique r such that $Holds(At(r), \tau)$. Hence, as $Up \neq Down$, only one condition applies in each instance of the state update axiom. Let

     $$\tau' = \tau - At(r) + At(r \pm 1)$$

     The first two domain constraints are satisfied in τ' since they are satisfied in τ and since the precondition for $Go(d)$ ensures that $1 \leq r \pm 1 \leq n$. As τ satisfies the other domain constraints, so does τ'. □

Deterministic domains obey a property which is essential for agent programming based on special fluent calculus: If an agent starts off with complete state information and performs possible actions only, then the axiomatization entails complete knowledge in every situation reached along the way.

Corollary 1.18 *Let Σ be a deterministic domain axiomatization. Consider a ground situation $\sigma = Do([\alpha_1, \ldots, \alpha_n], S_0)$ with $\Sigma \models Poss([\alpha_1, \ldots, \alpha_n], S_0)$, then there exists a ground state τ such that $\Sigma \models State(\sigma) = \tau$.*

Proof: By induction on n. If $n = 0$, then $\sigma = S_0$ and the claim follows from initial state axiom $\Sigma_{init} = \{State(S_0) = \tau_0\}$ in Σ. Suppose the claim holds for n, and consider a situation $Do(\alpha_{n+1}, Do([\alpha_1, \ldots, \alpha_n], S_0))$. The induction hypothesis implies the existence of a ground state τ such that $\Sigma \models State(Do([\alpha_1, \ldots, \alpha_n], S_0)) = \tau$; hence, Definition 1.16 and assumption $Poss(\alpha_{n+1}, Do([\alpha_1, \ldots, \alpha_n], S_0))$ imply the claim. ■

Domain axiomatizations using fluent calculus set the stage for agent programs. Precondition and state update axioms provide the background knowledge which allows agents to maintain an internal model of their environment. The foundational axioms of fluent calculus define how conditional statements in agent programs are evaluated against the internal model. In the upcoming chapter, we cast both the foundational axioms and the domain-dependent ones into a logic program, and we show how to write agent programs on top of it.

1.4 Bibliographical Notes

The roots of formal logic trace back to an analysis of the structure of logical arguments by the Greek philosopher Aristotle in his treatise *Organon* (that is, "instrument"). The foundations for modern predicate logic have been laid in [Frege, 1879]. There are numerous textbooks providing gentle introductions to first-order logic, among which is the classical [Quine, 1982]. Most of the logic notation used in this book has been taken from [Davis, 1993], where it has also been shown that the use of sorts and the special treatment of equality can in fact be reduced to standard predicate logic.

The idea to build intelligent agents using logic has been pursued ever since the emergence of Artificial Intelligence, beginning with [McCarthy, 1958]. The oldest formalism for axiomatizing actions and effects is **situation calculus**, which has introduced the basic notions of fluents, actions, and situations [McCarthy, 1963]. The fundamental frame problem has been uncovered in [McCarthy and Hayes, 1969] in the context of situation calculus. In [Green, 1969], situation calculus is applied to problem solving using a system for automated theorem proving. Lacking a good solution to the frame problem, this system is limited to domains with a small state space and short action sequences.

Trading the expressivity of situation calculus for efficiency, **STRIPS** (for: Stanford Research Institute Problem Solver) [Fikes and Nilsson, 1971] is restricted to **complete states** in using set operations to calculate the result of actions. With the aim of combining the expressivity of logic with the efficiency of STRIPS, several **non-classical** logics have been proposed as solutions to the frame problem [Bibel, 1986; Masseron *et al.*, 1993; Bibel, 1998]. Combining the fundamental notion of states with that of situations traces back to the logic programming-based approach of [Hölldobler and Schneeberger, 1990]. Characterizing states as multisets of fluents (where, as opposed to sets, multiple occurrences of elements count), this formalism has been shown equivalent to the aforementioned non-classical approaches for a particular kind of planning problems [Große *et al.*, 1996]. The paper [Hölldobler and Kuske, 2000] contains some decidability results for this approach.

The foundations for fluent calculus have been laid in [Thielscher, 1999] with the introduction of state update axioms as a combination of a state-based representation with elements of situation calculus. The signature of fluent calculus is an amalgamation of the axiomatization of states as in [Hölldobler and Schneeberger, 1990] and the variant of situation calculus presented in [Pirri and Reiter, 1999]. The full axiomatic theory of fluent calculus has first been presented in [Thielscher, 2001c], based on investigations carried out in [Störr and Thielscher, 2000]. As long as fluent calculus is applied to complete states only, it is essentially a reformulation of STRIPS in classical logic.

1.5 Exercises

1.1. Prove that foundational axioms Σ_{sstate} entail

(a) $z \circ \emptyset = z$;

(b) $z \circ z = z$.

1.2. Prove that foundational axioms Σ_{sstate} are mutually independent.

Hint: Find a model for the negation of each axiom with all other axioms satisfied.

1.3. Let $\tau = f_1 \circ \ldots \circ f_n$ be a finite state. Prove that foundational axioms Σ_{sstate} entail

(a) $\bigwedge_{i=1}^{n} Holds(f_i, z) \supset z \circ \tau = z$;

(b) $\bigwedge_{i=1}^{n} \neg Holds(f_i, z) \supset z - \tau = z$.

1.4. Suppose our mailbot has just one bag, and consider the initial situation where the robot is at room 2 with an empty bag and two requests, from 1 to 5 and from 3 to 4, respectively. Find the situation with the fewest *Go* actions in which the two packages have been delivered, and prove that in this situation no requests are left and all bags are empty.

1.5. Suppose that initially there are only two requests in the mail delivery world with a robot that has three mail bags, and all mail bags are initially empty. Show that there is no ground situation $\sigma = Do([\alpha_1, \ldots, \alpha_n], S_0)$ such that the mailbot domain axiomatization entails $(\forall b) \neg Holds(Empty(b), \sigma)$.

1.6. Suppose the mailbot can temporarily park packages at offices different from their destination.

(a) Replace the binary fluent $Request(r_1, r_2)$ by the ternary variant $Request(r_1, r_2, r_3)$ with the meaning that in room r_1 there lies a package addressed from room r_2 to r_3. Rewrite the domain constraints and the precondition and state update axioms accordingly.

(b) Design precondition and state update axioms for a new action called $Putdown(b)$ of putting down the contents of mail bag b at the current location. Show that the modified domain axiomatization, too, is deterministic.

(c) Solve Exercise 1.4 in the modified domain.

1.7. Design the environment of a software agent which transfers files to remote locations.

(a) Define domain constraints for the fluents $Request(addr, file, size)$, indicating that the file named *file* of size *size* should be moved to the remote location *addr*; and $TarFile(addr, size)$, indicating that an archive file of total size *size* exists, to be sent to the remote site *addr*.

(b) Define precondition axioms and state update axioms for the actions $Tar(addr, file)$, for adding *file* to the archive file for remote site *addr*; and $Ftp(addr)$, for transferring the archive file for *addr*. The latter action shall have the side effect of removing the file that has been sent. Assume that the size of the archive file is simply the sum of the sizes of the files it contains. Furthermore, assume that an archive file can be safely ftp'ed only if it has less than one megabyte $(1,048,576$ bytes).

(c) Consider the initial state

$$
\begin{aligned}
State(S_0) \;=\; & Request(\text{"ftp://photo.de"}, \text{"img1.jpg"}, 510000) \circ \\
& Request(\text{"ftp://photo.de"}, \text{"img2.jpg"}, 480000) \circ \\
& Request(\text{"ftp://photo.de"}, \text{"img3.jpg"}, 410000) \circ \\
& Request(\text{"ftp://photo.de"}, \text{"img4.jpg"}, 370000) \circ \\
& Request(\text{"ftp://photo.de"}, \text{"img5.jpg"}, 150000) \circ \\
& Request(\text{"ftp://photo.de"}, \text{"img6.jpg"}, 120000)
\end{aligned}
$$

Find a situation σ such that the domain axiomatization entails $State(\sigma) = \emptyset$, and prove this.

(d) Show that if an initial request concerns a file larger than 1 MB, then there is no ground situation $\sigma = Do([\alpha_1, \ldots, \alpha_n], S_0)$ such that $State(\sigma) = \emptyset$ is entailed.

Chapter 2

Special FLUX

FLUX, the FLUent eXecutor, is a programming language based on the ideas of fluent calculus. It allows to write concise and modular programs for agents which base their decisions on an explicit world model. In order to keep this model up to date during the execution of the program, every agent program requires an encoding of the underlying domain axiomatization, in particular the state update axioms. Due to the declarative nature of FLUX, domain axioms in fluent calculus translate directly into executable program statements.

FLUX agents are implemented on top of a kernel, which endows robotic agents with general reasoning capabilities for maintaining and using their internal model. As a **logic program**, the kernel can be easily verified against the semantics of fluent calculus. In a similar fashion, formal methods can be applied to the agent programs themselves in order to prove their correctness, which is of particular interest if agents handle secure information or are used for safety-critical control tasks.

In this chapter, we introduce the **special** FLUX language and system, which is suitable for programming agents with complete state information. As we will see, high-level agent control programs can be divided into three layers, which we introduce in a bottom-up manner; the last section deals with the interaction with other agents and in particular with users:

2.1. A domain-independent **kernel program** endows the agent with basic reasoning facilities.

2.2. On top of the kernel, the **domain encoding** provides the agent with the necessary knowledge of its actions and environment.

2.3. Finally, the **control part** defines the intended, goal-oriented behavior of the agent.

2.4. If the state of its environment is not under the complete control of an agent, then its program needs to be extended to allow for **exogenous actions**, such as the addition and cancellation of requests to the mailbot.

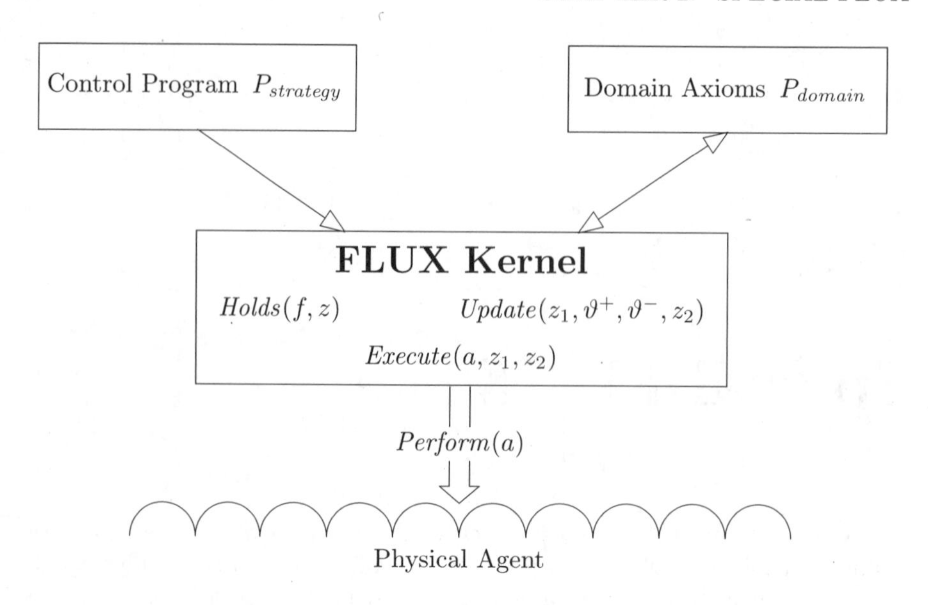

Figure 2.1: Architecture of agent programs written in special FLUX.

2.1 The Kernel

The general architecture of a FLUX agent is depicted in Figure 2.1. The program $P_{strategy}$ implements the behavior of the agent with the help of two basic commands defined in the kernel program: The expression $Holds(f, z)$ is used to condition the behavior of the agent on the current state of the environment. The predicate means exactly what it does in fluent calculus: It is true if and only if condition (i.e., fluent) f is true in the current world model (i.e., state) z. Secondly, the control program $P_{strategy}$ uses the statement $Execute(a, z_1, z_2)$ to trigger the agent to actually perform an action a in the outside world and, simultaneously, to update the current state z_1 to z_2 according to the changes effected by action a. The actual performance of an action needs to be programmed via $Perform(a)$.

The update computation itself relies on the domain specification P_{domain}. In particular, the effects of actions are defined by state update axioms. To facilitate these specifications, the FLUX kernel provides a definition of the auxiliary predicate $Update(z_1, \vartheta^+, \vartheta^-, z_2)$. Its intuitive meaning is that state z_2 is the result of positive and negative effects ϑ^+ and ϑ^-, respectively, in state z_1. In other words, the predicate encodes the update equation $z_2 = z_1 - \vartheta^- + \vartheta^+$.

Syntax and semantics of both agent programs in FLUX and the kernel derive from the paradigm of logic programming. We appeal to the standard completion semantics and negation-as-failure computation rule. Agent programs frequently utilize the "cut," whose operational semantics is given by a reduction of the computation tree. The programs in this book are all implemented using the logic programming language Eclipse.[1]

A **logic program** is a finite sequence of **clauses** of the form $p(\vec{t}) \leftarrow L_1, \ldots, L_n$ where $p(\vec{t})$ is an atom and $L_1, \ldots, L_n$ are **literals**, that is, atoms or negated atoms ($n \geq 0$). The semantics of a program is to regard all clauses with the same leading predicate p as the definition of p. Formally, let p be any predicate symbol of the underlying signature such that

$$\begin{array}{lcl} p(\vec{t}_1) & \leftarrow & L_{11}, \ldots, L_{1n_1} \\ \vdots & & \\ p(\vec{t}_m) & \leftarrow & L_{m1}, \ldots, L_{mn_m} \end{array}$$

are the clauses for p in a program P $(m \geq 0)$. Take a sequence $\vec{x}$ of pairwise different variables not occurring in any of the clauses, then the **definition** for p in P is given by the formula

$$p(\vec{x}) \equiv \bigvee_{i=1}^{m} (\exists \vec{y}_i)\, (\vec{x} = \vec{t}_i \wedge L_{i1} \wedge \ldots \wedge L_{in_i})$$

where the $\vec{y}_i$'s are the variables of the respective clause. The **completion** of a program P, written $COMP[P]$, is the definitions of all predicates in P along with unique-name axioms for all function symbols.

Special FLUX is suitable for the programming of agents which have complete information about the state of their environment, so that all computations concern ground states as in special fluent calculus. A ground state is encoded as a **list** of ground fluents such that no fluent occurs more than once. While states are (semantically) equal whenever they contain the same fluents, two lists are different if they contain the same fluents but in a different order. Hence, there is a one-to-many correspondence between states in fluent calculus and lists of fluents in FLUX.

Definition 2.1 Let $\langle \mathcal{F}, \circ, \emptyset \rangle$ be a fluent calculus state signature. A **ground FLUX state** is a list

$$[f_1, \ldots, f_n] \tag{2.1}$$

of pairwise different ground fluents $(n \geq 0)$. If τ is a ground state, then by $[\![\tau]\!]$ we denote any ground FLUX state containing the same fluents as τ. □

In the following we will distinguish between a fluent calculus state and its FLUX representation only when necessary. The initial state in our mailbot scenario

[1] The Eclipse© system was originally developed by the former European Computer-Industry Research Center (ECRC) in Munich, Germany, and is now maintained and distributed by the Center for Planning and Resource Control at Imperial College (IcParc) in London, U.K.

```
holds(F, [F|_]).
holds(F, Z) :- Z=[F1|Z1], F\==F1, holds(F, Z1).

holds(F, [F|Z], Z).
holds(F, Z, [F1|Zp]) :- Z=[F1|Z1], F\==F1, holds(F, Z1, Zp).

minus(Z, [], Z).
minus(Z, [F|Fs], Zp) :-
   (\+ holds(F, Z) -> Z1=Z ; holds(F, Z, Z1)),
   minus(Z1, Fs, Zp).

plus(Z, [], Z).
plus(Z, [F|Fs], Zp) :-
   (\+ holds(F, Z) -> Z1=[F|Z] ; holds(F, Z), Z1=Z),
   plus(Z1, Fs, Zp).

update(Z1, ThetaP, ThetaN, Z2) :-
   minus(Z1, ThetaN, Z), plus(Z, ThetaP, Z2).

execute(A, Z1, Z2) :- perform(A), state_update(Z1, A, Z2).
```

Figure 2.2: $P_{skernel}$: The kernel of special FLUX.

depicted in Figure 1.1, for example, may be encoded by the following clause, which defines an initial ground FLUX state:

```
init(Z0) :-
  Z0 = [ at(1),empty(1),empty(2),empty(3),                    (2.2)
         request(1,2),request(1,3),...,request(6,4)].
```

The completion semantics of logic programs entails that any two syntactically different ground terms are unequal. Hence, FLUX makes the implicit assumption of uniqueness of names for all fluent functions of a signature, so that the list just given is indeed free of multiple occurrences, as stipulated by Definition 2.1.

Based on the encoding of states as lists, the complete kernel of special FLUX is depicted in Figure 2.2. It consists of three parts: a definition for fluents to hold in states, a bunch of definitions culminating in a clause for updating a state by a list of positive and a list of negative effects, and a definition for executing an elementary action. Let this program be denoted by $P_{skernel}$. Its declarative semantics is given by the completion $COMP[P_{skernel}]$. In the following, we will discuss in detail each part of the kernel program.

2.1.1 Holds/2

The topmost two clauses in Figure 2.2 encode the assertion $Holds(f, z)$, using the standard decomposition $[Head \,|\, Tail]$ of a list into its first element and the rest list. The first clause says that if f occurs as the head of FLUX state z, then it holds in that state, and the second clause says that it holds in the state if it occurs in the tail of list z. The inequality condition $f \neq f_1$ in the second clause has been added for the sake of efficiency: Given that FLUX states contain no multiple occurrences, a fluent which is identical to the head cannot reappear in the tail.

Logic programs are executed by **queries**. A query A, Q starting with an atom can be **resolved** with a program clause $A' \leftarrow Q'$ iff there exists a most general variable substitution θ such that $A\theta = A'\theta$. As the **resolvent** one obtains the query $(Q', Q)\theta$.
A **computation tree** for a program and query $P \cup \{\leftarrow Q_0\}$ is obtained as the limit of a sequence of trees, starting with query Q_0.

1. If the current query Q starts with an atom A, then Q has a finite number of resolvents as ordered children, one for each program clause whose head unifies with the atom A, and the new current query is the leftmost child. Q **fails** if no clause is applicable in this way, and the new current query is the next unsolved query in the tree.

2. If Q is of the form $\neg Q_1, Q'$, then the negated first sub-goal is treated by the principle of **negation-as-failure**: Q is linked to a subsidiary tree consisting of just the root Q_1, which becomes the new current query. If later on all branches of the subsidiary tree are failure nodes, then Q itself has Q' as the only child. Otherwise, if the subsidiary tree contains a finite branch ending with the empty query, then Q itself fails.

The process of extending a tree may go on forever, in which case the limit is an infinite computation tree. A **successful** derivation in the main tree ends with the empty query. The **answer** computed by a successful derivation consists of a substitution for the variables in the query which is obtained by combining all substitutions used for resolving the queries along the branch.

As an example, recall the clause for the initial state in the mail delivery scenario given earlier, (2.2), and consider the following query, which determines all initial requests issued from the office where the robot is currently located:

```
?- init(Z), holds(at(R),Z), holds(request(R,R1),Z).

R  = 1
R1 = 2
Z  = [at(1),empty(1),...]  More?

R  = 1
R1 = 3
```

```
Z  = [at(1),empty(1),...]  More?

R  = 1
R1 = 5
Z  = [at(1),empty(1),...]  More?

no (more) solution.
```

Notice that the inequality in the second clause for *Holds* of the kernel does not apply when resolving the query $Holds(Request(1, r_1), [At(1), \ldots])$, since the term $Request(1, r_1)$, by virtue of containing a variable, is not identical to any element of the ground FLUX state. The computed answers are obviously correct under the semantics of fluent calculus, as Σ_{sstate} along with $UNA[At, \ldots, Request]$ entails

$$Holds(At(r), Z) \wedge Holds(Request(r, r_1), Z) \equiv \\ r = 1 \wedge r_1 = 2 \vee r = 1 \wedge r_1 = 3 \vee r = 1 \wedge r_1 = 5$$

if Z is the term (1.1) at the beginning of Chapter 1 (page 5).

Using negation (which is written "\+" in Prolog) the next sample query computes the nearest destination of a package waiting at the current location of the robot:

```
?- init(Z), holds(at(R),Z), holds(request(R,R1),Z),
   \+ ( holds(request(R,R2),Z), abs(R-R2) < abs(R-R1) ).

R  = 1
R1 = 2
Z  = [at(1),empty(1),...]  More?

no (more) solution.
```

Put in words, r_1 is the nearest requested destination if there is no requested destination r_2 strictly closer to the current location, r. Again, it is easy to see that the resulting answer is supported by fluent calculus.

The following theorem shows that the correspondence is not coincidental between the computed answers in the two examples and the underlying fluent calculus semantics. The clauses of the kernel which define $Holds(f, z)$ are provably correct in general.

Theorem 2.2 *Consider a fluent calculus state signature with fluent functions $\mathcal{F}$, and let φ be a fluent and τ a ground state, then for any $[\![\tau]\!]$,*

$$COMP[P_{skernel}] \models Holds(\varphi, [\![\tau]\!]) \text{ iff } \Sigma_{sstate} \cup UNA[\mathcal{F}] \models Holds(\varphi, \tau)$$

and

$$COMP[P_{skernel}] \models \neg Holds(\varphi, [\![\tau]\!]) \text{ iff } \Sigma_{sstate} \cup UNA[\mathcal{F}] \models \neg Holds(\varphi, \tau)$$

Proof: The completion $COMP[P_{skernel}]$ entails the definition

$$Holds(f, z) \equiv (\exists z')\, z = [f \mid z'] \\ \vee \\ (\exists f_1, z_1)\,(z = [f_1 \mid z_1] \wedge f \neq f_1 \wedge Holds(f, z_1)) \tag{2.3}$$

Let $[\![\tau]\!] = [f_1, f_2, \ldots, f_n]$ $(n \geq 0)$, then $\Sigma_{sstate} \models \tau = f_1 \circ f_2 \circ \ldots \circ f_n$ according to the foundational axioms of associativity and commutativity, and $\neg Holds(f_1, f_2 \circ \ldots \circ f_n)$ according to Definition 2.1 and Proposition 1.5. Decomposition and irreducibility imply that $\Sigma_{sstate} \cup UNA[\mathcal{F}]$ entails

$$Holds(\varphi, \tau) \equiv \varphi = f_1 \vee \varphi \neq f_1 \wedge Holds(\varphi, f_2 \circ \ldots \circ f_n)$$

which, along with the instance $\{f/\varphi, z/[f_1, f_2, \ldots, f_n]\}$ of (2.3), proves the claim. ■

2.1.2 Update/4

Predicate $Update(z_1, \vartheta^+, \vartheta^-, z_2)$ encodes the equation $z_2 = z_1 - \vartheta^- + \vartheta^+$. Its definition is based on the predicates $Minus(z, \vartheta^-, z')$ and $Plus(z, \vartheta^+, z')$, which encode, respectively, the macros $z' = z - \vartheta^-$ and $z' = z + \vartheta^+$. The former in turn uses the ternary $Holds(f, z, z')$, whose intended semantics is $Holds(f, z) \wedge z' = z - f$. The definition of *Minus* follows closely the corresponding macro expansion as given in Chapter 1. The encoding of *Plus* uses a case distinction, too, making sure that no multiple occurrences are introduced when adding a fluent to a list.

Clauses of logic programs, and hence resolvents, may include the **cut**, written "!". The unique resolvent of a query Q of the form $!, Q'$ is Q', but the side effect of this resolution step is that in the computation tree all branches to the right of the one leading to Q are cut off. We will frequently use the standard syntax $Q_1 \rightarrow Q_2$ in clause bodies, which stands for $Q_1, !, Q_2$.
Syntactic sugar allows for nested negation by defining $\leftarrow Q_1, \neg(Q), Q_2$ as $\leftarrow Q_1, \neg\diamond, Q_2$ along with the clause $\diamond \leftarrow Q$ where "$\diamond$" is a predicate symbol not occurring elsewhere. We will also use **disjunction** in clauses, written $p(\vec{t}) \leftarrow Q, (Q_1; Q_2), Q'$, which is defined as $p(\vec{t}) \leftarrow Q, \diamond$ along with the two clauses $\diamond \leftarrow Q_1, Q'$ and $\diamond \leftarrow Q_2, Q'$.

As an example, consider the update of the initial state in our mail delivery scenario by the positive effects $At(2) \circ Empty(1)$ (the latter of which happens to be true) and the negative effects $At(1) \circ Request(6, 5)$ (the latter of which happens to be false):

```
?- init(Z0), update(Z0,[at(2),empty(1)],
                       [at(1),request(6,5)],Z1).
```

```
Z0 = [at(1),empty(1),empty(2),empty(3),
      request(1,2),request(1,3),...,request(6,4)]
Z1 = [at(2),empty(1),empty(2),empty(3),
      request(1,2),request(1,3),...,request(6,4)]
```

It is easy to verify that the second FLUX state is indeed a list $[\![\tau_1]\!]$ such that $\tau_1 = \tau_0 - At(1) \circ Request(2,1) + At(2) \circ Empty(1)$, where $[\![\tau_0]\!]$ is the initial FLUX state of clause (2.2).

Correctness of the FLUX definition of update is established by a series of results concerning the auxiliary predicates, showing that each is defined in $P_{skernel}$ according to the intended meaning. To begin with, the following lemma shows that the ternary $Holds(f, z, z')$ is true just in case f holds in z and z' is z without f.

Lemma 2.3 *Consider a fluent calculus state signature with fluent functions $\mathcal{F}$, and let φ be a fluent and τ, τ' be ground states, then for any $[\![\tau]\!]$ there exists $[\![\tau']\!]$ such that*

$$COMP[P_{skernel}] \models Holds(\varphi, [\![\tau]\!], [\![\tau']\!]) \quad \textit{iff} \quad \Sigma_{sstate} \cup UNA[\mathcal{F}] \models Holds(\varphi, \tau) \wedge \tau' = \tau - \varphi$$

Proof: The completion $COMP[P_{skernel}]$ entails the definition

$$\begin{array}{rl} Holds(f,z,z') \equiv & z = [f \mid z'] \\ & \vee \\ & (\exists f_1, z_1, z'')\,(\, z' = [f_1 \mid z''] \wedge z = [f_1 \mid z_1] \wedge \\ & \qquad f \neq f_1 \wedge Holds(f, z_1, z'')) \end{array} \tag{2.4}$$

Let $[\![\tau]\!] = [f_1, f_2, \ldots, f_n]$ $(n \geq 0)$, then $UNA[\mathcal{F}] \models f_i \neq f_j$ for any $i \neq j$, and $\Sigma_{sstate} \models \tau = f_1 \circ f_2 \circ \ldots \circ f_n$ according to the foundational axioms of associativity and commutativity. Decomposition, irreducibility, and Proposition 1.6 imply that $\Sigma_{sstate} \cup UNA[\mathcal{F}]$ entails

$$\begin{array}{rl} Holds(\varphi, \tau) \wedge \tau' = \tau - \varphi \equiv & \varphi = f_1 \wedge \tau' = f_2 \circ \ldots \circ f_n \\ & \vee \\ & (\exists z'')\,(\,\tau' = f_1 \circ z'' \wedge \varphi \neq f_1 \wedge \\ & \qquad Holds(\varphi, f_2 \circ \ldots \circ f_n) \wedge \\ & \qquad z'' = f_2 \circ \ldots \circ f_n - \varphi) \end{array}$$

which, along with the instance $\{f/\varphi, z/[f_1, f_2, \ldots, f_n], z'/\tau'\}$ of (2.4), proves the claim. ■

The conditions of this lemma should be carefully observed: Among all FLUX states $[\![\tau']\!]$ which encode τ' only one satisfies $Holds(\varphi, [\![\tau]\!], [\![\tau']\!])$. This is due to the fact that the fluents in the list $[\![\tau']\!]$ occur in the very same order as in the list $[\![\tau]\!]$. Thus, for example,

```
?- holds(f2,[f1,f2,f3],[f1,f3]).

yes.

?- holds(f2,[f1,f2,f3],[f3,f1]).

no (more) solution.
```

This restriction propagates to the definition of predicate *Minus*, which is verified next.

Lemma 2.4 *Consider a fluent calculus state signature with fluent functions* $\mathcal{F}$, *and let* τ, ϑ^-, τ' *be ground states, then for any* $[\![\tau]\!]$ *and* $[\![\vartheta^-]\!]$ *there exists* $[\![\tau']\!]$ *such that*

$$COMP[P_{skernel}] \models Minus([\![\tau]\!], [\![\vartheta^-]\!], [\![\tau']\!]) \\ \text{iff} \\ \Sigma_{sstate} \cup UNA[\mathcal{F}] \models \tau' = \tau - \vartheta^-$$

Proof: The completion $COMP[P_{skernel}]$ entails the definition

$$\begin{aligned} & Minus(z, z^-, z') \equiv \\ & \quad z^- = [\,] \wedge z' = z \\ & \quad \vee \\ & \quad (\exists f, z_1^-, z_1)\,(z^- = [f \mid z_1^-] \wedge \\ & \qquad\qquad (\neg Holds(f, z) \wedge z_1 = z \vee Holds(f, z, z_1)\,) \wedge \\ & \qquad\qquad Minus(z_1, z_1^-, z')\,) \end{aligned} \tag{2.5}$$

Let $[\![\vartheta^-]\!] = [f_1, \ldots, f_n]$, then $\Sigma_{sstate} \models \vartheta^- = f_1 \circ f_2 \circ \ldots \circ f_n$ according to the foundational axioms of associativity and commutativity. The proof is by induction on n.

If $n = 0$, then $\vartheta^- = \emptyset$ and $[\![\vartheta^-]\!] = [\,]$. The instance $\{z/[\![\tau]\!], z^-/[\,]\}$ of (2.5) implies

$$Minus([\![\tau]\!], [\,], z') \equiv z' = [\![\tau]\!]$$

which proves the claim since $z - \emptyset = z'$ is equivalent to $z = z'$ by definition.

In case $n > 0$, the instance $\{z/[\![\tau]\!], z^-/[\![\vartheta^-]\!]\}$ of (2.5) implies

$$\begin{aligned} & Minus([\![\tau]\!], [f_1, f_2, \ldots, f_n], z') \equiv \\ & \quad (\exists z_1)\,(\,(\neg Holds(f_1, [\![\tau]\!]) \wedge z_1 = [\![\tau]\!] \vee Holds(f_1, [\![\tau]\!], z_1)\,) \wedge \\ & \qquad Minus(z_1, [f_2, \ldots, f_n], z')\,) \end{aligned}$$

which proves the claim according to Theorem 2.2, Lemma 2.3, the macro definition for "$-$", and the induction hypothesis. ■

Again, the order in which the fluents occur in the resulting FLUX state is determined by their order in the original list. For example,

```
?- minus([f1,f2,f3],[f2,f4],[f1,f3]).

yes.

?- minus([f1,f2,f3],[f2,f4],[f3,f1]).

no (more) solution.
```

In a similar fashion, the clauses for *Plus* are correct, as the following lemma shows.

Lemma 2.5 *Consider a fluent calculus state signature with fluent functions $\mathcal{F}$, and let τ, ϑ^+, τ' be ground states, then for any $[\![\tau]\!]$ and $[\![\vartheta^+]\!]$ there exists $[\![\tau']\!]$ such that*

$$COMP[P_{skernel}] \models Plus([\![\tau]\!], [\![\vartheta^+]\!], [\![\tau']\!])$$
$$\textit{iff}$$
$$\Sigma_{sstate} \cup UNA[\mathcal{F}] \models \tau' = \tau \circ \vartheta^+$$

Proof: Exercise 2.2 ■

Obviously, the order in which the fluents occur in the resulting state is determined by their order in both the original list and the list of fluents to be added. For example,

```
?- plus([f1,f2,f3],[g1,g2],Z).

Z = [g2,g1,f1,f2,f3]

no (more) solution.
```

The intermediate results culminate in the following theorem, which shows that the encoding of update is correct. Naturally, the definition inherits the restriction which applies to the auxiliary predicates so that just one particular representative of the updated state is inferred. That is to say, the definition is correct for computing the result of a state update rather than for verifying that two states satisfy an update equation. This is in accordance with the intended use of $Update(z_1, \vartheta^+, \vartheta^-, z_2)$ in agent programs, namely, to call it with the first three arguments instantiated and the last argument left variable. (See Exercise 2.3 on a more general definition of *Update*, which allows to infer all representatives of the resulting state.)

Theorem 2.6 *Consider a fluent calculus state signature with fluent functions $\mathcal{F}$, and let τ_1, ϑ^+, ϑ^-, and τ_2 be ground states, then for any $[\![\tau_1]\!]$, $[\![\vartheta^+]\!]$, and $[\![\vartheta^-]\!]$, there exists $[\![\tau_2]\!]$ such that*

$$COMP[P_{skernel}] \models Update([\![\tau_1]\!], [\![\vartheta^+]\!], [\![\vartheta^-]\!], [\![\tau_2]\!])$$
$$\textit{iff}$$
$$\Sigma_{sstate} \cup UNA[\mathcal{F}] \models \tau_2 = \tau_1 - \vartheta^- + \vartheta^+$$

Proof: The completion $COMP[P_{skernel}]$ entails the definition

$$Update(z_1, z^-, z^+, z_2) \equiv (\exists z)\,(Minus(z_1, z^-, z) \wedge Plus(z, z^+, z_2))$$

The claim follows from Lemma 2.4 and 2.5 along with the macro definition for update equations. ■

2.1.3 Execute/3

The predicate $Execute(a, z_1, z_2)$ is defined so as to trigger the actual performance of elementary action a and, simultaneously, to update the current state z_1 to state z_2. The kernel program presupposes the definition of a predicate $Perform(a)$ to cause the physical agent to carry out action a in the environment. Furthermore, the programmer needs to define the predicate $StateUpdate(z_1, a, z_2)$ in such a way that it encodes the state update axioms for each action the agent may execute.

2.2 Specifying a Domain

The domain specification is an essential component of every FLUX agent program. In order to decide which of its actions are possible at a certain stage, the agent must know about the preconditions of actions. In order to set up and maintain a model of the environment, the agent needs to be told the initial conditions and the effects of every action. A domain specification in FLUX is obtained by a direct translation of a domain axiomatization in fluent calculus into logic programming clauses.

Precondition Axioms

An agent program should trigger the actual performance of an action only if the action is possible. A precondition axiom $Poss(A(\vec{x}), z) \equiv \Pi(z)$ translates into the logic programming clause

$$Poss(A(\vec{x}), z) \;\leftarrow\; \widehat{\Pi}(z).$$

where $\widehat{\Pi}$ is a re-formulation of Π using the syntax of logic programs. Provided that there is no other program clause defining $Poss$ with first argument $A(\vec{x})$, the "only-if" direction of the precondition axiom is implicit in the completion semantics. The encoding of the precondition axioms (1.10) (page 16) for the mail delivery domain, for example, are included in Figure 2.3, which depicts the full background theory for mailbot agents.

State Update Axioms

For the execution of an action, the FLUX kernel assumes that the predicate $StateUpdate(z_1, A(\vec{x}), z_2)$ has been defined in such a way as to reflect the update

```
poss(pickup(B,R),Z) :- holds(empty(B), Z),
                       holds(at(R1), Z),
                       holds(request(R1,R), Z).

poss(deliver(B), Z) :- holds(at(R), Z),
                       holds(carries(B,R), Z).

poss(go(D), Z) :- holds(at(R), Z),
                  ( D=up, R<6 ; D=down, R>1 ).

state_update(Z1, pickup(B,R), Z2) :-
   holds(at(R1), Z1),
   update(Z1, [carries(B,R)], [empty(B),request(R1,R)], Z2).

state_update(Z1, deliver(B), Z2) :-
   holds(at(R), Z1),
   update(Z1, [empty(B)], [carries(B,R)], Z2).

state_update(Z1, go(D), Z2) :-
   holds(at(R), Z1),
   ( D = up -> R1 is R + 1
            ; R1 is R - 1 ),
   update(Z1, [at(R1)], [at(R)], Z2).
```

Figure 2.3: Background theory for the mailbot program.

---

of state z_1 to z_2 according to the effects of action $A(\vec{x})$. Appealing to the kernel predicates *Holds* and *Update*, the encoding of a state update axiom of the form (1.11) (page 16) is straightforward using the following scheme:

$$\begin{aligned} \mathit{StateUpdate}(z_1, A(\vec{x}), z_2) \leftarrow\ & \widehat{\Delta}_1(z_1),\ \mathit{Update}(z_1, \vartheta_1^+, \vartheta_1^-, z_2)\ ; \\ & \ldots\ ; \\ & \widehat{\Delta}_n(z_1),\ \mathit{Update}(z_1, \vartheta_n^+, \vartheta_n^-, z_2). \end{aligned} \tag{2.6}$$

where the $\widehat{\Delta}_i$'s are logic programming encodings of the conditions Δ_i for the respective effects $\vartheta_i^+, \vartheta_i^-$. For the sake of efficiency, action preconditions are not verified when inferring the state update, since actions are assumed to be performed only in case the current state entails that they are possible. The actual verification of precondition is left to the control program for an agent. The second part of Figure 2.3 shows the encoding of the state update axioms (1.12) (page 17) for the mailbot domain.

Initial State

Agents initialize their internal world model by a list representing the initial state. A standard way of encoding a domain axiom $\Sigma_{init} = \{\mathit{State}(S_0) = \tau\}$ is by the clause

$$Init(z_0) \leftarrow z_0 = [\![\tau]\!].$$

for some FLUX state $[\![\tau]\!]$ representing τ. An example is clause (2.2) in Section 2.1, which encodes the initial initial situation of the mailbot as depicted in Figure 1.1.

Let P_{mail} be the FLUX domain specification of Figure 2.3, and consider the program $P_{mail} \cup \{(2.2)\} \cup P_{skernel}$. The following sample query infers the state in situation $Do(Deliver(1), Do(Go(Up), Do(Pickup(1,2), S_0)))$, making sure that each action in the sequence is indeed possible at the time of performance:

```
?- init(Z0),
   poss(pickup(1,2),Z0), state_update(Z0,pickup(1,2),Z1),
   poss(go(up),Z1),      state_update(Z1,go(up),Z2),
   poss(deliver(1),Z2),  state_update(Z2,deliver(1),Z3).

Z0 = [at(1),empty(1),empty(2),empty(3),
      request(1,2),request(1,3),...,request(6,4)]
Z1 = [carries(1,2),at(1),empty(2),empty(3),
      request(1,3),request(1,5),...,request(6,4)]
Z2 = [at(2),carries(1,2),empty(2),empty(3),
      request(1,3),request(1,5),...,request(6,4)]
Z3 = [empty(1),at(2),empty(2),empty(3),
      request(1,3),request(1,5),...,request(6,4)]
```

The reader may compare the result with the corresponding derivation in fluent calculus (page 17).

Domain Constraints

A set of domain constraints Σ_{dc} can be encoded in FLUX by a clause defining the predicate $Consistent(z)$ so as to be true if state z is consistent wrt. the constraints. Let $\Sigma_{dc} = \{(\forall s)\, \Delta_1(s), \ldots, (\forall s)\Delta_n(s)\}$, then the FLUX translation is

$$Consistent(z) \leftarrow \widehat{\Delta}_1(z),\ \ldots,\ \widehat{\Delta}_n(z).$$

where the $\widehat{\Delta}_i$'s are logic programming encodings of the Δ_i's. Domain constraints (1.9) (page 15) of the mail delivery domain, for example, can be verified by the following clause, assuming a hallway with $n = 6$ offices and a robot with $k = 3$ mail bags:

```
consistent(Z) :-
   holds(at(R),Z), 1=<R, R=<6,
   \+ ( holds(at(R1),Z), holds(at(R2),Z), R1\=R2 ),
   \+ ( holds(empty(B),Z), holds(carries(B,_),Z) ),
   \+ ( holds(empty(B),Z), (B<1 ; B>3) ),
   \+ ( holds(carries(B,R),Z), (B<1 ; B>3 ; R<1 ; R>6) ),
   \+ ( holds(carries(B,R1),Z),
```

```
        holds(carries(B,R2),Z), R1=\=R2 ),
   \+ ( holds(request(R1,R2),Z), (R1<1 ; R1>6 ;
                                   R2<1 ; R2>6 ; R1=R2) ).
```

Our sample initial state (2.2) can then be proved consistent:

```
?- init(Z0), consistent(Z0).

Z0 = [at(1),empty(1),empty(2),empty(3),request(1,2),...]
```

For the sake of efficiency, domain constraints need not be included in an agent program if they are enforced by other means. In particular, the FLUX specification of a domain which satisfies the assumptions of Theorem 1.17 does not require the consistency check after each action.

Auxiliary Axioms

Using standard logic programming, the universal assumption of uniqueness-of-names is built into FLUX so that these axioms need not be encoded separately. Other auxiliary axioms need to be encoded only if they are required for the precondition or state update axioms.

2.3 Control Programs

Control programs $P_{strategy}$ guide robotic agents over time. A typical program starts with the initialization of the internal model via the predicate $Init(z)$. Executability of actions and other properties of the current state are determined using the kernel predicate $Holds(f, z)$, and the actual execution of an action is triggered via the kernel predicate $Execute(a, z_1, z_2)$.

Example 1 (continued) The control program depicted in Figure 2.4 implements the simple strategy for our mailbot which we have proposed at the very beginning, cf. Figure 1.2 (page 3). After initializing the internal world model, the program enters a main loop by which possible actions are systematically selected and executed. The order in which the actions are selected implies that the mailbot delivers packages as long as possible, then collects new packages as long as possible, followed by going up or down the hallway towards some office where it can pick up mail or for which it has mail. If the robot can neither deliver nor collect a package, and if there is no office to be visited, then the program terminates since all requests have been carried out. The reader may notice how the "→" operator, i.e., the Prolog cut, is employed to avoid backtracking once an action has been executed. This is necessary since actions which have actually been performed in the outside world cannot be "undone."

Our simple mailbot agent does not care as to which concrete instance of a $Deliver(b)$ or $Pickup(b, r)$ action gets selected. Choosing whether to move up or down the hallway is also arbitrary if both directions are possible. Running the agent program with the initial state of Figure 1.1 (page 2), encoded as

```
main :- init(Z), main_loop(Z).

main_loop(Z) :-
   poss(deliver(B), Z)       -> execute(deliver(B), Z, Z1),
                                main_loop(Z1) ;
   poss(pickup(B,R), Z)      -> execute(pickup(B,R), Z, Z1),
                                main_loop(Z1) ;
   continue_delivery(D, Z) -> execute(go(D), Z, Z1),
                                main_loop(Z1) ;
   true.

continue_delivery(D, Z) :-
   ( holds(empty(_), Z), holds(request(R1,_), Z)
     ;
     holds(carries(_,R1), Z) ),
   holds(at(R), Z),
   ( R < R1 -> D = up
            ; D = down ).
```

Figure 2.4: A simple mailbot agent in FLUX.

clause (2.2), the robot starts with collecting at office 1 the package for room 2 and ends after exactly 80 actions with delivering to room 3 the package sent from room 6.

To give an impression of the general performance of the program, the following tables show the lengths of the generated action sequence for different scenarios of increasing size. The results are for **maximal** mail delivery problems, that is, where requests are issued from each office to each other one. Hence, a scenario with n offices consists of $(n-1) \cdot (n-1)$ initial requests. The two tables contrast the solution lengths for a mailbot with $k = 3$ bags to one which can carry $k = 8$ packages at a time.

$k = 3$

n	# act	n	# act
11	640	21	3880
12	814	22	4424
13	1016	23	5016
14	1248	24	5658
15	1512	25	6352
16	1810	26	7100
17	2144	27	7904
18	2516	28	8766
19	2928	29	9688
20	3382	30	10672

$k = 8$

n	# act	n	# act
11	580	21	3720
12	744	22	4254
13	936	23	4836
14	1158	24	5468
15	1412	25	6152
16	1700	26	6890
17	2024	27	7684
18	2386	28	8536
19	2788	29	9448
20	3232	30	10422

Figure 2.5 gives a detailed analysis of the runtime behavior of the program.

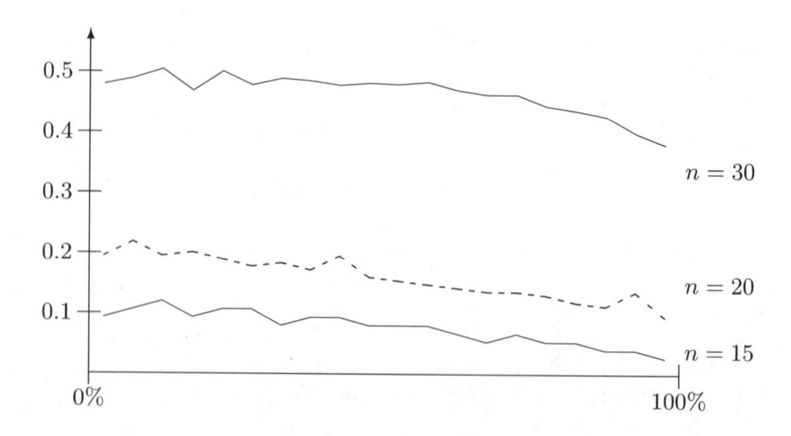

Figure 2.5: The computational behavior of the simple program for the mail delivery robot in the course of its execution. The horizontal axis depicts the degree to which the run is completed while the vertical scale is in seconds per 100 actions. The graphs are for maximal mail delivery problems with $k = 3$ mail bags.

Depicted are, for three selected problem sizes, the average time for action selection and state update computation in the course of the execution of the program.[2] The curves show that the computational effort remains essentially constant throughout. A slight general descent is discernible, which is due to the fact that the state size decreases in the course of time as fewer and fewer requests remain. □

Having implemented the simplest possible strategy for the mailbot, there are many ways of improving the performance. First, however, we will use our program to illustrate how one can establish important formal properties of FLUX agents, such as whether the generated sequence of actions is always possible.

The results of logic programs are determined by their computation tree. Putting aside the question of computational efficiency, the value of an agent program is judged by its generated sequence of actions. Most of the interesting properties of agent programs are therefore concerned with how the **situation** of the robot unfolds. From this perspective, the crucial nodes in a computation tree are those by which actions are executed, that is, queries where the atom *Execute* is resolved. In what follows, by agent program we mean a program including the background theory and the kernel $P_{skernel}$.

Definition 2.7 An **execution node** in a computation tree is a query starting with atom $Execute(\alpha, \tau_1, \tau_2)$.

Consider an agent program P and query Q_0, and let T be the computation tree for $P \cup \{\leftarrow Q_0\}$. For each node Q in T, the situation **associated**

[2] All runtimes in this book were measured on a PC with a 500 MHz processor.

with Q is $Do([\alpha_1, \ldots, \alpha_n], S_0)$ where $Execute(\alpha_1, \_, \_), \ldots, Execute(\alpha_n, \_, \_)$ is the ordered sequence of execution nodes to the left and above Q in T $(n \geq 0)$. □

It is worth stressing that associated situations are determined by all execution nodes that have been resolved, including—if any—those that lie on failed branches. This is because executed actions have actually been performed once and for all in the outside world, no matter whether backtracking occurs afterwards in the agent program.

Figure 2.6 depicts an excerpt of the computation tree for the mailbot program if applied to initial state (2.2). Starting with the query *Main*, the first branching occurs when resolving the query *MainLoop*, resulting in four alternative continuations. The leftmost branch eventually fails as no instance of $Carries(b, 1)$ holds in the initial state and hence no delivery action is possible. The second branch ramifies because multiple instances of $Empty(b)$ hold in the initial state, and later since multiple instances of $Request(1, r)$ hold. However, all open alternatives are discarded when resolving the cut prior to arriving at the first execution node. In Figure 2.6, this derivation step is indicated by an arrow pointing downwards. In the depicted excerpt of the tree, situation S_0 is associated with all but the bottommost node, which is associated with $Do(Pickup(1, 2), S_0)$.

A property which agent programs should always exhibit is to execute actions only if they are possible according to the underlying domain theory. This includes the requirement that the first argument of $Execute(a, z_1, z_2)$ is fully instantiated whenever this atom is resolved, for otherwise the underlying execution model of the agent could not actually perform the (partially specified) action. The following notion of soundness captures these intuitions.

Definition 2.8 Agent program P and query Q are **sound** wrt. a domain axiomatization Σ if for every execution node with atom $Execute(\alpha, \_, \_)$ in the computation tree for $P \cup \{\leftarrow Q\}$, α is ground and $\Sigma \models Poss(\alpha, \sigma)$, where σ is the situation associated with the node. □

The following theorem provides a strategy for writing sound programs. One of the conditions is that the underlying domain satisfies the assumptions of Theorem 1.17 (page 19), and hence is deterministic. Further aspects are, informally speaking, to

1. use *Execute* solely in the main derivation, that is, not in a derivation for a negated sub-goal;

2. make sure that no backtracking occurs over execution nodes;

3. systematically link the execution nodes thus:

$$Init(z_0),\ Execute(\alpha_1, z_0, z_1),\ Execute(\alpha_2, z_1, z_2), \ldots$$

making sure that at the time the execution node is resolved, every α_i is ground and $Poss(\alpha_i, z_{i-1})$ holds.

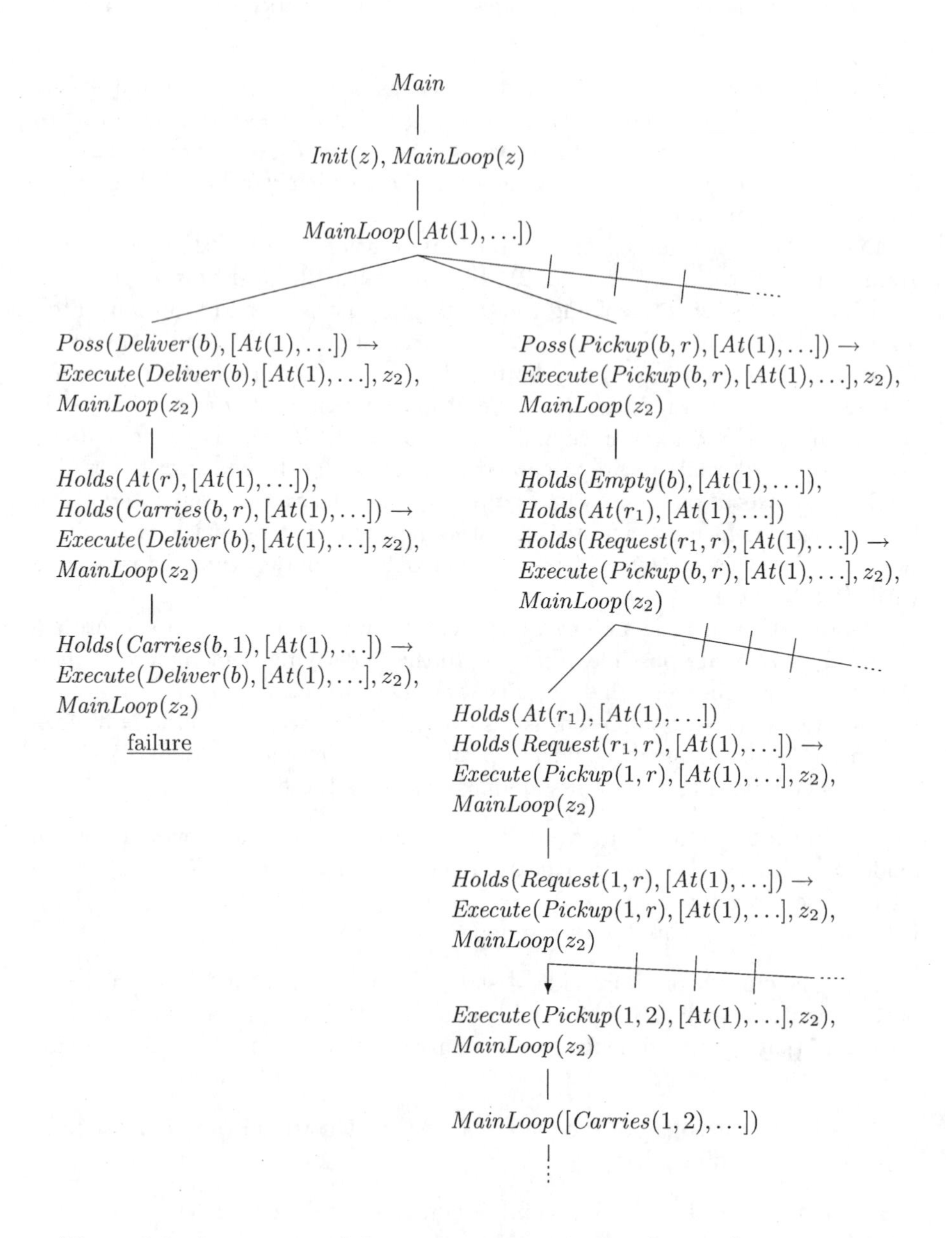

Figure 2.6: An excerpt of the computation tree for the mailbot program.

Formally, we say that an execution node Q_2 starting with $Execute(a_2, z_2, \_)$ is **linked** to another execution node Q_1 starting with $Execute(a_1, \_, z_1)$ if z_2 is $z_1\theta$ where θ is the substitution used when resolving Q_1.

Theorem 2.9 *Let Σ be a domain axiomatization satisfying the assumptions of Theorem 1.17. Let P be an agent program and Q_0 a query such that every execution node Q in the computation tree for $P \cup \{\leftarrow Q_0\}$ satisfies the following: Let $Execute(a, z_1, z_2)$ be the leading atom of Q, σ the associated situation, and τ_1 a state such that $[\![\tau_1]\!] = z_1$, then*

1. *Q is in the main tree and there is no branch to the right of Q;*

2. *a and z_1 are ground while z_2 is a variable;*

3. *if $\sigma = S_0$, then $\Sigma \models State(S_0) = \tau_1$, else Q is linked to its preceding execution node;*

4. *$\Sigma \models Poss(a, \tau_1)$.*

Then P and Q_0 are sound wrt. Σ.

Proof: Consider an execution node Q starting with $Execute(a, z_1, z_2)$ and whose associated situation is $\sigma = Do([\alpha_1, \ldots, \alpha_n], S_0)$. By condition 1, all execution nodes lie on one branch; hence, unless $\sigma = S_0$, Q is linked to an execution node whose associated situation is $\sigma_{n-1} = Do([\alpha_1, \ldots, \alpha_{n-1}], S_0)$. By induction on n, we prove that $\Sigma \models State(\sigma) = \tau_1$, which implies the claim according to condition 4. The base case $n = 0$ follows by condition 3. In case $n > 0$, the induction hypothesis along with condition 4 imply that $\Sigma \models Poss(\alpha_n, \sigma_{n-1})$. Since Σ is deterministic according to Theorem 1.17, there is a state τ such that $\Sigma \models State(Do(\alpha_n, \sigma_{n-1})) = \tau$. From the encoding of the state update axioms, (2.6), along with Theorem 2.6 and the fact that Q is linked to the execution node with associated situation σ_{n-1} it follows that $\Sigma \models \tau = \tau_1$. ■

Example 1 (continued) With the help of the theorem, our agent program for the mailbot can be proved sound wrt. the domain axiomatization of Chapter 1 for any initial state which satisfies the domain constraints.

1. There is no negation in the program; hence, all execution nodes are in the main tree. Moreover, since every occurrence of $Execute(A(\vec{x}), z, z_1)$ is immediately preceded by "$\rightarrow$", there are no branches to the right of any execution node.

2. Resolving $Init(z)$ grounds z. Given that z is ground, resolving any of $Poss(Deliver(b), z)$, $Poss(Pickup(b, r), z)$, and $ContinueDelivery(d, z)$ grounds the respective first argument. Hence, the first two arguments in every execution node are ground. On the other hand, the third argument in every occurrence of $Execute(a, z_1, z_2)$ remains variable until the atom is resolved.

3. It is assumed that resolving $Init(z)$ instantiates the argument with a FLUX state representing the initial state. This state is used when first executing an action. The construction of the clause for $MainLoop$ implies that all further execution nodes are linked to their predecessors.

4. Actions $Deliver(b)$ and $Pickup(b, r)$ are verified to be possible prior to executing them. According to precondition axiom (1.10) on page 16, action $Go(Up)$ is possible in z if $Holds(At(r), z)$ for some $r < n$. This is indeed the case whenever there is a request from or mail to some $r_1 > r$, given that the initial state satisfies the domain constraints (1.9) on page 15. Likewise, action $Go(Down)$ is possible in z if $Holds(At(r), z)$ for $r > 1$, which is true whenever there is a request from or mail to some $r_1 < r$. □

It is worth mentioning that all which has been said applies unconditionally to programs that continue controlling agents without ever terminating. The property of a program to be sound is independent of whether or not it contains a finite branch ending with success. Hence, infinite computation trees which satisfy the conditions of Theorem 2.9 are guaranteed to generate executable, albeit infinite, sequences of actions.

Our program for the mailbot, however, can be proved to terminate for any consistent initial state. Moreover, we can show that upon termination all requested packages will have been picked up and delivered.

Proposition 2.10 *Let Σ_{mail} be the axiomatization of the mailbot domain, including an initial state axiom $State(S_0) = \tau$ such that τ is consistent wrt. the domain constraints in Σ_{mail}. Then the computation tree for the program $P_{mail} \cup \{Init(z) \leftarrow z = [\![\tau]\!]\} \cup P_{skernel}$ and the query $\{\leftarrow Main\}$ contains a single successful branch. Moreover, if σ is the situation associated with the success node, then*

$$\begin{aligned}\Sigma_{mail} \models\ & \neg(\exists r_1, r_2)\, Holds(Request(r_1, r_2), \sigma) \\ & \wedge \neg(\exists b, r)\, Holds(Carries(b, r), \sigma)\end{aligned}$$

Proof: We already know that all actions are possible in the situation in which they are executed. By virtue of being finite, the initial state contains a finite number of instances of $Request$ and a finite number of instances of $Carries$. Each $Pickup$ action reduces the number of instances of $Request$ which hold. Each $Deliver$ action reduces the number of instances of $Carries$ which hold while preserving the number of instances of $Request$. Each Go action brings the robot closer to a selected office where it can pick up or deliver a package, so that after finitely many Go actions a $Pickup$ or a $Deliver$ follows. Consequently, after a finite number of actions a situation σ is reached in which no instance of $Request$ and no instance of $Carries$ holds. Let τ be a state such that $\Sigma_{mail} \models State(\sigma) = \tau$, then each of $Poss(Deliver(b), [\![\tau]\!])$, $Poss(Pickup(b, r), [\![\tau]\!])$, and $ContinueDelivery(d, [\![\tau]\!])$ fails. Hence, the rightmost branch of the overall derivation tree ends with resolving $MainLoop([\![\tau]\!])$ to the empty query. ■

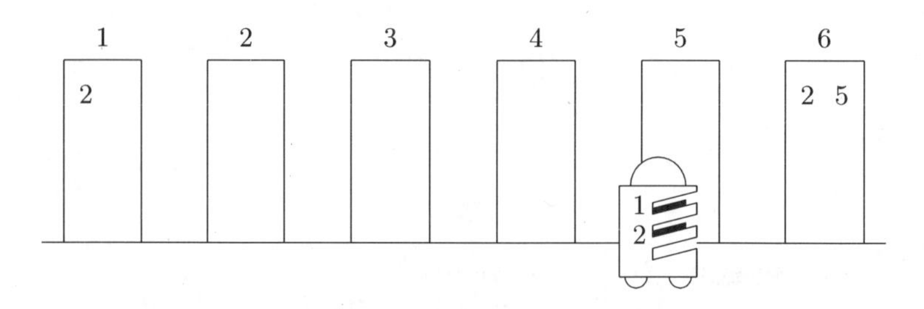

Figure 2.7: Carrying two packages with three more waiting to be picked up, the mailbot has to choose whether to move up or down the hallway, and if it decides to go to room 6 first, then it has to choose which of the two packages to put into the last free bag.

While the simple agent program for the mailbot solves the mail delivery problem, it does so rather inefficiently. This should be clear from the observation that a robot endowed with more mail bags is not significantly faster, as the figures in Section 2.3 indicate. One of the key advantages of high-level agent programming is to support concise and modular implementations of complex strategies. The mail delivery program, for example, can easily be improved by having the robot making better choices when selecting packages to be picked up or when heading towards another office.

Example 1 (continued) Figure 2.7 depicts a typical situation of choice for the mailbot. Since there are no packages to be delivered or collected at its current location, the robot must decide whether it should move down to room 2 (to deliver mail) or further up the hallway (to pick up a package). If it goes to room 6 first, then there is another choice to be made: Since there is only one free bag, the robot has to decide which of the two packages to handle first. Obviously, the smart robot decides to walk up to room 6, grabs and delivers the package for 5, goes back to 6 to pick up the second package, and then moves down to the other end of the hallway to finish the job. The agent program in the previous section, however, makes just arbitrary choices, which most of the time are not the optimal ones.

A more sophisticated strategy is to have the mailbot always take into account the distance to its current location when deciding which request to deal with next. This concerns both comparing the destination of the packages to choose from when filling the mail bag, and calculating the nearest office where something is to be done when moving along the hallway. The program depicted in Figure 2.8 implements this strategy. As before, packages are delivered in no prioritized order, because a specific order would make no difference. However, a mail bag is filled with a package only if there is no other package at the current location whose addressee is closer. Likewise, to pick up mail elsewhere, the mailbot goes to the nearest office with a request, and it goes to another office to deliver a package only if there is no closer office for which it carries mail, too.

```
main_loop(Z) :-
   poss(deliver(B), Z)      -> execute(deliver(B), Z, Z1),
                               main_loop(Z1) ;
   fill_a_bag(B, R, Z)      -> execute(pickup(B,R), Z, Z1),
                               main_loop(Z1) ;
   continue_delivery(D, Z) -> execute(go(D), Z, Z1),
                               main_loop(Z1) ;
   true.

fill_a_bag(B, R1, Z) :-
   poss(pickup(B,_), Z) ->
     holds(at(R), Z),
     holds(request(R,R1), Z),
     \+ ( holds(request(R,R2), Z), closer(R2, R1, R) ).

continue_delivery(D, Z) :-
   holds(at(R), Z),
   ( holds(empty(_), Z),
     holds(request(_,_), Z)
     -> ( holds(request(R1,_), Z),
          \+ ( holds(request(R2,_), Z), closer(R2, R1, R)) )
      ; ( holds(carries(_,R1), Z),
          \+ ( holds(carries(_,R2), Z), closer(R2, R1, R)) )
   ),
   ( R < R1 -> D = up
             ; D = down ).

closer(R1, R2, R) :- abs(R1-R) < abs(R2-R).
```

Figure 2.8: A smart mailbot agent in FLUX.

The improved FLUX program is sound, too (see Exercise 2.4). Running this program with the initial state of Figure 1.1, the mailbot needs 74 instead of 80 actions to carry out all requests. The following tables show the lengths of the action sequences applied to the same set of maximal mail delivery problems as in Section 2.3.

$k = 3$

n	# act	n	# act
9	264	20	2020
10	344	21	2274
11	430	22	2600
12	534	23	2930
13	656	24	3254
14	798	25	3676
15	956	26	4098
16	1100	27	4470
17	1286	28	4996
18	1516	29	5478
19	1760	30	5980

$k = 8$

n	# act	n	# act
9	206	20	1350
10	272	21	1498
11	342	22	1686
12	416	23	1890
13	502	24	2098
14	592	25	2320
15	694	26	2546
16	794	27	2792
17	908	28	3054
18	1054	29	3306
19	1180	30	3590

A comparison with the results for the naïve program shows that for $k = 3$ the improved strategy leads to solutions which are almost just half in length. Endowing the robot with five more slots reduces the solution length up to an additional 40%.

A comparative analysis of the runtime behavior of the two agent programs brings to light an interesting phenomenon. Not unexpectedly, the improved algorithm solves each individual problem much faster, as can be seen by the following figures (in seconds CPU time) for a robot with $k = 3$ mail bags:

simple strategy

n	time	n	time
11	0.21	21	5.23
12	0.32	22	6.57
13	0.45	23	8.21
14	0.68	24	10.18
15	0.95	25	12.50
16	1.31	26	15.14
17	1.83	27	18.15
18	2.42	28	21.99
19	3.14	29	26.26
20	4.04	30	31.14

improved strategy

n	time	n	time
11	0.15	21	3.04
12	0.21	22	3.79
13	0.31	23	4.59
14	0.43	24	5.63
15	0.60	25	6.91
16	0.80	26	8.32
17	1.09	27	9.92
18	1.44	28	11.93
19	1.88	29	14.09
20	2.38	30	16.51

On the other hand, a closer examination reveals that the time for action selection and update computation tends to be higher for the second program. This can be seen by comparing the initial segments of the graphs in Figure 2.9 with those of Figure 2.5. In particular with the graph for $n = 30$ it is easy to see that at the beginning more time is needed, on average, to select an action. This

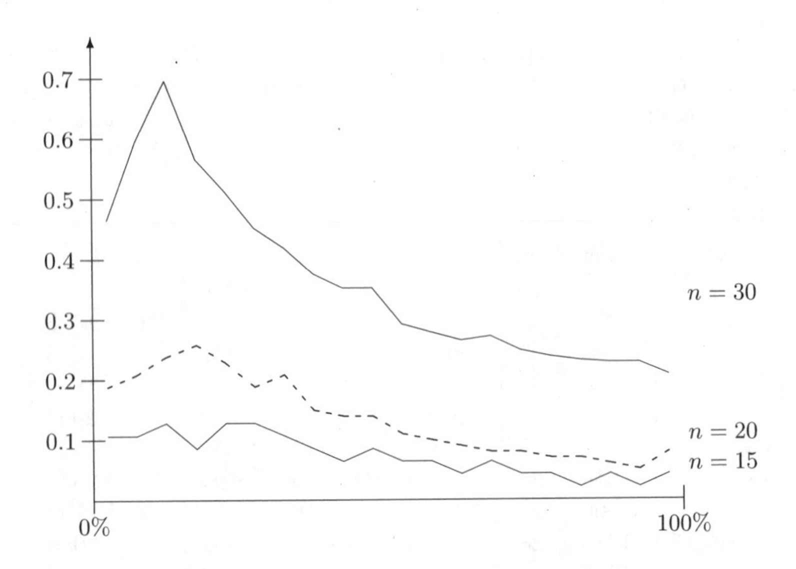

Figure 2.9: The computational behavior of the improved program for the mail delivery robot in the course of its execution. The horizontal axis depicts the degree to which the run is completed while the vertical scale is in seconds per 100 actions. The graphs are for maximal mail delivery problems with $k = 3$ mail bags.

---

is not surprising since the sophisticated strategy requires to compare several instances before making a decision, as opposed to just making an arbitrary choice. The general principle behind this observation is that improving the quality of the generated action sequence is usually payed for by an increased computation time per action. In the extreme case, there might be too much of a tradeoff in order for a highly sophisticated action selection strategy to be of practical value (see also Exercise 2.6).

In our particular example, however, the smart robot turns out to be even faster on average in the long run: The graphs in Figure 2.9 show a steeper general descent as the program proceeds. This is because the state size decreases faster due to generally shorter delivery times. As a consequence, the average computation time for the problem with $n = 30$ offices, say, is $31.14/10672 = 2.92 \cdot 10^{-3}$ seconds per action under the simple strategy and just $16.51/5980 = 2.76 \cdot 10^{-3}$ if we employ the sophisticated program. □

2.4 Exogenous Actions

The programs for the mailbot have been written under the assumption that all delivery requests are given initially and that the work is done once all requests have been carried out. In a practical setting, however, it should be possible

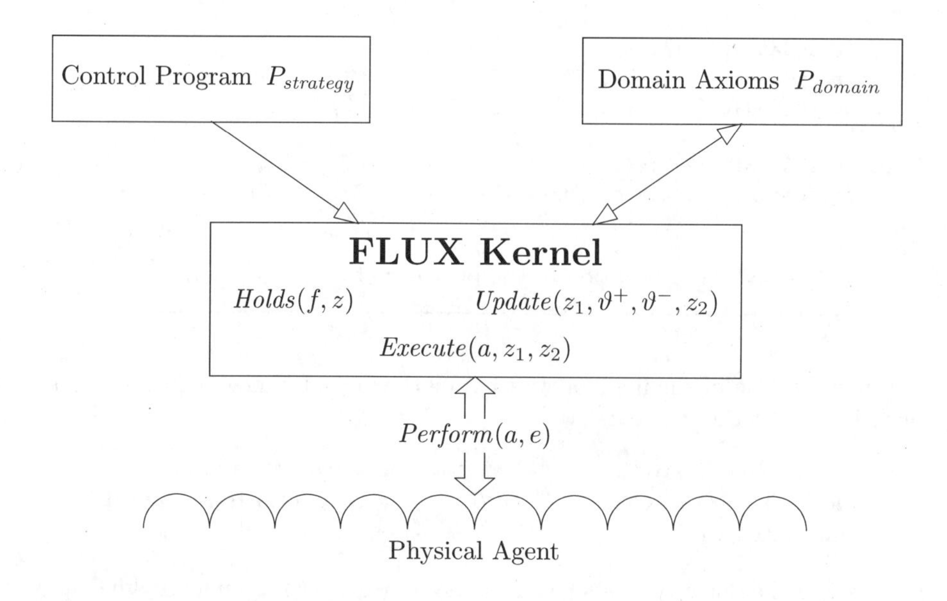

Figure 2.10: Architecture of FLUX programs for exogenous actions.

---

at any time to dynamically issue new requests (or, for that matter, to cancel a request). The automatic post boy must then be on the alert at any time and react sensibly to changes as soon as they occur. Most agents are actually embedded in an environment which includes other active entities, be it humans or fellow agents. As a consequence, some state properties may not be under the sole control of one agent. Agents must therefore take into account actions besides their own when maintaining their internal world model.

From the subjective perspective of an agent, an action is **exogenous** if it is not performed by the agent itself but does affect one or more of the relevant fluents. In the presence of exogenous actions, the interaction between an agent and its environment is no longer unidirectional. Figure 2.10 depicts the modified architecture of FLUX agent programs in which exogenous actions are accounted for. The standard FLUX predicate $Perform(a)$ is extended into $Perform(a, e)$, where e is a (possibly empty) list of actions that happen as the agent performs action a. These exogenous actions are taken into account upon updating the internal world model. This is achieved by an extension of the kernel predicate $Execute(a, z_1, z_2)$ as shown in Figure 2.11. According to this definition, a successor state is obtained by inferring the effect of the agent's action a and of all exogenous actions that have additionally occurred. So doing requires the agent to be equipped with state update axioms also for the possible exogenous actions in a domain.

Example 1 (continued) The mail delivery scenario shall be extended by the exogenous actions of adding and cancelling requests. Furthermore, we introduce

```
state_updates(Z, [], Z).
state_updates(Z1, [A|S], Z2) :-
   state_update(Z1, A, Z), state_updates(Z, S, Z2).

execute(A, Z1, Z2) :-
   perform(A, E), state_updates(Z1, [A|E], Z2).
```

Figure 2.11: Execution in the presence of exogenous actions.

---

the simple action of the mailbot waiting for the arrival of new requests in case there is none at the moment:

$$\begin{array}{rll} Add: & \mathbb{N} \times \mathbb{N} \mapsto \text{ACTION} & Add(r_1, r_2) \hat{=} \text{ new request from } r_1 \text{ to } r_2 \\ Cancel: & \mathbb{N} \times \mathbb{N} \mapsto \text{ACTION} & Cancel(r_1, r_2) \hat{=} \text{ request } r_1 \text{ to } r_2 \text{ cancelled} \\ Idle: & \text{ACTION} & \end{array}$$

We assume that new requests can only be issued if the arguments differ and are in the allowed range. Requests can be cancelled just in case they exist, so that it is too late for cancellation if the robot has already picked up the package in question. Idling is, of course, always possible (though it should not be the preferred action of the robot). Hence the following precondition axioms:

$$\begin{aligned} Poss(Add(r_1, r_2), z) &\equiv r_1 \neq r_2 \wedge 1 \leq r_1, r_2 \leq n \\ Poss(Cancel(r_1, r_2), z) &\equiv Holds(Request(r_1, r_2), z) \\ Poss(Idle, z) &\equiv \top \end{aligned}$$

The state update axioms for the new actions are straightforward:

$$\begin{aligned} & Poss(Add(r_1, r_2), s) \supset \\ & \quad State(Do(Add(r_1, r_2), s)) = State(s) + Request(r_1, r_2) \end{aligned}$$

$$\begin{aligned} & Poss(Cancel(r_1, r_2), s) \supset \\ & \quad State(Do(Cancel(r_1, r_2), s)) = State(s) - Request(r_1, r_2) \end{aligned}$$

$$Poss(Idle, s) \supset State(Do(Idle, s)) = State(s)$$

Figure 2.12 shows the FLUX encoding of these state update axioms.[3] Based on this extended background theory, our FLUX programs for the mailbot can be directly applied to the new setting, with just a simple modification by which the robot is triggered to idle in case there is nothing to do; see Figure 2.12. The extended control program enables the robot to react immediately and sensibly whenever requests are added or cancelled. A sample scenario is illustrated in Figure 2.13, whereby it is assumed that the robot employs the improved selection

[3] Exogenous actions are necessarily possible when they happen. Hence, there is no need to encode the underlying precondition axioms. These will be needed, however, for stating and proving correctness of FLUX programs in the presence of exogenous actions.

```
state_update(Z1, add(R1,R2), Z2) :-
   update(Z1, [request(R1,R2)], [], Z2).

state_update(Z1, cancel(R1,R2), Z2) :-
   update(Z1, [], [request(R1,R2)], Z2).

state_update(Z, idle, Z).

main_loop(Z) :-
   ( poss(deliver(B), Z)       -> execute(deliver(B), Z, Z1) ;
     fill_a_bag(B, R, Z)       -> execute(pickup(B,R), Z, Z1) ;
     continue_delivery(D, Z) -> execute(go(D), Z, Z1) ;
     execute(idle, Z, Z1)
   ),
   main_loop(Z1).
```

Figure 2.12: Additional state update axioms and slightly modified main loop of a mailbot controller which allows for adding and cancelling requests during program execution.

---

strategy: After picking up the initial requests in rooms 1 and 2, the robot heads towards office 4 with the intention to collect a third package. However, the corresponding request is cancelled while the robot goes from room 3 to 4. At the same time, a new request arises in office 3. This causes the robot to turn around immediately and go back to this room. In the meantime, a second request from office 6 arises. At the end, after dropping the package that came from office 6 the robot idles in front of room 2 awaiting further exogenous actions by which new requests are issued. □

Exogenous actions can be viewed as **input** to a FLUX program. This input has obviously significant influence on the actual computation. In particular, the generated sequence of actions depends on the exogenous actions that occur during the run of a program. As a consequence, it is no longer appropriate to speak of *the* computation tree of a program where there can be many different ones. As the computation trees for a program are each determined by the input, they can be characterized by the substitutions for e in every resolved $Perform(a, e)$ node:

Definition 2.11 Let T be a computation tree for an agent program and query. Let $Execute(\alpha_1, \_, \_), Execute(\alpha_2, \_, \_), \ldots$ be the ordered sequence of execution nodes in T. Tree T is said to **process** exogenous actions $\vec{\varepsilon}_1, \vec{\varepsilon}_2, \ldots$ if $Perform(\alpha_1, e_1), Perform(\alpha_2, e_2), \ldots$ is the ordered sequence of child nodes of the execution nodes in T and $\{e_1/\vec{\varepsilon}_1\}, \{e_2/\vec{\varepsilon}_2\}, \ldots$ are the substitutions used to resolve these child nodes. Tree T is said to **generate** the sequence of actions $\alpha_1, \alpha_2, \ldots$. □

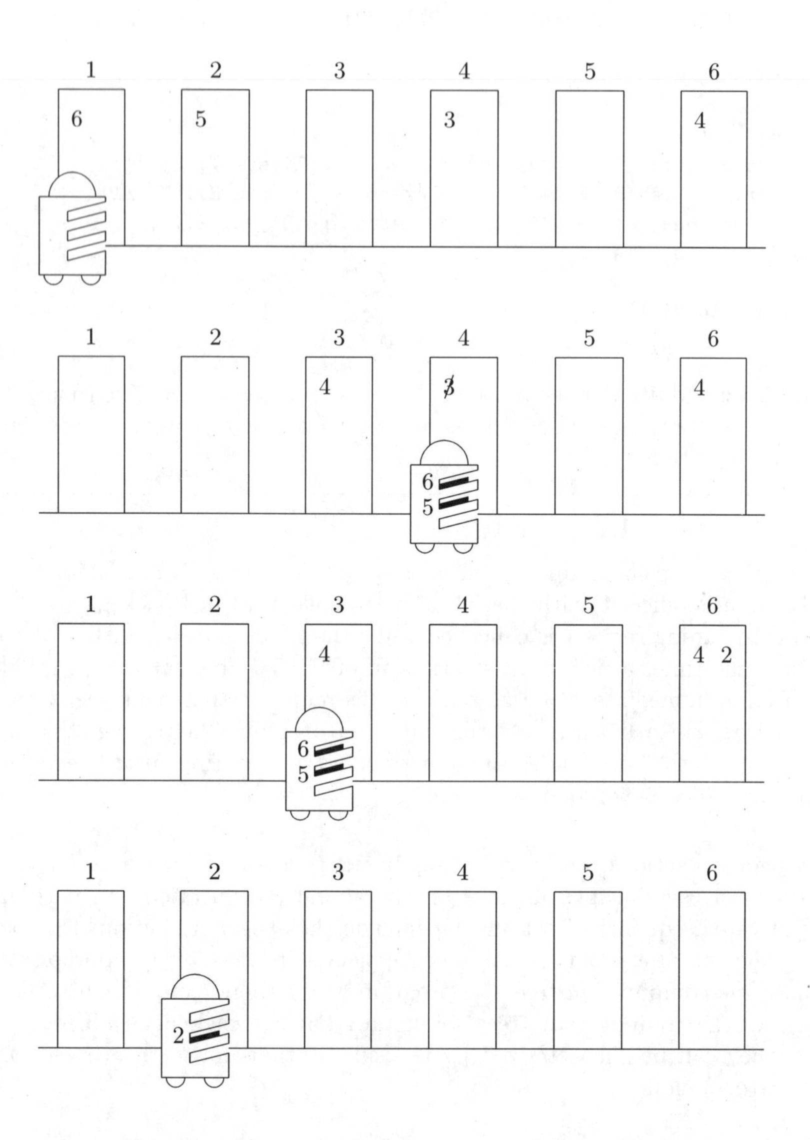

Figure 2.13: Sample run where the requests change dynamically.

As an example, Figure 2.14 depicts skeletons of two different computation trees for the mailbot program. It is assumed that the initial state contains no requests and all mail bags are empty. The tree sketched on the left hand side processes the exogenous actions

$$[], [Add(1,2)], [], [], [], [], \ldots$$

and generates actions *Idle*, *Idle*, *Pickup*(1, 2), *Go*(*Up*), *Deliver*(1), *Idle*, ... The tree on the right hand side processes the exogenous actions

$$[Add(2,1)], [], [], [], [], [], \ldots$$

and generates *Idle*, *Go*(*Up*), *Pickup*(1, 1), *Go*(*Down*), *Deliver*(1), *Idle*, ...

The exogenous actions along with the actions of the agent determine the situation the agent is in at each stage of the program. This is crucial when it comes to proving that a program is sound, which requires all executed actions to be possible. Suppose, for example, the mailbot starts in a state without any initial request and eventually executes a *Pickup* action. This action can only be justified by a preceding exogenous action by which a new request has been issued. Thus it is necessary to generalize the notion of an associated situation as follows:

Definition 2.12 Consider an agent program P and query Q_0, and let T be a computation tree for $P \cup \{\leftarrow Q_0\}$. Let Q be a node in T and $Execute(\alpha_1, \_, \_), \ldots, Execute(\alpha_n, \_, \_)$ the ordered sequence of execution nodes to the left and above Q in T $(n \geq 0)$. If $\vec{\varepsilon}_1, \vec{\varepsilon}_2, \ldots, \vec{\varepsilon}_n, \ldots$ are the exogenous actions processed by T, then the situation **associated** with Q is $Do([\alpha_1, \vec{\varepsilon}_1, \ldots, \alpha_n, \vec{\varepsilon}_n], S_0)$.[4] □

It is easy to see that this definition coincides with the previous one in the special case that no exogenous action at all happens. For example, in both trees of Figure 2.14 the topmost and the second node are associated with S_0, and, say, the sixth node in the tree on the left hand side is associated with $Do([Idle, Idle, Add(1,2)], S_0)$ whereas the sixth node in the tree on the right hand side is associated with $Do([Idle, Add(2,1), Go(Up)], S_0)$.

We are now in a position to generalize to exogenous actions the notion of soundness of programs. Informally speaking, an agent program is sound if for any sequence of exogenous actions, all generated actions are possible in the situation in which they are executed—provided that the exogenous actions themselves are possible. The latter restriction allows to ignore impossible inputs, for which a program need not be prepared.

Definition 2.13 Agent program P and query Q are **sound** wrt. a domain axiomatization Σ if for every computation tree T for $P \cup \{\leftarrow Q\}$ the following

[4] By a slight abuse of notation, we identify a list $\vec{\varepsilon}_i$ with a (possibly empty) sequence of actions.

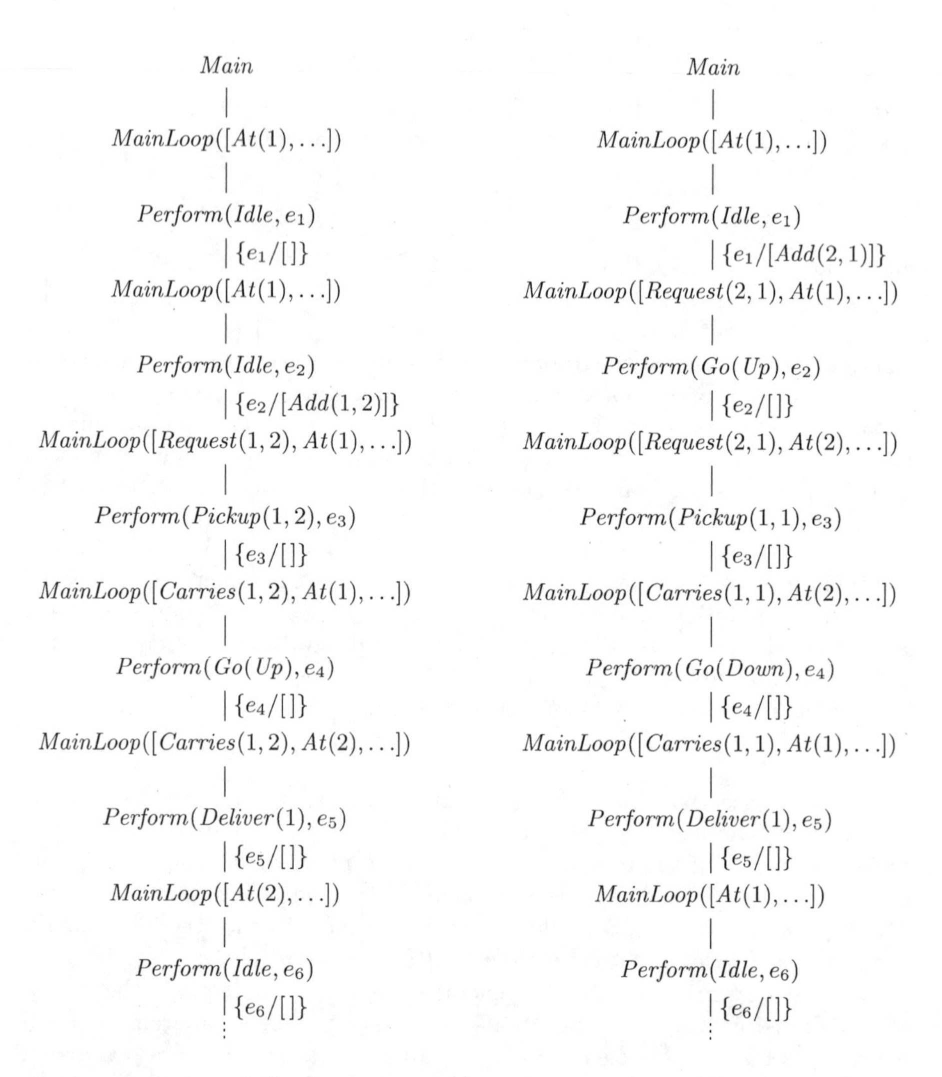

Figure 2.14: Two different computation trees for the mailbot program with exogenous actions.

holds: Let $\vec{\varepsilon}_1, \vec{\varepsilon}_2, \ldots$ be the exogenous actions processed by T and $\alpha_1, \alpha_2, \ldots$ be the actions generated by T. If for all $i = 1, 2, \ldots$,

$$\Sigma \models Poss(\vec{\varepsilon}_i, Do(\alpha_i, \sigma_i))$$

then for all $i = 1, 2, \ldots$,

$$\Sigma \models Poss(\alpha_i, \sigma_i)$$

where σ_i is the situation associated with the i-th execution node in T. □

That is to say, a program is sound if in every computation tree in which all exogenous actions are possible, all generated actions are possible, too.

Theorem 2.9, which provides a general strategy for writing sound programs, carries over to FLUX with exogenous actions (see Exercise 2.10). This is so because the definition of execution of Figure 2.11 ensures that the internal world model is updated according to the actions of both the agent and the environment. The FLUX programs for our mailbot can thus be proved sound also in case requests are dynamically added or cancelled (see Exercise 2.11). The possibility to issue requests at runtime raises, however, a novel challenge. If constantly new requests occur, then both of our control programs risk to never handle some requests if the robot is all the time busy with new ones. For practical purposes it is therefore necessary to employ a refined selection strategy that guarantees that every request is handled eventually. Exercise 2.12 is concerned with developing such a strategy.

2.5 Bibliographical Notes

The use of logic as a programming paradigm has been introduced in [Kowalski, 1974]. Around the same time, the logic programming language Prolog (for: **Pro***grammation en* **log***ique*) has been developed [Colmerauer *et al.*, 1972]. There are many textbooks on the theory of logic programming, among which are the classical [Lloyd, 1987] and the more recent [Apt, 1997]. Prolog companions such as [Clocksin and Mellish, 1994] focus on the practical side of programming. The negation-as-failure rule and the completion semantics have been developed in [Clark, 1978]. Our definition of computation trees with negation and cut follows [Apt and Bol, 1994; Apt, 1997].

General-purpose languages and systems for reasoning agents and robots have not been developed until recently, beginning with **GOLOG** (for: Al**gol** in **logic**) [Levesque *et al.*, 1997]. The underlying action formalism is situation calculus. A typical encoding in GOLOG of how a fluent changes upon the performance of actions is

```
holds(at(R),do(A,S)) :- A=go(D), holds(at(R1),S),
                        (D=up, R is R1+1 ; D=down, R is R1-1)
                        ;
                        \+ A=go(_), holds(at(R),S).
```

Put in words, either $At(r)$ becomes true by performing a Go action with the corresponding $At(r_1)$ being true in the preceding situation, or the location of the mailbot remains unchanged because a different action is performed. The logic programs for GOLOG described in the literature all apply the principle of **regression** to evaluate conditions in agent programs: The question whether, for example, $At(r)$ holds after a sequence of actions is solved by recursively applying the clause just mentioned until the initial situation is reached. A consequence of this is that the computational effort to evaluate a condition increases with the number of actions that have been performed. Control strategies are encoded in GOLOG programs using a special syntax for control structures such as conditional branching and loops. GOLOG has been extended into various directions, such as the incorporation of sensing actions, concurrency, or decision theoretic elements [Giacomo and Levesque, 1999; Lakemeyer, 1999; Giacomo and Levesque, 2000; Boutilier *et al.*, 2000; Grosskreutz and Lakemeyer, 2000]. The textbook [Reiter, 2001a] provides a detailed account of the mathematical foundations of GOLOG.

FLUX as a high-level programming language and system has been introduced in [Thielscher, 2002a]. The computational behaviors of GOLOG and FLUX have been compared in [Thielscher, 2004], where it is shown how the principle of **progression** allows to evaluate conditions directly against the updated (i.e., progressed) world model. Other systems for automated reasoning about actions are described in [Shanahan and Witkowski, 2000; Kvarnström and Doherty, 2000; Giunchiglia *et al.*, 2004].

2.6 Exercises

2.1. Formulate queries to the FLUX background theory for the mailbot domain to compute the following:

(a) the initial requests with the farthest distance between sender and recipient;

(b) the initially possible actions;

(c) an executable sequence consisting of five actions which ends with a delivery in office 4.

2.2. Prove Lemma 2.5.

2.3. (a) Extend the FLUX kernel of Figure 2.2 by clauses P_{equal} defining a predicate $Equal(z_1, z_2)$ with the intended meaning that the two states z_1 and z_2 are equal under the semantics of fluent calculus.

(b) Construct the completion of the program thus extended. Prove that $COMP[P_{skernel} \cup P_{equal}] \models Equal([\![\tau_1]\!], [\![\tau_2]\!])$ iff $\Sigma_{sstate} \models \tau_1 = \tau_2$ for any two ground states τ_1, τ_2 and any of their FLUX encodings $[\![\tau_1]\!], [\![\tau_2]\!]$.

(c) Use the new predicate to generalize the definition of *Update* so as to allow for verifying that two ground states satisfy an update equation.

2.4. Show that the improved agent program of Figure 2.8 for the mailbot is sound.

2.5. Find and implement a refined strategy for the mailbot by which the solution lengths for maximal mail delivery problems are further optimized in comparison with the program of Figure 2.8. Show that the program is sound wrt. the axiomatization for the mail delivery domain.

2.6. Write a program for the mailbot which always generates the optimal solution. Compare its runtime behavior with that of the programs of Figure 2.8 and of the previous exercise.

2.7. Implement Exercise 1.6:

(a) Modify the FLUX encoding of the background theory for the mail delivery world by the 3-place *Request* predicate and the *Putdown*(b) action.

(b) Find and implement a strategy to reduce the number of *Go* actions using the new action. Show that the program is sound wrt. the domain axiomatization developed in Exercise 1.6.

2.8. Implement the domain of Exercise 1.7 and write an agent program with the aim to minimize the number of transfer actions. Show that the program is sound wrt. your domain axiomatization.

2.9. Consider a robot working in an office building with two elevators E_1 and E_2 next to each other. At each floor there is a single call button, which can be activated if no elevator happens to be at this floor. The activation has the nondeterministic effect that either of the two elevators arrives.

(a) Axiomatize this domain using the fluents $At(x, n)$, representing that $x \in \{E_1, E_2, Robot\}$ is at floor $n \in \mathbb{N}$, and $In(x)$, representing that the robot is inside of elevator x. The actions shall be $Call(n)$, pressing the call button at floor n, and $Enter(x)$, entering the elevator x. Suppose that the effect of calling an elevator be specified by this **nondeterministic** state update axiom:

$$
\begin{aligned}
& Poss(Call(n), s) \supset \\
& \quad (\exists n')\,(\,Holds(At(E_1, n'), s) \wedge \\
& \qquad\qquad State(Do(Call(n), s)) = \\
& \qquad\qquad\quad State(s) - At(E_1, n') + At(E_1, n)\,) \\
& \quad \vee \\
& \quad (\exists n')\,(\,Holds(At(E_2, n'), s) \wedge \\
& \qquad\qquad State(Do(Call(n), s)) = \\
& \qquad\qquad\quad State(s) - At(E_2, n') + At(E_2, n)\,)
\end{aligned}
$$

Define suitable precondition axioms, a state update axiom for $Enter$, and appropriate domain constraints.

(b) Encode the axioms in FLUX. Show that there is an *unsound* agent program which satisfies all conditions of Theorem 2.9 except that the underlying domain is not deterministic.

2.10. Prove Theorem 2.9 in the presence of exogenous actions and with the generalized notion of soundness of Definition 2.13.

2.11. Prove soundness of the mailbot program of Section 2.4, which allows for dynamically adding and cancelling requests.

2.12. (a) With regard to the mailbot program for exogenous actions, find an initial state and an (infinite) sequence of exogenous Add actions such that one of the initial requests is never handled.

(b) A general solution to this problem is to give requests increasing priority the longer they have been around. One way of implementing this is to extend the mailbot domain by a fluent $Time(t)$ such that t indicates the number of Go actions the robot has performed since the beginning. Redefine, to this end, the fluents $Request(r_1, r_2)$ and $Carries(b, r)$ so as to include the information at what time the request in question has been issued. Modify the state update axioms accordingly. Program a refined control strategy by which every request is guaranteed to be carried out eventually. (Take care also that the robot does not carry around some package forever!)

Chapter 3

General Fluent Calculus

Imagine a cleaning robot in an office building whose task is to empty all waste bins of a floor. As this is a rather noisy procedure, the "cleanbot" works after hours. Still, it is supposed to not burst into any office which is occupied by people staying in late. At the beginning of its route, the robot does not know which offices are still in use. Fortunately, the cleanbot is equipped with a light sensor which is activated whenever inside of or adjacent to a room that is occupied. However, the sensor does not enable the robot to tell which direction the light comes from. Figure 3.1 depicts a sample office layout and scenario.[1] The elementary actions of the robot are to go forward to the adjacent square, to turn clockwise by 90 degrees, and to empty the waste bin at the current position. The challenge is to write a control program by which the cleanbot empties as many bins as possible without the risk to disturb anyone.

The essential difference to the mailbot of Chapter 1 is that the cleanbot acts under uncertainty as it does not have complete state knowledge. Its sensor allows the robot to acquire information in the course of time, yet there is no guarantee that it ever attains the full picture. Uncertainty requires the robot to be cautious: If it does not know whether a certain office is free, it should not enter. Consider, for example, the situation of the robot of Figure 3.1 after having successively emptied the waste bins at $(1,1)$, $(1,2)$, and $(1,3)$. There, it senses light. Since it cannot be decided whether the light comes from office $(1,4)$ or maybe $(2,3)$, the robot should avoid both of them for the moment. So it goes back and continues with cleaning the bin in the hallway at location $(2,2)$. Sensing no light there, it follows that after all office $(2,3)$ cannot be occupied. Moreover, if the cleanbot is smart enough to recall that it already knew that one of $(1,4)$ or $(2,3)$ *must* be occupied, then it can now indirectly conclude the state of the remote office $(1,4)$. This conclusion relies on the ability to interpret and logically combine sensor information that has been acquired over time.

This chapter and the following two are devoted to the question of how to

[1] For the moment, it is assumed that people do not leave while the robot performs its cleaning task. A dynamic variant of this domain is considered in Chapter 7.

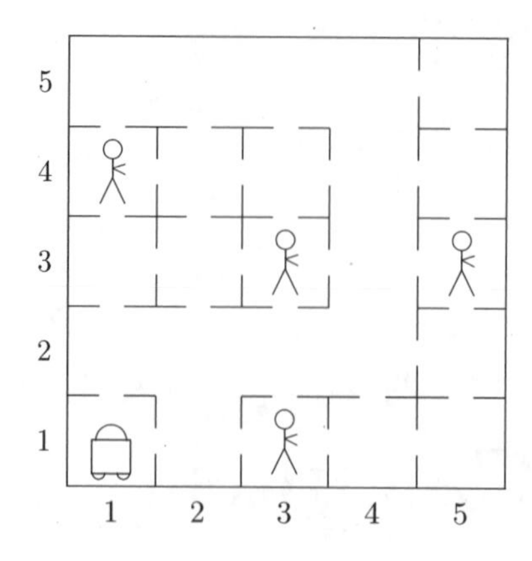

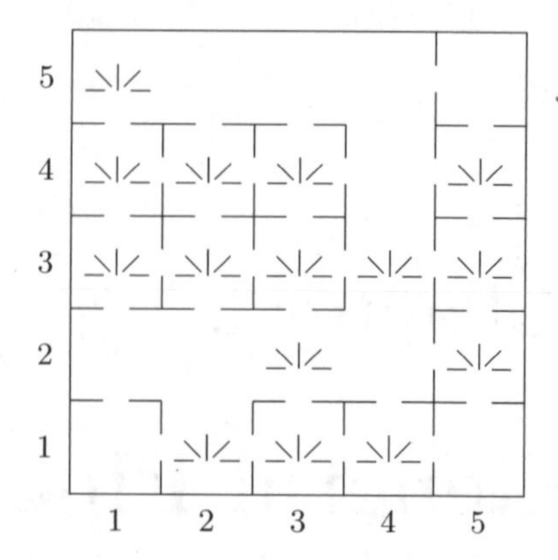

Figure 3.1: Layout of an office floor. It is assumed that waste bins are in every office and every square of the corridor. Room $(1, 1)$ is the home of the cleanbot. In the scenario depicted in the left diagram, four offices are still in use. The diagram on the right hand side shows the locations in which a light sensor would be activated that evening.

---

program autonomous agents that are capable of reasoning and acting with incomplete information. We begin by introducing the **general fluent calculus**, which allows to

3.1. specify and reason with incomplete states;

3.2. update incomplete states.

3.1 Incomplete States

General fluent calculus is a theory of reasoning about actions in the presence of incomplete states. The syntax of the general theory is that of special fluent calculus, whose signature can be readily used to formulate partial state knowledge.

Example 2 The relevant state information for the cleaning robot are its position, the direction it faces, whether a location is occupied, and whether the waste bin at a location is cleaned. Accordingly, we will describe the states by these four fluents:

$$\begin{array}{rll}
At: & \mathbb{N} \times \mathbb{N} \mapsto \text{FLUENT} & At(x,y) \hat{=} \text{ robot is at square } (x,y) \\
Facing: & \mathbb{N} \mapsto \text{FLUENT} & Facing(d) \hat{=} \text{ robot faces direction } d \\
Occupied: & \mathbb{N} \times \mathbb{N} \mapsto \text{FLUENT} & Occupied(x,y) \hat{=} \text{ square } (x,y) \text{ occupied} \\
Cleaned: & \mathbb{N} \times \mathbb{N} \mapsto \text{FLUENT} & Cleaned(x,y) \hat{=} \text{ bin at } (x,y) \text{ cleaned}
\end{array}$$

The four possible directions shall be encoded by the numbers 1 (north), 2 (east), 3 (south), and 4 (west), respectively. As usual, we introduce an axiom of uniqueness of names, here: $UNA[At, Facing, Occupied, Cleaned]$.

Suppose that initially the robot faces north, then we may say that the robot knows the following facts about the initial situation:

$$\begin{array}{l} Holds(At(1,1), S_0) \wedge Holds(Facing(1), S_0) \wedge \\ \neg Holds(Occupied(1,1), S_0) \wedge \\ \neg Holds(Occupied(2,1), S_0) \wedge \ldots \wedge \neg Holds(Occupied(4,5), S_0) \wedge \\ (\forall x, y) \neg Holds(Cleaned(x,y), S_0) \end{array} \quad (3.1)$$

That is, given are the robot's location and orientation, the fact that neither the cleaning room nor any of the squares in the corridor is occupied, and that no waste bin has been cleaned thus far. This is an example of an incomplete state specification. Instead of equating the expression $State(S_0)$ with a ground term, formula (3.1) merely puts some constraints on this state (in the disguise of the *Holds* macro).

Further information about the initial state can be derived from the general domain constraints that apply to the cleaning robot world:

$$\begin{array}{l} (\exists x, y)\,(Holds(At(x,y), s) \wedge 1 \leq x, y \leq 5) \\ Holds(At(x_1,y_1), s) \wedge Holds(At(x_2,y_2), s) \supset x_1 = x_2 \wedge y_1 = y_2 \\ (\exists d)\,(Holds(Facing(d), s) \wedge 1 \leq d \leq 4) \\ Holds(Facing(d_1), s) \wedge Holds(Facing(d_2), s) \supset d_1 = d_2 \\ (\forall x) \neg Holds(Occupied(x,0), s) \wedge (\forall x) \neg Holds(Occupied(x,6), s) \\ (\forall y) \neg Holds(Occupied(0,y), s) \wedge (\forall y) \neg Holds(Occupied(6,y), s) \end{array} \quad (3.2)$$

Put in words, in any situation the robot is somewhere unique (first and second constraint) facing a unique direction (third and fourth constraint), and offices outside of the boundaries of the floor cannot be occupied (fifth and sixth constraint).

When it comes to logical reasoning, there is no essential difference between formulas which merely constrain states and those that fully specify a state. Consider, for instance, the following formula, which is central to the cleanbot's understanding of its environment. It defines the meaning of light at a certain location (x, y) in a particular state z:

$$\begin{array}{l} Light(x,y,z) \equiv \\ \quad Holds(Occupied(x,y), z) \vee \\ \quad Holds(Occupied(x+1,y), z) \vee Holds(Occupied(x,y+1), z) \vee \\ \quad Holds(Occupied(x-1,y), z) \vee Holds(Occupied(x,y-1), z) \end{array} \quad (3.3)$$

Now, for the sake of argument suppose it is given that initially there is no light at square $(1,2)$ but at square $(1,3)$, that is,

$$\neg Light(1,2,State(S_0)) \wedge Light(1,3,State(S_0))$$

From axiom (3.3) and $\neg Light(1,2,State(S_0))$ it follows

$$\neg Holds(Occupied(1,3), S_0) \quad (3.4)$$

On the other hand, from $Light(1, 3, State(S_0))$ it follows

$$\begin{array}{l} Holds(Occupied(1,3), S_0) \vee \\ Holds(Occupied(2,3), S_0) \vee Holds(Occupied(1,4), S_0) \vee \\ Holds(Occupied(0,3), S_0) \vee Holds(Occupied(1,2), S_0) \end{array}$$

The penultimate domain constraint in (3.2) implies that there is no room with coordinates $(0, 3)$ which can be occupied, nor can the hallway square $(1, 2)$ be occupied according to the initial specification, axiom (3.1); hence, with (3.4) it follows

$$Holds(Occupied(2,3), S_0) \vee Holds(Occupied(1,4), S_0)$$

This disjunction cannot be resolved further. If it were additionally given that $\neg Light(2, 2, State(S_0))$, then by (3.3), $\neg Holds(Occupied(2, 3), S_0)$, which implies $Holds(Occupied(1, 4), S_0)$. □

From a semantic point of view, a crucial difference between complete and incomplete states is that the former, according to Definition 1.4, always contain finitely many fluents. In contrast, it is straightforward to specify an incomplete state in which an infinite number of fluents hold. A simple example is the formula

$$(\forall n)\, Holds(F(n), S_0)$$

where $F : \mathbb{N} \mapsto$ FLUENT. The fact that states can be infinite is, however, not adequately captured by the foundational axioms of special fluent calculus. A particular deficiency arises with the notion of state difference, which is central to fluent calculus when it comes to the representation of actions and their effects. The problem is that the following formula is satisfiable under the axiomatic foundation of special fluent calculus:

$$(\exists z, f)(\forall z')\, z' \neq z - f$$

That is to say, there are (infinite) states z for which fluent removal is not defined (see Exercise 3.1). As a consequence, the axiomatic foundation of special fluent calculus is too weak to define update for actions with negative effects in the presence of incomplete states.

In order to guarantee that state difference is always defined, an additional foundational axiom stipulates the existence of a state for every combination of fluents. Since first-order logic is too weak for this purpose, general fluent calculus appeals to second-order logic.

Second-order logic adds sorted variables for predicates and functions to the language of first-order logic. Interpretations for second-order formulas assign relations (of the right arity and sort) to predicate variables, and mappings (of the right arity and sort) to function variables. The second-order formula $(\forall P)(\exists x)\, P(x)$, for example, is contradictory, because there exists an assignment for variable P, viz. the empty relation, which is false for any x. Substitutions for predicate and function variables in formulas use **λ-expressions**. These are of the form $\lambda x_1, \ldots, x_n.\tau$ with n being the arity of the variable and where τ is a first-order formula or term, respectively. The result of the substitution is that the variable expression $P(t_1, \ldots, t_n)$ (or $f(t_1, \ldots, t_n)$, respectively) is replaced by $\tau\{x_1/t_1, \ldots, x_n/t_n\}$. E.g., applying the substitution $\{P/\lambda d.\, \mathit{Facing}(d) \wedge \neg \mathit{Facing}(d)\}$ to the formula $(\exists x)\, P(x)$ results in the (inconsistent) first-order formula $(\exists x)\,(\mathit{Facing}(x) \wedge \neg \mathit{Facing}(x))$.

Definition 3.1 The **foundational axioms** Σ_{state} **of fluent calculus** are Σ_{sstate} augmented with the axiom of existence of states:

$$(\forall P)(\exists z)(\forall f)\,(\mathit{Holds}(f, z) \equiv P(f)) \tag{3.5}$$

where P is a second-order predicate variable of sort FLUENT. □

The new axiom is very powerful. It implies the existence of a state for any intensional specification one might think of. If, for example, we wish to construct a state in which the fluent $F : \mathbb{N} \mapsto$ FLUENT holds just for all multiples of 7 between 14 and 98, and the fluent $G : \mathbb{N} \mapsto$ FLUENT holds for all even numbers and nothing else holds, then the instance of (3.5),

$$\{P/\lambda f.\ (\exists n : \mathbb{N})\,(f = F(7 \cdot n) \wedge 1 < n < 15)\ \vee\ (\exists n : \mathbb{N})\, f = G(2 \cdot n)\}$$

guarantees that such a state exists.

The axiom of state existence can be safely added, unlike its counterpart in general set theory: Stipulating the existence of a set for any intensional definition gives rise to the infamous Russell paradox of the set of all sets which do not contain themselves. No such contradiction exists for Σ_{state} since the second-order variable in axiom (3.5) ranges over fluents only. If P were replaced by the analogue to the paradox, that is, the expression $\lambda f.\, \neg \mathit{Holds}(f, f)$, then the axiom just stipulates the existence of the empty state. Thus the axiomatic foundation of general fluent calculus is consistent, too.

Proposition 3.2 Σ_{state} *is consistent.*

Proof: Let ι be an interpretation whose domain includes all sets of ground fluents and such that

- $\emptyset^{\iota} = \{\}$;

- $f^\iota = \{f\}$ for each ground fluent f;
- $(z_1 \circ z_2)^\iota = z_1^\iota \cup z_2^\iota$.

Then ι is a model for Σ_{sstate} (cf. the proof of Proposition 1.3 on page 7). It is also a model for (3.5): Let P^ι be any unary relation of ground fluents. If we define $Z = \{f : (f) \in P^\iota\}$, then for any f there exists a set Z' satisfying $Z = \{f\} \cup Z'$ just in case $(f) \in P^\iota$. ■

The following result shows that under the axiom of state existence, for any state z any fluent f, there exists a state which is as z but without f. This implies that the definition of fluent removal extends to infinite states in general fluent calculus. Incidentally, the proof of this proposition illustrates how the second-order axiom is used in logical derivations.

Proposition 3.3 **(Rewrite Law)**

$$\Sigma_{state} \models Holds(g, z) \equiv (\exists z')\,(z = g \circ z' \wedge \neg Holds(g, z'))$$

Proof: Suppose $Holds(g, z)$. Consider the instance

$$\{P/\lambda f.\ Holds(f, z) \wedge f \neq g\}$$

of the axiom of state existence. This instance implies that there is a state, call it z', such that

$$(\forall f)\,(Holds(f, z') \equiv Holds(f, z) \wedge f \neq g) \tag{3.6}$$

From $(\forall f)\,(Holds(f, z') \supset f \neq g)$ it follows that $\neg Holds(g, z')$. Decomposition and irreducibility imply $Holds(f, g \circ z') \equiv [f = g \vee Holds(f, z')]$; hence, according to (3.6), $Holds(f, g \circ z') \equiv [f = g \vee Holds(f, z)]$. From the assumption $Holds(g, z)$ it follows that $[f = g \vee Holds(f, z)] \equiv Holds(f, z)$. Consequently, $Holds(f, g \circ z') \equiv Holds(f, z)$. The axiom of state equivalence implies that $g \circ z' = z$.

For the converse, suppose $(\exists z')\, z = g \circ z'$, then $Holds(g, z)$. ■

As we will see in the following section, Proposition 3.3 lays the foundation for inferring the update of incomplete states. Therefore, the axiom of state existence is indispensable in general fluent calculus.

3.2 Updating Incomplete States

Agents with partial state knowledge face the challenge to maintain a world model which remains incomplete throughout the execution of the program. The concept of a state update axiom as a way to formalize the effects of actions applies in the presence of incomplete states, too. However, in special fluent calculus the result of an action is easily inferred by instantiating an update equation with the term that describes the current state, followed by some straightforward rewriting steps. This inference rule is not directly applicable to incomplete state specifications such as (3.1), because there is no explicit compound state term which can be readily inserted into the update equation at hand.

Proposition 3.3 helps here. It suggests a (fully mechanic) way of combining pieces of state information into one equation, which can then be used in combination with an update equation. More precisely, the goal of this procedure is to obtain from an arbitrary incomplete state specification for a situation s a single equation $State(s) = \tau$, possibly accompanied by further axioms constraining the variables in τ.

Example 2 (continued) To begin with, the following is obviously true:

$$(\exists z)\, State(S_0) = z$$

Next, we integrate into this equation all pieces of information given in state specification (3.1). Consider the first atom, $Holds(At(1,1), S_0)$. Combining this with the above equation implies

$$(\exists z)\,(State(S_0) = z \wedge Holds(At(1,1), z))$$

According to Proposition 3.3, this can be equivalently written as

$$(\exists z')\,(\, State(S_0) = At(1,1) \circ z' \wedge \neg Holds(At(1,1), z')\,) \tag{3.7}$$

Simple macro expansion of $Holds(At(1,1), S_0)$ would of course yield the same equation for $State(S_0)$. However, the additional requirement that z' does not contain $At(1,1)$ will prove helpful later when it comes to inferring negative effects of actions.

Continuing with the atom $Holds(Facing(1), S_0)$ in (3.1), from (3.7) it follows

$$\begin{aligned}(\exists z')\,(\, & State(S_0) = At(1,1) \circ z' \wedge \neg Holds(At(1,1), z') \\ & \wedge Holds(Facing(1), At(1,1) \circ z')\,)\end{aligned} \tag{3.8}$$

The conjunct in the second line is equivalent to $Holds(Facing(1), z')$ according to the foundational axioms of decomposition and irreducibility along with uniqueness-of-names. Hence, Proposition 3.3 can be applied again and implies that there exists z'' such that $z' = Facing(1) \circ z''$ and $\neg Holds(Facing(1), z'')$. Combining this with (3.8) yields

$$\begin{aligned}(\exists z'')\,(\, & State(S_0) = At(1,1) \circ Facing(1) \circ z'' \\ & \wedge \neg Holds(At(1,1), Facing(1) \circ z'') \wedge \neg Holds(Facing(1), z'')\,)\end{aligned}$$

The conjunct $\neg Holds(At(1,1), Facing(1) \circ z'')$ can be simplified according to uniqueness-of-names along with decomposition and irreducibility, yielding

$$\begin{aligned}(\exists z'')\,(\, & State(S_0) = At(1,1) \circ Facing(1) \circ z'' \\ & \wedge \neg Holds(At(1,1), z'') \wedge \neg Holds(Facing(1), z'')\,)\end{aligned} \tag{3.9}$$

Having cast into a single equation all positive *Holds* statements of (3.1), next we incorporate the negative state knowledge. This will result in further constraints on the state variable z'' in (3.9). Consider, to begin with, the literal $\neg Holds(Occupied(1,1), S_0)$. With regard to formula (3.9), it follows that $Occupied(1,1)$ does not hold in the term to the right of the equation sign. Decomposition, irreducibility, and uniqueness-of-names imply that this is equivalent to $Occupied(1,1)$ not occurring in z'', that is,

$$\begin{aligned}(\exists z'')\,(\,&State(S_0) = At(1,1) \circ Facing(1) \circ z'' \\ &\wedge \neg Holds(At(1,1), z'') \wedge \neg Holds(Facing(1), z'') \\ &\wedge \neg Holds(Occupied(1,1), z''))\end{aligned}$$

Adding the other pieces of information in (3.1) in a similar fashion yields

$$\begin{aligned}(\exists z'')\,(\,&State(S_0) = At(1,1) \circ Facing(1) \circ z'' \\ &\wedge \neg Holds(At(1,1), z'') \wedge \neg Holds(Facing(1), z'') \\ &\wedge \neg Holds(Occupied(1,1), z'') \\ &\wedge \ldots \wedge \neg Holds(Occupied(4,5), z'') \\ &\wedge (\forall x, y)\,\neg Holds(Cleaned(x,y), z''))\end{aligned} \tag{3.10}$$

This incomplete state specification is logically equivalent to (3.1) under the foundational axioms and the axiom of uniqueness-of-names for the fluents. The advantage of the modified representation is to provide a descriptive specification of $State(S_0)$, which can be readily inserted into an update equation. Prior to this, the underlying domain constraints should be incorporated into the equational state specification, too, in order to be able to deduce as much as possible about the updated state. Combining (3.10) with the constraints of the cleaning robot world, (3.2), results in further constraints on the state variable z'':

$$\begin{aligned}(\exists z'')\,(\,&State(S_0) = At(1,1) \circ Facing(1) \circ z'' \\ &\wedge \neg Holds(Occupied(1,1), z'') \\ &\wedge \ldots \wedge \neg Holds(Occupied(4,5), z'') \\ &\wedge (\forall x, y)\,\neg Holds(Cleaned(x,y), z'') \\ &\wedge (\forall x, y)\,\neg Holds(At(x,y), z'') \\ &\wedge (\forall d)\,\neg Holds(Facing(d), z'') \\ &\wedge (\forall x)\,\neg Holds(Occupied(x,0), z'') \\ &\wedge (\forall x)\,\neg Holds(Occupied(x,6), z'') \\ &\wedge (\forall y)\,\neg Holds(Occupied(0,y), z'') \\ &\wedge (\forall y)\,\neg Holds(Occupied(6,y), z''))\end{aligned} \tag{3.11}$$

Thereby, the two conjuncts $\neg Holds(At(1,1), z'')$ and $\neg Holds(Facing(1), z'')$ of formula (3.10) became redundant in the light of the additional constraints $(\forall x, y)\,\neg Holds(At(x,y), z'')$ and $(\forall d)\,\neg Holds(Facing(d), z'')$. □

The general form of an incomplete equational state specification, suitable for insertion into an update equation, is $(\exists z)\,(State(s) = \tau \circ z \wedge \Phi)$. Combining this with an equation $State(Do(a,s)) = State(s) - \vartheta^- + \vartheta^+$ deriving from a state update axiom yields

$$(\exists z)\,(\,State(Do(a,s)) = \tau \circ z - \vartheta^- + \vartheta^+ \,\wedge\, \Phi\,)$$

This formula provides a specification of the successor situation. If possible, rewriting steps may be applied to simplify the term to the right of the equation sign. In particular, negative effects may be "cancelled out" as in special fluent calculus with the help of the cancellation law (Proposition 1.7, page 10).

Example 2 (continued) The three actions of the cleaning robot shall be represented by the following constants:

Go :	ACTION	go forward to adjacent square
Turn :	ACTION	turn clockwise by 90°
Clean :	ACTION	empty waste bin at current location

Let the precondition axioms be

$$\begin{aligned}
Poss(Go,z) \;&\equiv\; (\forall d,x,y)\,(\,Holds(At(x,y),z) \wedge \\
&\qquad\qquad Holds(Facing(d),z) \supset \\
&\qquad\qquad\quad (\exists x',y')\,Adjacent(x,y,d,x',y')) \qquad (3.12)\\
Poss(Turn,z) \;&\equiv\; \top \\
Poss(Clean,z) \;&\equiv\; \top
\end{aligned}$$

in conjunction with the auxiliary axiom

$$\begin{aligned}
Adjacent(x,y,d,x',y') \;\equiv\;& 1 \le d \le 4 \,\wedge\, 1 \le x,x',y,y' \le 5 \,\wedge \\
& [\,d=1 \wedge x'=x \wedge y'=y+1 \,\vee \\
& \;\; d=2 \wedge x'=x+1 \wedge y'=y \,\vee \qquad (3.13)\\
& \;\; d=3 \wedge x'=x \wedge y'=y-1 \,\vee \\
& \;\; d=4 \wedge x'=x-1 \wedge y'=y\,]
\end{aligned}$$

That is to say, going forward requires the robot not to face the wall of the building while making a quarter turn clockwise and emptying a waste bin is always possible. The state update axioms for the three actions are as follows:

$$\begin{aligned}
& Poss(Go,s) \supset \\
& \quad (\exists d,x,y,x',y')\,(\,Holds(At(x,y),s) \wedge Holds(Facing(d),s) \wedge \\
& \qquad\qquad Adjacent(x,y,d,x',y') \wedge \\
& \qquad State(Do(Go,s)) = State(s) - At(x,y) + At(x',y')\,) \\
& \\
& Poss(Turn,s) \supset \\
& \quad (\exists d)\,(Holds(Facing(d),s) \wedge \qquad (3.14)\\
& \qquad State(Do(Turn,s)) = \\
& \qquad\quad State(s) - Facing(d) + Facing(d \bmod 4 + 1) \\
& \\
& Poss(Clean,s) \supset \\
& \quad (\exists x,y)\,(Holds(At(x,y),s) \wedge \\
& \qquad State(Do(Clean,s)) = State(s) + Cleaned(x,y)\,)
\end{aligned}$$

Put in words, going forward brings the robot to the adjacent location. Turning has the effect that the cleanbot faces the direction which is next on the compass. Cleaning causes the waste bin in the current square to be cleaned.

Recall the reformulation (3.11) of our sample initial state specification, and consider the action Go. From the first precondition axiom in (3.12) it follows that $Poss(Go, S_0)$ since square $(1, 2)$ is adjacent to the initial position $(1, 1)$ in the direction the robot faces initially, viz. $d = 1$. Thus the state update axiom in (3.14) for Go implies

$$State(Do(Go, S_0)) = State(S_0) - At(1,1) + At(1,2)$$

Combining this update equation with formula (3.11), it follows that there exists z'' such that

$$State(Do(Go, S_0)) = At(1,1) \circ Facing(1) \circ z'' - At(1,1) + At(1,2)$$

along with all sub-formulas in (3.11) which constrain state variable z''. Macro expansion implies the existence of a state z such that $\neg Holds(At(1,1), z)$ and

$$z \circ At(1,1) = At(1,1) \circ Facing(1) \circ z'' \wedge State(Do(Go, S_0)) = z \circ At(1,2)$$

The equation to the left allows the application of the cancellation law thanks to $\neg Holds(At(1,1), z'')$ according to (3.11), and due to the foundational axioms of irreducibility and decomposition along with uniqueness-of-names. It follows that $z = Facing(1) \circ z''$. Hence, along with all constraints on z'' known from axiom (3.11) we obtain

$$\begin{array}{l}(\exists z'')\,(\, State(Do(Go, S_0)) = At(1,2) \circ Facing(1) \circ z'' \\ \quad \wedge \neg Holds(Occupied(1,1), z'') \\ \quad \wedge \ldots \wedge \neg Holds(Occupied(4,5), z'') \\ \quad \wedge (\forall x, y)\, \neg Holds(Cleaned(x,y), z'') \\ \quad \wedge (\forall x, y)\, \neg Holds(At(x,y), z'') \\ \quad \wedge (\forall d)\, \neg Holds(Facing(d), z'') \\ \quad \wedge (\forall x)\, \neg Holds(Occupied(x,0), z'') \\ \quad \wedge (\forall x)\, \neg Holds(Occupied(x,6), z'') \\ \quad \wedge (\forall y)\, \neg Holds(Occupied(0,y), z'') \\ \quad \wedge (\forall y)\, \neg Holds(Occupied(6,y), z''))\end{array}$$

Thus we have derived an incomplete equational state specification of the successor situation. □

In general, an incomplete state may not entail whether a fluent to be added or removed currently holds or does not. In this case, partial information on this fluent entailed by the state may no longer be entailed by the updated state. Consider, for example, the partial state specification

$$(\exists z)\,(State(s) = F(y) \circ z \wedge [Holds(F(A), z) \vee Holds(F(B), z)]\,) \tag{3.15}$$

along with the update equation

$$State(Do(a,s)) = State(s) - F(A)$$

With the usual inference scheme it follows that

$$\begin{aligned}(\exists z)\,(\,&State(Do(a,s)) = (F(y) \circ z) - F(A) \\ &\wedge [\,Holds(F(A),z) \vee Holds(F(B),z)\,]\,)\end{aligned} \tag{3.16}$$

This formula does not entail whether $F(A)$ holds in the state it is subtracted from. By macro expansion of "$-$" it follows that $\Sigma_{state} \cup \{(3.16)\}$ implies $\neg Holds(F(A), Do(a,s))$. But it cannot be decided whether $F(y)$ holds in $State(Do(a,s))$ or whether $F(B)$ does, since $\Sigma_{state} \cup \{(3.16)\}$ entails

$$\begin{aligned}&[\,y = A \supset \neg Holds(F(y), Do(a,s))\,] \wedge \\ &[\,y \neq A \supset Holds(F(y), Do(a,s))\,] \wedge \\ &[\,\neg Holds(F(B),z) \supset \neg Holds(F(B), Do(a,s))\,] \wedge \\ &[\,Holds(F(B),z) \supset Holds(F(B), Do(a,s))\,]\end{aligned}$$

Hence, (3.16) just entails that

$$(\exists z)\,(State(Do(a,s)) = z \wedge \neg Holds(F(A),z))$$

so that partial knowledge in (3.15) concerning the fluent that has been removed is not implied by the resulting state specification.

Partial information is lost in the same way when adding a fluent wrt. an incomplete state specification which does not entail the status of this fluent. Consider, for example, the specification

$$\begin{aligned}(\exists z)\,(\,&State(s) = z \\ &\wedge [\,Holds(F(x),z) \vee Holds(F(A),z)\,] \\ &\wedge \neg Holds(F(B),z)\,)\end{aligned} \tag{3.17}$$

along with the update equation

$$State(Do(a,s)) = State(s) + F(x)$$

With the usual inference scheme it follows that

$$\begin{aligned}(\exists z)\,(\,&State(Do(a,s)) = z + F(x) \\ &\wedge [Holds(F(x),z) \vee Holds(F(A),z)] \\ &\wedge \neg Holds(F(B),z)\,)\end{aligned} \tag{3.18}$$

This formula does not entail whether $F(A)$ holds, nor whether $F(B)$ does, since $\Sigma_{state} \cup \{(3.18)\}$ implies

$$\begin{aligned}&[\,Holds(F(A),z) \vee x = A \supset Holds(F(A), Do(a,s))\,] \wedge \\ &[\,\neg Holds(F(A),z) \wedge x \neq A \supset \neg Holds(F(A), Do(a,s))\,] \wedge \\ &[\,x = B \supset Holds(F(B), Do(a,s))\,] \wedge \\ &[\,x \neq B \supset \neg Holds(F(B), Do(a,s))\,]\end{aligned}$$

Hence, (3.18) just entails that

$$(\exists z)\,(State(Do(a,s)) = F(x) \circ z \wedge \neg Holds(F(x), z))$$

so that partial knowledge in (3.17) concerning the fluent that has been added is not implied by the resulting state specification.

The fundamental Theorem 1.14 of Chapter 1, which says that state update axioms solve the frame problem, carries over to general fluent calculus as it makes no assumption about states being completely specified. Hence, all parts of a state specification are still entailed wrt. the updated state if they are not potentially affected by the update.

While states may be infinite in general fluent calculus, state update axioms are still restricted to finite positive and negative effects by definition. It is not too difficult to generalize the notion of an update equation to infinite collections of effects (see Exercise 3.6). However, this necessitates the intensional description of resulting states, which would spoil the computational merits of the extensional definition. For the moment, therefore, agent programs rely on the assumption that actions always have a bounded number of effects. Later, in Chapter 9, we will consider a different way of generalizing the concept of a state update axiom by which an unbounded number of effects can be specified in an extensional manner.

3.3 Bibliographical Notes

Reasoning about actions in the presence of incomplete states is intimately connected to the frame problem [McCarthy and Hayes, 1969]. The early approach of [Green, 1969] has simply circumvented this problem by introducing additional axioms for all non-effects of actions. Tracing back to a proposal of [Sandewall, 1972], the search for more concise solutions to the frame problem has stimulated the development of **nonmonotonic** logics [Bobrow, 1980]. A prominent approach based on **circumscription** [McCarthy, 1986] has proved to be erroneous by an infamous counterexample known as the "Yale Shooting scenario" [Hanks and McDermott, 1987]. This has led to a variety of refined attempts to solve the frame problem in situation calculus, such as [Shoham, 1988; Kautz, 1986; Lifschitz, 1987; Baker, 1991; Turner, 1997; Kartha and Lifschitz, 1995]. Some of these have turned out to cause other problems [Kartha, 1994; Stein and Morgenstern, 1994]. These difficulties with nonmonotonic approaches have inspired the search for concise solutions to the frame problem in classical logic [Kowalski, 1979; Haas, 1987; Schubert, 1990; Elkan, 1992], culminating in the concept of **successor state axioms** [Reiter, 1991]. State update axioms in fluent calculus can be viewed as a development of successor state axioms by introducing the notion of a state into situation calculus [Thielscher, 1999].

As an alternative to the situation-based representation of actions, the **event calculus** [Kowalski and Sergot, 1986] uses time points as a fundamental sort, which defines a linear time structure. Originally formulated as a logic program, a more general variant appeals to circumscription as a solution to the frame problem [Shanahan, 1995]. Differences and common grounds of situation calculus and event calculus have been extensively studied, e.g., [Kowalski and Sadri, 1994; Miller and Shanahan, 1994].

Formal methods for assessing the range of applicability of action formalisms aim at comparing different approaches to the frame problem [Lifschitz, 1987; Lin and Shoham, 1991]. The two most prominent frameworks today originate in the articles [Sandewall, 1993a] and, introducing the **action description language**, [Gelfond and Lifschitz, 1993]. The former is topic of the textbook [Sandewall, 1994], in which a number of nonmonotonic solutions to the frame problem have been formally assessed. This method itself has led to an extensive action representation formalism [Doherty *et al.*, 1998] based on a linear time structure as in event calculus. The action description language, on the other hand, has been used, for example, in [Kartha, 1993] to prove correctness of nonmonotonic action formalisms for well-defined problem classes. The language itself has been developed into various directions [Baral, 1995; Giunchiglia *et al.*, 1997; Kakas and Miller, 1997b] and has led to successful logic programming-based systems [McCain and Turner, 1998; Kakas *et al.*, 2001].

Mainly around the time where many attempts to solve the frame problem have proved erroneous, the problem has been subject to philosophical reflections [Dennet, 1984; Pylyshyn, 1987] questioning the feasibility of Artificial Intelligence in general.

3.4 Exercises

3.1. Find a model for $\Sigma_{sstate} \cup \{(\exists z, f)(\forall z')\, z' \neq z - f\}$.
Hint: Construct a model for sort STATE which includes some but not all infinite sets of fluents.

3.2. Suppose the cleanbot is initially at square $(1, 1)$ while it is unknown whether it faces north or west, that is,

$$Holds(At(1,1), S_0) \wedge [Holds(Facing(1), S_0) \vee Holds(Facing(4), S_0)]$$

Use the domain axioms for the cleaning robot to prove that

(a) the robot does not face east initially;

(b) the robot faces north or east after turning;

(c) action *Go* is possible after turning;

(d) if the robot is in square $(1, 2)$ after turning and going forward, then it faces north.

3.3. Use the axiom of state existence to prove that

(a) there is a state z in which the cleanbot is at square $(1, 1)$ facing north and in which all instances of $Cleaned(x, y)$ hold;

(b) for any state z_1 there is a state z_2 which is as z_1 except that no instance of $Cleaned(x, y)$ holds.

Show that if z_1 in Exercise (b) is substituted by z of Exercise (a), then z_2 of Exercise (b) satisfies $\Sigma_{state} \models z_2 = At(1, 1) \circ Facing(1)$.

3.4. Prove that the axiomatization of the cleaning robot domain is deterministic.

3.5. (a) Define a macro "$\ominus$" which generalizes "$-$" to the subtraction of infinite states.

(b) Let z_1 be the state which consists of all instances of fluent $G(n)$ where $n \in \mathbb{N}$, and let z_2 be the state which consists of all instances of fluent $G(n)$ where n is an even natural number. Show that if $z_3 = z_1 \ominus z_2$, then z_3 consists of all instances of $G(n)$ where n is an odd number.

(c) Prove that "$\ominus$" is indeed a generalization of "$-$", that is,

$$z_2 = z_1 - (f_1 \circ \ldots \circ f_n) \equiv z_2 = z_1 \ominus (f_1 \circ \ldots \circ f_n)$$

(d) Prove that foundational axioms Σ_{state} entail

$$(\forall z_1, z_2)(\exists z_3)\, z_3 = z_1 \ominus z_2$$

3.6. (a) Use the previous exercise to define the macro for a generalized state equation $z_2 = (z_1 \ominus z^-) \oplus z^+$ for arbitrary states z^+, z^-.

(b) Generalize Theorem 1.14 to "$\ominus, \oplus$" and prove it.

(c) Consider the fluent *File*(*dir*, *file*) with the intended meaning that *file* is in directory *dir*. Use "$\ominus, \oplus$" to define an update axiom for the action $Move{*}(dir_1, dir_2)$ which moves all files from directory dir_1 to dir_2. Show that after *Move*∗("/tmp/", "home/") all files that were initially in */tmp/* are in *home/* and that directory */tmp/* is empty.

3.7. Design the environment of an agent to buy books at a given selection of Internet stores.

(a) Consider the fluents *Price*(*book*, *store*, *price*), indicating that *store* has *book* in stock and sells it at *price*; *GiftCertificate*(*store*, *price*), indicating that the agent possesses electronic gift certificates for *store* of total value *price*; and, finally, *InCart*(*book*, *store*), indicating that the agent has put *book* into its virtual cart at *store*. Define suitable domain constraints for these fluents.

(b) Define precondition axioms and state update axioms for the actions *Add*(*book*, *store*) and *Remove*(*book*, *store*), for adding and removing, respectively, *book* from the cart at *store*.

(c) Define the auxiliary predicate *Total*(*price*, *s*) such that *price* is the total amount to be paid for the contents of the carts of all shops in the state of situation *s*.

(d) Axiomatize the following information concerning the initial situation: No store sells $B_1 = \langle$"Gottlob Frege","Begriffsschrift"$\rangle$; Store *nile.com* sells book $B_2 = \langle$"Jonathan Franzen","The Corrections"$\rangle$ at \$20.95 while *brownwell.com* sells this book at \$18.20; the agent has a gift certificate of \$5 for *nile.com*; and all carts are empty. Show that no ground situation $\sigma = Do([\alpha_1, \ldots, \alpha_n], S_0)$ exists such that $(\exists x)\, Holds(InCart(B_1, x), \sigma)$ is entailed. Prove that a ground situation σ exists which satisfies

$$Holds(InCart(B_2), \sigma) \wedge [\, Total(p, \sigma) \supset p \leq \$16\,]$$

Chapter 4

General FLUX

In this chapter, we use the ideas of general fluent calculus to extend the special FLUX system. The goal is to provide agents with the general ability to set up and maintain an incomplete world model and to reason about sensor information. To this end, we will present

4.1. an encoding of incomplete states using **constraints** in logic programs;

4.2. a method for handling these constraints;

4.3. a proof that this method is correct wrt. the foundational axioms of fluent calculus;

4.4. an extension of the FLUX kernel for the update of incomplete states.

4.1 Incomplete FLUX States

General FLUX is based on a particular technique for the encoding of incomplete state specifications. Incomplete collections of fluents are modeled by lists with a **variable tail**. This is a very natural way of representing compound states in fluent calculus which contain a state variable. For example, the term $At(1,1) \circ Facing(1) \circ z$ is encoded by this list:

$$[At(1,1), Facing(1) \mid z]$$

Incomplete lists cannot be subjected to the principle of negation-as-failure as employed in special FLUX for the evaluation of negated *Holds* expressions. This is because an assertion like, for example,

$$\neg Holds(Occupied(1,1),\ [At(1,1), Facing(1) \mid z])$$

would simply fail: The corresponding affirmative *Holds* statement can always be satisfied via substituting the tail variable, z, by a list starting with the fluent in question. General FLUX, therefore, appeals to the paradigm of **constraint**

logic programming (CLP, for short). Certain atoms, the so-called constraints, which can occur in clause bodies and queries, may be partially evaluated only and kept in store for later consideration.

Constraints add a great deal of flexibility and efficiency to logic programs. A standard constraint system is **FD** (for: *finite domains*), which includes arithmetic constraints over integers. Based on the standard functions $+$, $-$, and $*$, constraints are built using the usual equality, inequality, and ordering predicates. **Range constraints** of the form $x :: [m..n]$ define an interval of natural numbers for a variable. All of these constraints can be logically combined using conjunction and disjunction, respectively. Constraints may occur in the body of program clauses as well as in queries. In the course of a derivation, the encountered constraints are put into a **constraint store**. The contents of the store is constantly evaluated in the background, and a derivation fails as soon as an inconsistency among the constraints is detected. If a derivation succeeds with one or more constraints still unsolved, then these are part of the computed answer. For example, the query $x :: [1..4],\ x > 2 * y,\ y \geq 1$ results in the answer $\{y/1\}$ along with the unsolved constraint $x :: [3..4]$. If the sample query included, say, $x < 3$, it would fail.

FLUX employs two kinds of constraints. First, special **state constraints** encode both negative and disjunctive state knowledge by restricting the tail variable of an incomplete list. Second, standard **arithmetic constraints** are used to define restrictions on unknown arguments of fluents. In so doing, FLUX assumes the arguments of fluents to be encoded by natural or rational numbers. This enables the use of a standard high-speed constraint solver.

The available basic state constraints determine the set of state specifications that can be directly expressed in FLUX.

Definition 4.1 A state formula $\Phi(z)$ is **FLUX-expressible** if it is of the form

$$(\exists \vec{x}, z')\,(z = \varphi_1 \circ \ldots \circ \varphi_m \circ z' \wedge \Psi_1(\vec{x}, z') \wedge \ldots \wedge \Psi_n(\vec{x}, z')) \qquad (4.1)$$

where $m, n \geq 0$, each φ_i $(1 \leq i \leq m)$ is a fluent with variables among $\vec{x}$, and each $\Psi_j(\vec{x}, z')$ $(1 \leq j \leq n)$ is of one of the following forms:

1. $\neg Holds(\varphi, z')$, where φ is a fluent with variables among $\vec{x}$;

2. $(\forall \vec{y})\neg Holds(\varphi, z')$, where φ is a fluent with variables $\vec{y}$;

3. $Holds(\varphi'_1, z') \vee \ldots \vee Holds(\varphi'_k, z')$, where $k \geq 1$ and φ'_i $(1 \leq i \leq k)$ is a fluent with variables among $\vec{x}$;

4. quantifier-free arithmetic formula over integers with variables among $\vec{x}$.

□

Thus, FLUX-expressible are state specifications which consist of existentially quantified positive and negated fluents, universally quantified negated fluents, positive disjunctions, and auxiliary axioms expressing properties of the arguments of fluents. For example, the fact that a fluent F has a unique value in a state z is FLUX-expressible by the formula

$$(\exists \vec{x}, z')\,(z = F(\vec{x}) \circ z' \,\wedge\, (\forall \vec{y})\,\neg Holds(F(\vec{y}), z')\,) \qquad (4.2)$$

Likewise FLUX-expressible is the initial state specification in the cleaning robot domain, if we take the rewritten formula (3.11) on page 66.

Universally quantified positive statements, like $(\forall n : \mathbb{N})\, Holds(F(n), z)$, cannot be directly expressed with the state constraints considered in this book (but see Exercise 4.2 for how to encode universally quantified domain constraints by adding domain-specific constraints to a FLUX program). This ensures that only a finite number of fluents holds in a state. When designing a signature for FLUX agents, the programmer must therefore take care that the fluent functions are defined in such a way that all states can be encoded by finitely many positive fluents. In some cases this may be achieved by replacing a fluent F, of which infinitely many instances may be true in some state, by a new fluent $\overline{F}$ whose semantics is $\neg F$ and, hence, of which infinitely many instances would be false in the same state. For example, it is not advisable to use a fluent $NotInDirectory(x, y)$ to represent that a file x is *not* contained in a directory y; one should rather employ the semantic negation $InDirectory(x, y)$. Further state formulas that are not FLUX-expressible according to Definition 4.1 are disjunctions with negated literals or exclusive disjunctions, but see Exercises 4.3 and 4.5 for an extension of the expressiveness.

FLUX-expressible formulas are encoded as so-called FLUX states, which consist of an incomplete list of fluents along with a set of constraints. The constraints encode the sub-formulas Ψ_j of a state formula (4.1). An auxiliary constraint called $DuplicateFree(z)$ is used to ensure that the state list z is free of multiple occurrences of fluents.

Definition 4.2 Let $\langle \mathcal{F}, \circ, \emptyset \rangle$ be a fluent calculus state signature. A **FLUX state** $\Phi(z)$ consists of a list

$$z = [f_1, \ldots, f_n \mid z']$$

of pairwise different fluents ($n \geq 0$) along with the constraint $DuplicateFree(z)$ and a finite number of constraints of the form

1. $NotHolds(f, z')$;
2. $NotHoldsAll(f, z')$;
3. $OrHolds([f_1', \ldots, f_k'], z')$ where $k \geq 1$;
4. arithmetic constraints.

```
init(Z0) :- Z0 = [at(1,1),facing(1) | Z],
            not_holds(occupied(1,1), Z),
            not_holds(occupied(2,1), Z),           % hallway
            ..., not_holds(occupied(4,5), Z),      %
            not_holds_all(cleaned(_,_), Z),
            consistent(Z0).

consistent(Z) :-
   holds(at(X,Y),Z,Z1) -> [X,Y]::1..5, not_holds_all(at(_,_),Z1),
   holds(facing(D),Z,Z2) -> [D]::1..4, not_holds_all(facing(_),Z2),
   not_holds_all(occupied(_,0), Z),
   not_holds_all(occupied(_,6), Z),
   not_holds_all(occupied(0,_), Z),
   not_holds_all(occupied(6,_), Z),
   duplicate_free(Z).
```

Figure 4.1: Sample state specification in general FLUX.

If $\Phi(z)$ is a FLUX-expressible state formula (4.1), then by $[\![\Phi(z)]\!]$ we denote any FLUX state containing the same positive fluents and the respective set of constraints 1.–4. representing the sub-formulas Ψ_j. □

Again, we will distinguish between a FLUX-expressible formula and the actual FLUX state only when necessary.

Example 2 (continued) Figure 4.1 shows the FLUX encoding of the initial state specification (3.11) (page 66) in the cleaning robot scenario. The auxiliary predicate *Consistent*(z) defines state z to be consistent in view of the underlying domain constraints. A bit tricky is the encoding of the fact that there is always a unique instance of the two functional fluents, *At* and *Facing*. Inspired by formula (4.2), this is expressed via the ternary *Holds*(f, z, z_1) known from special FLUX, where state z_1 is z without fluent f. □

4.2 FLUX Constraint Solver*

The introduction of constraints requires to extend the FLUX kernel by a constraint solver. A standard constraint system for arithmetics is used to handle constraints on variable arguments of fluents. We follow the syntax of the FD-library for Eclipse, where the arithmetic predicates are preceded by the symbol "#" to distinguish them from the standard Prolog operations. Conjunctions and disjunctions are denoted by "#/\" and "#\/", respectively.

The second component of the constraint solver for general FLUX is a system of rules for handling the state constraints of Definition 4.2.

Constraint Handling Rules (**CHR**s, for short) are a general method of specifying, in a declarative way, rules for processing constraints. Consider an arbitrary signature for constraints which consists of predicate and function symbols, variables, and the special constraint *False*. Then a Constraint Handling Rule is an expression of the form

$$H_1, \ldots, H_m \Leftrightarrow G_1, \ldots, G_k \mid B_1, \ldots, B_n \tag{4.3}$$

where

- the **head** $H_1, \ldots, H_m$ is a sequence of constraints ($m \geq 1$);
- the **guard** $G_1, \ldots, G_k$ and the **body** $B_1, \ldots, B_n$ are queries ($k, n \geq 0$).

An empty guard is omitted; the empty body is denoted by *True*. The declarative interpretation of a CHR of the form (4.3) is given by the formula

$$(\forall \vec{x})\,(G_1 \wedge \ldots \wedge G_k \supset [H_1 \wedge \ldots \wedge H_m \equiv (\exists \vec{y})\,(B_1 \wedge \ldots \wedge B_n)]) \tag{4.4}$$

where $\vec{x}$ are the variables in both guard and head and $\vec{y}$ are the variables which additionally occur in the body. Consider, e.g., a constraint $Doublet(x)$ with the intended meaning that x is a list of doublets, i.e., pairs of identical elements, as in $[3,3,4,4]$. The following two CHRs can be used to handle instances of this constraint:

$$Doublet([x_1, x_2 \mid y]) \Leftrightarrow \neg x_1 = x_2 \mid False \tag{4.5}$$

$$Doublet([x \mid y]) \Leftrightarrow y = [x \mid y'],\ Doublet(y') \tag{4.6}$$

Their interpretation is $(\forall x_1, x_2, y)\,(x_1 \neq x_2 \supset (Doublet([x_1, x_2 \mid y]) \equiv \bot))$ and $(\forall x, y)\,(Doublet([x \mid y]) \equiv (\exists y')\,(y = [x \mid y'] \wedge Doublet(y')))$.
The procedural interpretation of a CHR is given by an extension of the notion of a resolvent in a computation tree. If the head of a CHR can be matched against elements $C_1, \ldots, C_m$ of the constraint store using a substitution θ and if the subsidiary computation tree for the guard $G_1, \ldots, G_k$ is successful, then the constraints $C_1, \ldots, C_m$ are removed from the constraint store and the instantiated body $(B_1, \ldots, B_n)\theta$ of the CHR is added to the current query. If the special constraint *False* is thus introduced, then the entire derivation fails. Consider, e.g., the query $Doublet([3, x \mid z])$, then the guard in CHR (4.5) fails while CHR (4.6) is applicable and results in $\{x/3\}$ along with the pending constraint $Doublet(z)$. The query $Doublet([3, x, x, 4])$, on the other hand, by CHR (4.6) and substitution $\{x/3\}$ yields $Doublet([3, 4])$, which in turn fails according to CHR (4.5).
The notation $H_1 \setminus H_2 \Leftrightarrow G \mid B$ abbreviates $H_1, H_2 \Leftrightarrow G \mid H_1, B$. It is commonly used to remove one or more constraints H_2 which are subsumed by constraints H_1.

```
not_holds(_,[])          <=> true.                                   %1
not_holds(F,[F1|Z]) <=> neq(F,F1), not_holds(F,Z).                   %2

not_holds_all(_,[])          <=> true.                               %3
not_holds_all(F,[F1|Z]) <=> neq_all(F,F1), not_holds_all(F,Z).       %4

not_holds_all(F,Z) \ not_holds(G,Z)        <=> inst(G,F) | true.     %5
not_holds_all(F,Z) \ not_holds_all(G,Z) <=> inst(G,F) | true.        %6

duplicate_free([])        <=> true.                                  %7
duplicate_free([F|Z]) <=> not_holds(F,Z), duplicate_free(Z).         %8

neq(F,F1)         :- or_neq(exists,F,F1).
neq_all(F,F1) :- or_neq(forall,F,F1).

or_neq(Q,Fx,Fy) :- functor(Fx,F,M), functor(Fy,G,N),
                   ( F=G, M=N -> Fx =.. [_|ArgX], Fy =.. [_|ArgY],
                                 or_neq(Q,ArgX,ArgY,D), call(D)
                               ; true ).

or_neq(_,[],[],(0#\=0)).
or_neq(Q,[X|X1],[Y|Y1],D) :-
  or_neq(Q,X1,Y1,D1),
  ( Q=forall, var(X), \+ is_domain(X)
    -> ( binding(X,X1,Y1,YE) -> D=((Y#\=YE)#\/D1)
                              ; D=D1 )
     ; D=((X#\=Y)#\/D1) ).

binding(X,[X1|ArgX],[Y1|ArgY],Y) :- X==X1 -> Y=Y1
                                            ; binding(X,ArgX,ArgY,Y).
```

Figure 4.2: CHRs for negation.

4.2.1 Negation Constraints

Figure 4.2 depicts the first part of the system of CHRs for FLUX, which handles the negation constraints as well as the auxiliary constraint on multiple occurrences. The topmost CHR defines $NotHolds(f, z)$ to be true if z is the empty state, which is a direct encoding of the empty state axiom of fluent calculus. The second CHR propagates a constraint $NotHolds(f, z)$ through a compound state, following the contraposition of the foundational axioms of decomposition and irreducibility. The auxiliary predicate $Neq(f, f_1)$ encodes the disequation $f \neq f_1$ and is defined by auxiliary clauses (see below). In a similar fashion, the next two CHRs are used to evaluate $NotHoldsAll(f, z)$, where the auxiliary predicate $NeqAll(f, f_1)$ encodes the disequation $(\forall \vec{x})\, f \neq f_1$ with $\vec{x}$ being the variables in f. By CHRs 5 and 6, a negation constraint of the form $\neg Holds(g, z)$ or $(\forall \vec{x})\, \neg Holds(g, z)$ is resolved if it is subsumed by some other negation constraint $(\forall \vec{y})\, \neg Holds(f, z)$. The auxiliary predicate $Inst(g, f)$ means that fluent g is an instance of fluent f. Finally, the two CHRs for the auxiliary constraint on multiple occurrences stipulate that $DuplicateFree(z)$ is true for the empty list while compound lists are free of duplicates if the head does not occur in the tail and the tail itself is free of multiple occurrences.

The auxiliary clauses in Figure 4.2 reduce the inequality of two fluents to an arithmetic constraint over the arguments. If the two fluent function symbols are unequal, then the clause for the ternary $OrNeq$ succeeds, thus encoding that $F(\vec{x}) \neq G(\vec{y})$ under uniqueness-of-names. Otherwise, the resulting arithmetic constraint is a *disjunction* of argument-wise disequations. The base case of the empty disjunction is encoded as the unsatisfiable constraint $0 \neq 0$. The distinction between an existential disequation $Neq(f, f_1)$ and a universal $NeqAll(f, f_1)$ is made in the penultimate clause: In case of universal quantification, the pure variable arguments[1] in f are discarded while possibly giving rise to dependencies among the arguments of the second fluent. These dependencies are inferred using the auxiliary predicate $Binding(x, \vec{x}_1, \vec{y}_1, y)$, which is true if x reappears in $\vec{x}_1$ and y occurs in $\vec{y}_1$ at the same position as x does in $\vec{x}_1$. For example, $NeqAll(F(x, x, x), F(u, v, w))$ reduces to the arithmetic constraint $u \neq v \vee v \neq w$ whereas $NeqAll(F(x, 2, y), F(u, v, w))$ reduces to $2 \neq v$. In case of existentially quantified variables, all arguments participate in the resulting constraint, so that, e.g., $Neq(F(x, 2, y), F(u, v, w))$ reduces to $x \neq u \vee 2 \neq v \vee y \neq w$.

The following derivation illustrates these CHRs at work, using a query which combines the information that the cleaning robot is at a unique location $(1, y)$ for $y \in \{2, 3\}$ with the fact that the robot is not in the second row. Not surprisingly, the conclusion is that the robot is at location $(1, 3)$. The elements in the constraint store are framed, and underlined are the atoms or constraints to which the next derivation step applies. The application of a CHR is indicated by its number according to Figure 4.2:

$$\begin{array}{ll}
& \underline{y :: [2..3]},\ z = [At(1, y) \mid z_1],\ NotHoldsAll(At(v, w), z_1), \\
& NotHoldsAll(At(x, 2), z),\ DuplicateFree(z) \\[1ex]
\Leftrightarrow & \boxed{y :: [2..3]},\ \underline{z = [At(1, y) \mid z_1]},\ NotHoldsAll(At(v, w), z_1), \\
& NotHoldsAll(At(x, 2), z),\ DuplicateFree(z) \\[1ex]
\Leftrightarrow & \boxed{y :: [2..3]},\ \underline{NotHoldsAll(At(v, w), z_1)}, \\
& NotHoldsAll(At(x, 2), [At(1, y) \mid z_1]),\ DuplicateFree([At(1, y) \mid z_1]) \\[1ex]
\Leftrightarrow & \boxed{y :: [2..3],\ NotHoldsAll(At(v, w), z_1)}, \\
& \underline{NotHoldsAll(At(x, 2), [At(1, y) \mid z_1])},\ DuplicateFree([At(1, y) \mid z_1])
\end{array}$$

[1] A variable is pure if it does not satisfy the standard FD-predicate $IsDomain(x)$, which is true if x is a constrained variable.

$$\Leftrightarrow \boxed{y :: [2..3],\ NotHoldsAll(At(v,w),z_1)},\ \boxed{\underline{NotHoldsAll(At(x,2),[At(1,y) \mid z_1])}},\ DuplicateFree([At(1,y) \mid z_1])$$

$$\overset{4}{\Leftrightarrow} \boxed{y :: [2..3],\ NotHoldsAll(At(v,w),z_1),\ \underline{NeqAll(At(x,2),At(1,y))}},\ \boxed{NotHoldsAll(At(x,2),z_1)},\ DuplicateFree([At(1,y) \mid z_1])$$

$$\Leftrightarrow \boxed{\underline{y :: [2..3]},\ NotHoldsAll(At(v,w),z_1),\ \underline{2 \neq y \vee 0 \neq 0}},\ \boxed{NotHoldsAll(At(x,2),z_1)},\ DuplicateFree([At(1,y) \mid z_1])$$

$$\Leftrightarrow \boxed{\underline{NotHoldsAll(At(v,w),z_1),\ NotHoldsAll(At(x,2),z_1)}},\ DuplicateFree([At(1,3) \mid z_1])$$

$$\overset{6}{\Leftrightarrow} \boxed{NotHoldsAll(At(v,w),z_1)},\ \underline{DuplicateFree([At(1,3) \mid z_1])}$$

$$\Leftrightarrow \boxed{NotHoldsAll(At(v,w),z_1),\ \underline{DuplicateFree([At(1,3) \mid z_1])}}$$

$$\overset{8}{\Leftrightarrow} \boxed{NotHoldsAll(At(v,w),z_1),\ \underline{NotHolds(At(1,3),z_1)},\ DuplicateFree(z_1)}$$

$$\overset{5}{\Leftrightarrow} \boxed{NotHoldsAll(At(v,w),z_1),\ DuplicateFree(z_1)}$$

Neither of the two resulting constraints can be resolved further. Hence, they are part of the computed answer, which otherwise consists of the substitution $\{y/3\}$ determined by the arithmetic constraint solver from $y :: [2..3]$ and $2 \neq y \vee 0 \neq 0$.

4.2.2 Disjunction Constraints

Figure 4.3 depicts the second part of the system of CHRs for FLUX, which handles disjunction constraints. The solver employs an extended notion of a disjunctive clause, where in addition to fluents each disjunction may include atoms of the form $Eq(\vec{x}, \vec{y})$. The meaning of such a general disjunctive constraint $OrHolds([\delta_1, \ldots, \delta_k], z)$ is

$$\bigvee_{i=1}^{k} \begin{cases} Holds(f, z) & \text{if } \delta_i \text{ is fluent } f \\ \vec{x} = \vec{y} & \text{if } \delta_i \text{ is } Eq(\vec{x}, \vec{y}) \end{cases} \tag{4.7}$$

This extension is needed for propagating disjunctions with variables through compound states. The constraint $OrHolds([F(x), F(1)], [F(y)|z])$, e.g., will be rewritten to $OrHolds([Eq([1],[y]), F(1), Eq([x],[y]), F(x)], z)$. This is in accordance with the fact that $\Sigma_{state} \cup UNA[F]$ entails

$$\begin{gathered} Holds(F(x), F(y) \circ z) \vee Holds(F(1), F(y) \circ z) \\ \equiv \\ x = y \vee Holds(F(x), z) \vee 1 = y \vee Holds(F(1), z) \end{gathered}$$

which follows by the foundational axioms of irreducibility and decomposition.

```
or_holds([F],Z) <=> F\=eq(_,_) | holds(F,Z).                        %9

or_holds(V,Z)    <=> \+(member(F,V), F\=eq(_,_))                    %10
                     | or_and_eq(V,D), call(D).

or_holds(V,[])   <=> member(F,V,W), F\=eq(_,_) | or_holds(W,[]).    %11

or_holds(V,Z)    <=> member(eq(X,Y),V),                             %12
                     or_neq(exists,X,Y,D), \+ call(D) | true.

or_holds(V,Z)    <=> member(eq(X,Y),V,W),                           %13
                     \+(and_eq(X,Y,D), call(D)) | or_holds(W,Z).

not_holds(F,Z)     \ or_holds(V,Z) <=> member(G,V,W), F==G          %14
                                       | or_holds(W,Z).

not_holds_all(F,Z) \ or_holds(V,Z) <=> member(G,V,W), inst(G,F)     %15
                                       | or_holds(W,Z).

or_holds(V,[F|Z])        <=> or_holds(V,[],[F|Z]).                  %16
or_holds([F1|V],W,[F|Z])<=> F1==F -> true ;                         %17
                            F1\=F -> or_holds(V,[F1|W],[F|Z]) ;
                            F1=..[_|ArgX], F=..[_|ArgY],
                            or_holds(V,[eq(ArgX,ArgY),F1|W],[F|Z]).
or_holds([],W,[_|Z])     <=> or_holds(W,Z).                         %18

and_eq([],[],(0#=0)).
and_eq([X|X1],[Y|Y1],D) :- and_eq(X1,Y1,D1), D=((X#=Y)#/\D1).

or_and_eq([],(0#\=0)).
or_and_eq([eq(X,Y)|Eq],(D1#\/D2)) :- and_eq(X,Y,D1), or_and_eq(Eq,D2).
```

Figure 4.3: CHRs for disjunction.

The first rule in Figure 4.3, CHR 9, handles a disjunction consisting of just a single fluent. Any such disjunction is simplified to a standard *Holds* statement. For example, the constraint $OrHolds([Occupied(1, y)], z)$ reduces to $Holds(Occupied(1, y), z)$.

Rule 10 simplifies a pure equality clause to a disjunctive arithmetic constraint. For instance, the constraint $OrHolds([Eq([x], [1]), Eq([2, y], [x, 3])], z)$ reduces to $x = 1 \vee (2 = x \wedge y = 3)$. The next rule, CHR 11, is used to simplify a disjunctive clause in the presence of the empty state. For example, $OrHolds([Occupied(x, 3), Eq([x], [3])], [\,])$ reduces to $OrHolds([Eq([x], [3])], [\,])$. Auxiliary predicate $Member(x, v)$ means that x is an element of v, and for $Member(x, v, w)$ to be true, w must additionally equal v without x.

CHRs 12 and 13 apply to disjunctions which include a decided equality. If the equality is true, then the entire disjunction is true, else if the equality is false, then it gets removed from the disjunction. For example, the constraint $OrHolds([Eq([1, 2], [1, 2]), Facing(3)], z)$ reduces to *True* as $1 \neq 1 \vee 2 \neq 2$ fails. Conversely, the constraint $OrHolds([Eq([1, 2], [x, 3]), Facing(x)], z)$ reduces to $OrHolds([Facing(3)], z)$ as $1 = x \wedge 2 = 3$ fails.

Rules 14–15 model a unit resolution step, by which a positive atom in a disjunction is resolved in the presence of its negation. For example, given the negative constraint $NotHoldsAll(Occupied(0, y), z)$, the disjunctive constraint $OrHolds([Occupied(1, 4), Occupied(0, 3)], z)$ reduces to the disjunction $OrHolds([Occupied(1, 4)], z)$ by CHR 15.

The last group of CHRs, 16–18, encode the propagation of a disjunction through a compound state. Informally speaking, each element in the disjunct is matched against the head of the state, if possible, and the respective equational constraint is introduced into the disjunction. More precisely, with the help of the auxiliary ternary constraint $OrHolds(v, w, [f|z])$, a disjunction is divided into two parts. List v contains the fluents that have not yet been evaluated against the head f of the state list, while list w contains those fluents that have been evaluated. In the special case that the disjunction contains a fluent f_1 which is identical to the head f of the state list, the constraint necessarily holds and, hence, is resolved to *True* by CHR 17. Otherwise, any fluent f_1 in the disjunction which does not unify with f is propagated without inducing an equality. Any fluent f_1 which does unify with f extends the disjunction by the equality of the arguments of f_1 and f. For example, the constraint

$$OrHolds([F(x), F(1)], [F(y)|z])$$

mentioned earlier is re-written by CHR 16 to the ternary disjunction

$$OrHolds([F(x), F(1)], [\,], [F(y)|z])$$

Rule 17 then yields

$$OrHolds([F(1)], [Eq([x], [y]), F(x)], [F(y)|z])$$

Applying the same rule again yields

$$OrHolds([\,], [Eq([1], [y]), F(1), Eq([x], [y]), F(x)], [F(y)|z])$$

Now CHR 18 applies since all unifiable fluents have been evaluated, and the propagation is carried out by combining the equations with the remaining atoms of the disjunction, resulting in $OrHolds([Eq([1],[y]), F(1), Eq([x],[y]), F(x)], z)$.

By appealing to the underlying arithmetic constraint solver, the auxiliary clauses in Figure 4.3 reduce equality of sequences of terms (and disjunctions of these equalities, respectively) to conjunctions of equations (disjunctions of these conjunctions, respectively). The clauses are similar to the encoding of disequations in case of negation constraints. The base case of an empty conjunction is encoded by the tautological constraint $0 = 0$ while an empty disjunction is encoded as before by the unsatisfiable constraint $0 \neq 0$.

The following derivation illustrates the CHRs for disjunction at work, using a query which encodes the state knowledge that there is exactly one occupied office, that this office is in row 3 and in a column higher than 1, and that there is an occupied office in column 1 or 2. The constraint solver arrives at the conclusion that office $(2,3)$ is occupied. As before, the atoms and constraints are underlined which are resolved next, and the application of a CHR is indicated by its number:[2]

$$\begin{aligned}
& \underline{z = [Occupied(x,3) \mid z_1]},\ x > 1,\ NotHoldsAll(Occupied(v,w), z_1), \\
& OrHolds([Occupied(1,y), Occupied(2,y)], z) \\
\Leftrightarrow\ & x > 1,\ NotHoldsAll(Occupied(v,w), z_1), \\
& \underline{OrHolds([Occupied(1,y), Occupied(2,y)], [Occupied(x,3) \mid z_1])} \\
\overset{16}{\Leftrightarrow}\ & x > 1,\ NotHoldsAll(Occupied(v,w), z_1), \\
& \underline{OrHolds([Occupied(1,y), Occupied(2,y)], [\,], [Occupied(x,3) \mid z_1])} \\
\overset{17}{\Leftrightarrow}\ & x > 1,\ NotHoldsAll(Occupied(v,w), z_1), \\
& \underline{OrHolds([Occupied(2,y)], [Eq([1,y],[x,3]), Occupied(1,y)],} \\
& \qquad\qquad\qquad\qquad \underline{[Occupied(x,3) \mid z_1])} \\
\overset{17}{\Leftrightarrow}\ & x > 1,\ NotHoldsAll(Occupied(v,w), z_1), \\
& \underline{OrHolds([\,], [Eq([2,y],[x,3]), Occupied(2,y),} \\
& \qquad \underline{Eq([1,y],[x,3]), Occupied(1,y)], [Occupied(x,3) \mid z_1])} \\
\overset{18}{\Leftrightarrow}\ & x > 1,\ \underline{NotHoldsAll(Occupied(v,w), z_1)}, \\
& \underline{OrHolds([Eq([2,y],[x,3]), Occupied(2,y),} \\
& \qquad \underline{Eq([1,y],[x,3]), Occupied(1,y)], z_1)} \\
\overset{15}{\Leftrightarrow}\ & x > 1,\ \underline{NotHoldsAll(Occupied(v,w), z_1)}, \\
& \underline{OrHolds([Eq([2,y],[x,3]), Eq([1,y],[x,3]), Occupied(1,y)], z_1)}
\end{aligned}$$

[2] Since all but the very first step take place inside of the constraint store, we refrain from making explicit the steps by which the constraints enter the store, and we omit the frames around the elements of the constraint store.

$$\overset{15}{\Leftrightarrow} \; x > 1,\; \mathit{NotHoldsAll}(\mathit{Occupied}(v,w), z_1), \\ \underline{\mathit{OrHolds}([\mathit{Eq}([2,y],[x,3]), \mathit{Eq}([1,y],[x,3])], z_1)}$$

$$\overset{10}{\Leftrightarrow} \; \underline{x > 1},\; \mathit{NotHoldsAll}(\mathit{Occupied}(v,w), z_1), \\ \underline{(2 = x \wedge y = 3 \wedge 0 = 0) \vee (1 = x \wedge y = 3 \wedge 0 = 0) \vee 0 \neq 0}$$

$$\Leftrightarrow \; \mathit{NotHoldsAll}(\mathit{Occupied}(v,w), z_1)$$

The computed answer substitution $\{x/2, y/3\}$ is determined by the arithmetic constraint solver in the last step.

The FLUX constraint solver endows agents with the general ability to logically combine and evaluate information about states. This enables agents to infer implicit knowledge from different observations, which means they are capable of **reasoning** about sensor information and incomplete states.

Example 2 (continued) The clause in Figure 4.4 encodes the specification of the light sensor in the world of the cleaning robot. Given the initial state specification of Figure 4.1, the FLUX constraint solver allows the cleanbot to infer implicit knowledge that follows from the perception of light at certain locations. The following two queries illustrate that the robot is able to evaluate given information about where and where not it would sense light:

```
?- init(Z0), light(1,2,false,Z0), light(1,3,true,Z0).

Z0 = [at(1,1),facing(1) | Z]

Constraints:
not_holds(occupied(1,3), Z)
or_holds([occupied(2,3),occupied(1,4)], Z)
...
```

Put in words, no light in square $(1,2)$ but in square $(1,3)$ implies that office $(1,3)$ is not occupied but either office $(2,3)$ or office $(1,4)$ is so (or both). If the robot is additionally told that there is no light at location $(2,2)$, then it concludes that office $(2,3)$ cannot be occupied and, hence, that office $(1,4)$ must be so:

```
?- init(Z0), light(1,2,false,Z0), light(1,3,true,Z0),
             light(2,2,false,Z0).

Z0 = [at(1,1), facing(1), occupied(1,4) | _Z]

Constraints:
not_holds(occupied(1,3), Z)
not_holds(occupied(2,3), Z)
...
```

□

```
light(X, Y, Percept, Z) :-
   X_east#=X+1, X_west#=X-1, Y_north#=Y+1, Y_south#=Y-1,
   ( Percept=false,
       not_holds(occupied(X,Y), Z),
       not_holds(occupied(X_east,Y), Z),
       not_holds(occupied(X_west,Y), Z),
       not_holds(occupied(X,Y_north), Z),
       not_holds(occupied(X,Y_south), Z) ;
     Percept=true,
       or_holds([occupied(X,Y),
                 occupied(X_east,Y),occupied(X,Y_north),
                 occupied(X_west,Y),occupied(X,Y_south)], Z) ).
```

Figure 4.4: A clause defining the functionality of the light sensor of the cleaning robot.

4.3 Correctness of the Constraint Solver *

The FLUX constraint solver can be proved correct under the semantics of fluent calculus. Thanks to their declarative nature, verification of Constraint Handling Rules against an axiomatic theory is not difficult.

4.3.1 Negation Handling

Let us begin with the auxiliary clauses of Figure 4.2, which define inequality of two fluents in terms of an arithmetic constraint over the arguments. This is justified by the assumption of uniqueness-of-names for fluents. The solver distinguishes two cases, depending on whether the variables in the first fluent are existentially or universally quantified. In view of the latter, we introduce the notion of a **schematic** fluent $f = h(\vec{x}, \vec{r})$ with $\vec{x}$ being the variable arguments in f and $\vec{r}$ the ground arguments. The following observation implies that the clauses defining Neq and $NeqAll$, respectively, are correct.

Observation 4.3 *Consider a fluent calculus signature with a set $\mathcal{F}$ of functions into sort* FLUENT. *Let $f_1 = g(r_1, \ldots, r_m)$ and $f = h(t_1, \ldots, t_n)$ be two fluents and $f_2 = g(x_1, \ldots, x_k, r_{k+1}, \ldots, r_m)$ a schematic fluent. Furthermore, let $Neq(f_1, f) \stackrel{\text{def}}{=} f_1 \neq f$ and $NeqAll(f_2, f) \stackrel{\text{def}}{=} (\forall x_1, \ldots, x_k)\, f_2 \neq f$, then*

1. *if $g \neq h$, then $UNA[\mathcal{F}] \models Neq(f_1, f)$ and $UNA[\mathcal{F}] \models NeqAll(f_2, f)$;*
2. *if $g = h$, then $m = n$, and $UNA[\mathcal{F}]$ entails*

$$Neq(f_1, f) \equiv r_1 \neq t_1 \vee \ldots \vee r_m \neq t_n \vee 0 \neq 0$$
$$NeqAll(f_2, f) \equiv [\textstyle\bigvee_{\substack{i \neq j \\ x_i = x_j}} t_i \neq t_j] \vee r_{k+1} \neq t_{k+1} \vee \ldots \vee r_m \neq t_n \vee 0 \neq 0$$

■

Based on this observation, the CHRs for negation constraints, 1–4, are justified by the following proposition.

Proposition 4.4 *Σ_{state} entails*

1. $\neg Holds(f, \emptyset)$; *and*

2. $\neg Holds(f, f_1 \circ z) \equiv f \neq f_1 \wedge \neg Holds(f, z)$.

Likewise, if $f = g(\vec{x}, \vec{r})$ *is a schematic fluent, then* Σ_{state} *entails*

3. $(\forall \vec{x}) \neg Holds(f, \emptyset)$; *and*

4. $(\forall \vec{x}) \neg Holds(f, f_1 \circ z) \equiv (\forall \vec{x})\, f \neq f_1 \wedge (\forall \vec{x}) \neg Holds(f, z)$.

Proof:

1. Follows by the empty state axiom.

2. We prove that $Holds(f, f_1 \circ z) \equiv f = f_1 \vee Holds(f, z)$:

   "$\Rightarrow$": Follows by the foundational axioms of decomposition and irreducibility.

   "$\Leftarrow$": If $f = f_1$, then $f_1 \circ z = f \circ z$, hence $Holds(f, f_1 \circ z)$. Likewise, if $Holds(f, z)$, then $z = f \circ z'$ for some z', hence $f_1 \circ z = f_1 \circ f \circ z'$, hence $Holds(f, f_1 \circ z)$.

The proof of the second part is similar. ■

Rules 5–6 of the FLUX solver, by which subsumed negation constraints are resolved, are justified since $(\forall \vec{x}) \neg Holds(f, z)$ implies $\neg Holds(g, z)$ for a schematic fluent $f = h(\vec{x}, \vec{r})$ and a fluent g such that $f\theta = g$ for some θ. Likewise, $(\forall \vec{x}) \neg Holds(f, z)$ implies $(\forall \vec{y}) \neg Holds(g, z)$ for any two schematic fluents $f = h_1(\vec{x}, \vec{r})$ and $g = h_2(\vec{y}, \vec{t})$ such that $f\theta = g$ for some θ.

Finally, CHRs 7–8 for the auxiliary constraint on multiple occurrences are correct since the empty list contains no duplicates while a non-empty list contains no duplicates iff the head does not occur in the tail and the tail itself is free of duplicates.

4.3.2 Disjunction Handling

Moving on to part two of the FLUX constraint solver, let us first consider the auxiliary clauses in Figure 4.3 which define equality of two sequences of terms, and disjunctions thereof. Their correctness is implied by the following observation.

Observation 4.5 *Let* $\vec{r} = r_1, \ldots, r_n$ *and* $\vec{t} = t_1, \ldots, t_n$ *be two sequences of terms of equal length* $n \geq 0$, *then*

$$\vec{r} = \vec{t} \equiv r_1 = t_1 \wedge \ldots \wedge r_n = t_n \wedge 0 = 0$$

Furthermore, if $\vec{r}_1, \vec{t}_1, \ldots, \vec{r}_m, \vec{t}_m$ all are sequences of terms such that $m \geq 0$ and for each $1 \leq i \leq m$, $\vec{r}_i, \vec{t}_i$ are of equal length, then

$$\bigvee_{i=1}^{m} [\vec{r}_i = \vec{t}_i] \;\equiv\; \vec{r}_1 = \vec{t}_1 \vee \ldots \vee \vec{r}_m = \vec{t}_m \vee 0 \neq 0$$

■

The first CHR in Figure 4.3 simplifies a singleton disjunction. Correctness of this rule follows by definition, because $OrHolds([f], z)$ means $Holds(f, z)$. CHR 10, by which a pure equational disjunction is reduced to an arithmetic constraint, is correct according to Observation 4.5. CHR 11, by which a disjunction is simplified in the presence of the empty state, is justified by the empty state axiom, which entails

$$[Holds(f, \emptyset) \vee \Psi] \;\equiv\; \Psi$$

for any formula Ψ.

Rules 12 and 13 are justified by Observation 4.5 along with the fact that for any formula Ψ,

$$\vec{r} = \vec{t} \supset [(\vec{r} = \vec{t} \vee \Psi) \;\equiv\; \top]$$
$$\vec{r} \neq \vec{t} \supset [(\vec{r} = \vec{t} \vee \Psi) \;\equiv\; \Psi]$$

Rules 14–15, which model unit resolution steps, are justified since for any formula Ψ,

$$[Holds(f, z) \vee \Psi] \wedge \neg Holds(f, z) \;\equiv\; \neg Holds(f, z) \wedge \Psi$$

and, given that $f_1\theta = f_2$ for some θ,

$$[Holds(f_2, z) \vee \Psi] \wedge (\forall \vec{x})\neg Holds(f_1, z) \;\equiv\; (\forall \vec{x})\neg Holds(f_1, z) \wedge \Psi$$

where $\vec{x}$ are the variables of f_1.

Finally, consider CHRs 16–18, by which a disjunction is propagated through a compound state. Given that the semantics of the ternary $OrHolds(\gamma, \delta, [f \mid z])$ is $OrHolds(\gamma, [f \mid z]) \vee OrHolds(\delta, z)$, the rules are justified by the following proposition.

Proposition 4.6 *Consider a fluent calculus signature with a set $\mathcal{F}$ of functions into sort* FLUENT. *Foundational axioms Σ_{state} and uniqueness-of-names $UNA[\mathcal{F}]$ entail each of the following:*

1. $\Psi \equiv [\Psi \vee \bigvee_{i=1}^{0} \Psi_i]$;
2. $[Holds(f, f \circ z) \vee \Psi_1] \vee \Psi_2 \equiv \top$;
3. $f_1 \neq f \supset ([Holds(f_1, f \circ z) \vee \Psi_1] \vee \Psi_2 \equiv \Psi_1 \vee [Holds(f_1, z) \vee \Psi_2])$;
4. $[Holds(F(\vec{x}), F(\vec{y}) \circ z) \vee \Psi_1] \vee \Psi_2 \equiv \Psi_1 \vee [\vec{x} = \vec{y} \vee Holds(F(\vec{x}), z) \vee \Psi_2]$;
5. $[\bigvee_{i=1}^{0} \Psi_i \vee \Psi] \equiv \Psi$.

```
holds(F,[F|_]).
holds(F,Z) :- nonvar(Z), Z=[F1|Z1], F\==F1, holds(F,Z1).

holds(F,[F|Z],Z).
holds(F,Z,[F1|Zp]) :- Z=[F1|Z1], F\==F1, holds(F,Z1,Zp).
```

Figure 4.5: The definition of *Holds* in general FLUX.

Proof: Claims 1 and 5 are obvious. Claim 2 follows by the definition of *Holds*. Claims 3 and 4 follow from the foundational axioms of decomposition and irreducibility along with $UNA[\mathcal{F}]$. ∎

Correctness of each single CHR implies that the constraint system for FLUX is correct, provided that the underlying arithmetic solver is correct.

Theorem 4.7 *Let P be the program of Figure 4.2 and 4.3 along with a correct arithmetic constraint solver. Consider a fluent calculus signature with a set $\mathcal{F}$ of functions into sort* FLUENT, *and let $[\![\Phi(z)]\!]$ be a FLUX state. If the derivation tree for $P \cup \{\leftarrow [\![\Phi(z)]\!]\}$ fails, then $\Sigma_{state} \cup \Phi(z) \models \bot$.*

Proof: Each CHR being an equivalence transformation on $\Phi(z)$, the claim follows by the correctness of the underlying arithmetic constraint solver. ∎

4.4 Updating Incomplete FLUX States

The constraint solver allows FLUX agents to reason about **static** incomplete state information. We now turn to the problem of reasoning about the effects of actions, that is, the computation of removal and addition of fluents to incomplete states. Generalizing the FLUX kernel of Chapter 2, Figure 4.5 depicts extended definitions for the *Holds* predicate. The only difference to special FLUX of Figure 2.2 (page 28) is the atom $Nonvar(z)$ occurring in the recursive clause. The reason for this addition is that otherwise a query $Holds(\varphi, z)$ with variable z would admit infinitely many answers of the form $z/[f_1, \ldots, f_n, \varphi \mid z']$ $(n \geq 0)$ with variables f_i. Under the semantics of fluent calculus, these substitutions are all subsumed by the answer $z/[\varphi \mid z']$ obtained by just applying the non-recursive clause.

Partially specified states may give rise to more than one way in which a fluent is derived to hold in a state. As an example, consider the derivation tree depicted in Figure 4.6. The leaves are labeled with the respective computed answer substitution along with the remaining elements of the constraint store.

The following result generalizes Theorem 2.2 to the correctness of the clauses in Figure 4.5 in the presence of incomplete states.

Theorem 4.8 *Let P be the program of Figure 4.5 along with the FLUX constraint solver. Consider a fluent calculus state signature with fluent functions $\mathcal{F}$, and let φ be a fluent and $[\![\Phi(z)]\!]$ a FLUX state.*

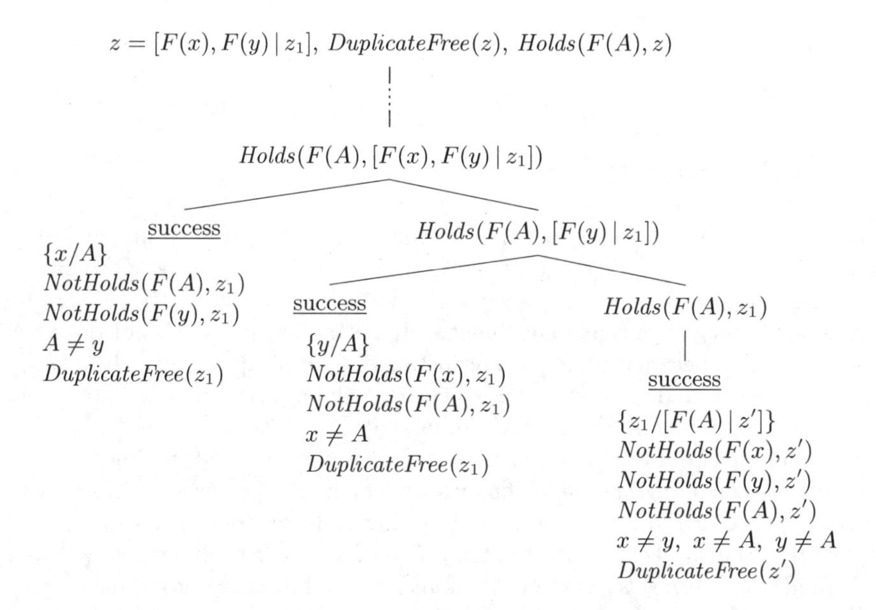

Figure 4.6: A derivation tree with three successful branches, each of which determines a different possibility for fluent $F(A)$ to hold in the partially specified state $[F(x), F(y) \mid z_1]$. The resulting negation constraints are all derived from the initial constraint *DuplicateFree*(z).

---

1. *If* $P \cup \{\leftarrow [\![\Phi(z)]\!], Holds(\varphi, z)\}$ *succeeds with answer* θ, *then*

$$\Sigma_{state} \models \Phi(z)\theta \supset Holds(\varphi, z)\theta$$

2. *If* $P \cup \{\leftarrow [\![\Phi(z)]\!], Holds(\varphi, z)\}$ *fails, then*

$$\Sigma_{state} \cup UNA[\mathcal{F}] \cup \{\Phi(z)\} \cup \{Holds(\varphi, z)\} \models \bot$$

Proof: Let $z = [f_1, \ldots, f_n \mid z']$ be the state equation in $[\![\Phi(z)]\!]$, then the foundational axioms of decomposition and irreducibility imply

$$\Sigma_{state} \cup \{\Phi(z)\} \models Holds(\varphi, z) \equiv \bigvee_{i=1}^{n} [\varphi = f_i] \vee (\exists z'')\, z' = \varphi \circ z'' \tag{4.8}$$

1. If $P \cup \{\leftarrow [\![\Phi(z)]\!], Holds(\varphi, z)\}$ succeeds with answer θ, then $\varphi\theta = f_i\theta$ for some $1 \leq i \leq n$, or $z'\theta = [\varphi \mid z'']$ for some z''. In either case the claim follows by (4.8).

2. If $P \cup \{\leftarrow [\![\Phi(z)]\!], Holds(\varphi, z)\}$ fails, then from the correctness of the constraint solver according to Theorem 4.7 it follows that

$$\Sigma_{state} \cup UNA[\mathcal{F}] \cup \{\Phi(z)\} \cup \{\varphi = f_i\} \models \bot$$

for all $i = 1, \ldots, n$, and also

$$\Sigma_{state} \cup UNA[\mathcal{F}] \cup \{\Phi(z)\{z'/[\varphi \mid z'']\}\} \models \bot$$

The claim follows by (4.8). ∎

This theorem implies that there is a derivation for $Holds(\varphi, z)$ under FLUX state $[\![\Phi(z)]\!]$ whenever there exists a state z satisfying the state specification $\Phi(z) \wedge Holds(\varphi, z)$. Hence, the program can be used to verify whether it can be **consistently asserted** that some fluent φ holds in a state. This is not the same as to say that φ necessarily holds. For instance, a simple query like $Holds(F, z)$ admits the computed answer $\{z/[F \mid z']\}$. But the opposite is satisfiable, too: The query $NotHolds(F, z)$ succeeds immediately. With general FLUX, it is therefore inappropriate to use *Holds* (or *NotHolds*) for conditioning in agent programs. Controlling the behavior of an agent in the presence of incomplete information requires a notion of **knowing** that a fluent does or does not hold. This will be the subject of the upcoming Chapter 5. For the moment, we note that *Holds* or *NotHolds* statements should be used in agent programs only to assert state information, as in the initial state specification of Figure 4.1 or the definition of light perception in Figure 4.4.

Updating a partially specified state by a possibly non-ground fluent requires an extension of the definition of update as used in special FLUX. If the incomplete state entails the status of a fluent φ, then the result of removing or adding φ is inferred as before. On the other hand, in Chapter 3 we have seen that if the status of the fluent is not entailed by the state specification $\Phi(z)$ at hand, then partial knowledge of φ in $\Phi(z)$ does not transfer to the resulting state. The loss of information when updating a state with a fluent whose status is unknown, is reflected in the extended definition of removal and addition of fluents in general FLUX; see Figure 4.7. The recursive clause for *Minus* defines $z - f$ as before in case $\{\Phi(z)\} \cup \{Holds(f, z)\}$ is unsatisfiable or in case $\{\Phi(z)\} \cup \{\neg Holds(f, z)\}$ is unsatisfiable. Otherwise, i.e., if the status of f wrt. $\Phi(z)$ is unknown, then all partial information about f in $\Phi(z)$ is cancelled prior to asserting that f does not hold in the resulting state. Similarly, the recursive clause for *Plus* defines $z + f$ as before in case $\{\Phi(z)\} \cup \{Holds(f, z)\}$ is unsatisfiable or in case $\{\Phi(z)\} \cup \{\neg Holds(f, z)\}$ is unsatisfiable. Otherwise, if the status of f wrt. $\Phi(z)$ is unknown, then all partial information about f in $\Phi(z)$ is cancelled prior to adding f to the state and asserting that f does not hold in the tail.

The definition of cancellation of a fluent f is given in Figure 4.8. In the base case, all negative and disjunctive state information affected by f is cancelled via the constraint $Cancel(f, z)$. The latter is resolved itself by the auxiliary constraint $Cancelled(f, z)$, indicating the termination of the cancellation procedure. In the recursive clause for $Cancel(f, z, z_1)$, each atomic, positive state information that unifies with f is cancelled.

This completes the kernel of general FLUX. To summarize, let P_{kernel} be the program of Figure 4.5, 4.7, and 4.8 along with the FLUX constraint solver. Much

```
holds(F,[F|Z],Z).
holds(F,Z,[F1|Zp]) :-
   nonvar(Z), Z=[F1|Z1], F\==F1, holds(F,Z1,Zp).

minus(Z,[],Z).
minus(Z,[F|Fs],Zp) :-
   ( \+ not_holds(F,Z) -> holds(F,Z,Z1) ;
     \+ holds(F,Z)     -> Z1 = Z ;
     cancel(F,Z,Z1), not_holds(F,Z1) ),
   minus(Z1,Fs,Zp).

plus(Z,[],Z).
plus(Z,[F|Fs],Zp) :-
   ( \+ holds(F,Z)     -> Z1=[F|Z] ;
     \+ not_holds(F,Z) -> Z1=Z ;
     cancel(F,Z,Z2), Z1=[F|Z2], not_holds(F,Z2) ),
   plus(Z1,Fs,Zp).

update(Z1,ThetaP,ThetaN,Z2) :-
   minus(Z1,ThetaN,Z), plus(Z,ThetaP,Z2).
```

Figure 4.7: FLUX clauses for updating incomplete states.

like in case of special FLUX, soundness of the definition of update in general FLUX is established by a series of results concerning the auxiliary predicates. We begin with the auxiliary ternary *Holds* predicate.

Lemma 4.9 *Consider a fluent calculus state signature with fluent functions* $\mathcal{F}$*, and let* φ *be a fluent and* $[\![\Phi(z)]\!]$ *a FLUX state.*

1. *If* $P_{kernel} \cup \{\leftarrow [\![\Phi(z)]\!], Holds(\varphi, z, z_1)\}$ *succeeds with answer* θ*, then*

$$\Sigma_{state} \models \Phi(z)\theta \supset Holds(\varphi, z)\theta \wedge (z - \varphi = z_1)\theta$$

2. *If* $P_{kernel} \cup \{\leftarrow [\![\Phi(z)]\!], Holds(\varphi, z, z_1)\}$ *fails, then*

$$\Sigma_{state} \cup UNA[\mathcal{F}] \cup \{\Phi(z)\} \cup \{Holds(\varphi, z)\} \models \bot$$

Proof: Let $z = [f_1, \ldots, f_n \,|\, z']$ be the state equation in $[\![\Phi(z)]\!]$, then the foundational axioms of decomposition and irreducibility imply

$$\Sigma_{state} \cup \{\Phi(z)\} \models Holds(\varphi, z) \equiv \bigvee_{i=1}^{n} [\varphi = f_i] \vee (\exists z'')\, z' = f \circ z'' \qquad (4.9)$$

1. If $P_{kernel} \cup \{\leftarrow [\![\Phi(z)]\!], Holds(\varphi, z, z_1)\}$ succeeds with answer θ, then either of two cases applies.

```
cancel(F,Z1,Z2) :-
   var(Z1)      -> cancel(F,Z1), cancelled(F,Z1), Z2=Z1 ;
   Z1 = [G|Z]  -> ( F\=G -> cancel(F,Z,Z3), Z2=[G|Z3]
                          ; cancel(F,Z,Z2) ) ;
   Z1 = []      -> Z2 = [].

cancel(F,Z) \ not_holds(G,Z)      <=> \+ F\=G | true.

cancel(F,Z) \ not_holds_all(G,Z) <=> \+ F\=G | true.

cancel(F,Z) \ or_holds(V,Z)       <=> member(G,V), \+ F\=G | true.

cancel(F,Z), cancelled(F,Z) <=> true.
```

Figure 4.8: Auxiliary clauses and CHRs for cancelling partial information about a fluent.

(a) There exists some $i = 1, \ldots, n$ such that $\varphi\theta = f_i\theta$ and where $z_1\theta = [f_1, \ldots, f_{i-1}, f_{i+1}, \ldots, f_n \,|\, z']\theta$. Thus, $(z_1 \circ \varphi = z)\theta$, and from $\mathit{DuplicateFree}(z)$ in $[\![\Phi(z)]\!]$ it follows that $\neg \mathit{Holds}(\varphi\theta, z_1\theta)$. This and (4.9) imply the claim.

(b) For some z'' we have that $z'\theta = [\varphi \,|\, z'']$ and $z_1\theta = [f_1, \ldots, f_n \,|\, z'']\theta$. Thus, $(z_1 \circ \varphi = z)\theta$, and from $\mathit{DuplicateFree}(z)$ in $[\![\Phi(z)]\!]$ it follows that $\neg \mathit{Holds}(\varphi\theta, z'')$ and $\neg \mathit{Holds}(\varphi\theta, f_1 \circ \ldots \circ f_n)$. This and (4.9) imply the claim.

2. Similar to the proof of Theorem 4.8. ■

Correctness of the definition for predicate *Minus* in case a fluent with unknown status is removed follows from the fundamental theorem of fluent calculus: Following Theorem 1.14 (page 18), properties which are unaffected by an update continue to hold in the updated state. This applies to all parts of a state specification which are not cancelled.

Lemma 4.10 *Consider a fluent calculus state signature with fluent functions $\mathcal{F}$, and let ϑ^- be a finite state and $[\![\Phi(z)]\!]$ a FLUX state, then for any $[\![\vartheta^-]\!]$, the query $P_{kernel} \cup \{\leftarrow [\![\Phi(z)]\!], \mathit{Minus}(z, [\![\vartheta^-]\!], z')\}$ succeeds with answer θ such that*

$$\Sigma_{state} \cup \mathit{UNA}[\mathcal{F}] \models \Phi(z)\theta \supset (z - \vartheta^- = z')\theta$$

Proof: Let $[\![\vartheta^-]\!] = [f_1, \ldots, f_n]$, then $\Sigma_{state} \models \vartheta^- = f_1 \circ f_2 \circ \ldots \circ f_n$ according to the foundational axioms of associativity and commutativity. The proof is by induction on n.

If $n = 0$, then $\vartheta^- = \emptyset$ and $[\![\vartheta^-]\!] = [\,]$. Hence, by the non-recursive clause the query $P_{kernel} \cup \{\leftarrow [\![\Phi(z)]\!], \mathit{Minus}(z, \vartheta^-, z')\}$ succeeds with answer θ such that $z\theta = z'\theta$, which proves the claim since $z - \emptyset = z'$ is equivalent to $z = z'$ by definition.

In case $n > 0$, we distinguish three cases.

1. If $P_{kernel} \cup \{\leftarrow [\![\Phi(z)]\!], NotHolds(f_1, z)\}$ fails, then by correctness of the constraint solver (Theorem 4.7), $\Sigma_{state} \cup \{\Phi(z)\} \cup \{\neg Holds(f_1, z)\} \models \bot$. Hence, $\Sigma_{state} \models \Phi(z) \supset Holds(f_1, z)$. From Lemma 4.9 it follows that $P_{kernel} \cup \{\leftarrow [\![\Phi(z)]\!], Holds(f_1, z, z_1)\}$ succeeds with answer substitution θ such that Σ_{state} and $\Phi(z)\theta$ entail $(z - \varphi = z_1)\theta$.

2. If $P_{kernel} \cup \{\leftarrow [\![\Phi(z)]\!], Holds(f_1, z)\}$ fails, then by correctness of the constraint solver (Theorem 4.7), $\Sigma_{state} \cup \{\Phi(z)\} \cup \{Holds(f_1, z)\} \models \bot$. Hence, $\Sigma_{state} \models \Phi(z) \supset \neg Holds(f_1, z)$. The macro definition for "$-$" implies that $z - f_1 = z_1$ is equivalent to $z_1 = z$.

3. Consider any fluent φ occurring in the state equation of $[\![\Phi(z)]\!]$ such that $\Sigma_{state} \cup UNA[\mathcal{F}] \models \varphi \neq f_1$. Then $\Sigma_{state} \cup \{\Phi(z)\} \models Holds(\varphi, z)$ and, hence, $\Sigma_{state} \cup \{\Phi(z)\} \cup UNA[\mathcal{F}] \models z - f_1 = z_1 \supset Holds(\varphi, z_1)$.

   Consider any constraint $NotHolds(\varphi, z')$ occurring in $[\![\Phi(z)]\!]$ such that $\Sigma_{state} \cup UNA[\mathcal{F}] \models \varphi \neq f_1$. Then $\Sigma_{state} \cup \{\Phi(z)\} \models \neg Holds(\varphi, z)$ and, hence, $\Sigma_{state} \cup \{\Phi(z)\} \cup UNA[\mathcal{F}] \models z - f_1 = z_1 \supset \neg Holds(\varphi, z_1)$ by Theorem 1.14.

   Consider any constraint $NotHoldsAll(\varphi, z')$ occurring in $[\![\Phi(z)]\!]$ such that $\Sigma_{state} \cup UNA[\mathcal{F}] \models \varphi \neq f_1$. Let $\vec{x}$ be the variables occurring in φ, then $\Sigma_{state} \cup \{\Phi(z)\} \models (\forall \vec{x})\, \neg Holds(\varphi, z)$ and, hence, $\Sigma_{state} \cup \{\Phi(z)\} \cup UNA[\mathcal{F}] \models z - f_1 = z_1 \supset (\forall \vec{x})\, \neg Holds(\varphi, z_1)$ by Theorem 1.14.

   Consider any disjunction constraint $OrHolds([\varphi_1, \ldots, \varphi_k], z')$ occurring in $[\![\Phi(z)]\!]$ (where $k \geq 1$) such that $\Sigma_{state} \cup UNA[\mathcal{F}] \models \varphi_i \neq f_1$ for all $1 \leq i \leq k$. Then $\Sigma_{state} \cup \{\Phi(z)\} \models \bigvee_{i=1}^{k} Holds(\varphi_i, z)$ and, hence, $\Sigma_{state} \cup \{\Phi(z)\} \cup UNA[\mathcal{F}] \models z - f_1 = z_1 \supset \bigvee_{i=1}^{k} Holds(\varphi_i, z_1)$ by Theorem 1.14.

The claim follows by the definition of the macro "$-$" and the induction hypothesis. ■

In a similar fashion, the clauses for *Plus* define a sound update wrt. a collection of positive effects.

Lemma 4.11 *Consider a fluent calculus state signature with fluent functions $\mathcal{F}$, and let ϑ^+ be a finite state and $[\![\Phi(z)]\!]$ a FLUX state, then for any $[\![\vartheta^+]\!]$, the query $P_{kernel} \cup \{\leftarrow [\![\Phi(z)]\!], Plus(z, [\![\vartheta^+]\!], z')\}$ succeeds with answer θ such that*

$$\Sigma_{state} \cup UNA[\mathcal{F}] \models \Phi(z)\theta \supset (z + \vartheta^+ = z')\theta$$

Proof: Exercise 4.7. ■

The results culminate in the following theorem, which shows that the encoding of update is sound.

```
adjacent(X, Y, D, X1, Y1) :-
    [X,Y,X1,Y1] :: 1..5,
    D :: 1..4,
    (D#=1) #/\ (X1#=X) #/\ (Y1#=Y+1)       % north
    #\/
    (D#=2) #/\ (X1#=X+1) #/\ (Y1#=Y)       % east
    #\/
    (D#=3) #/\ (X1#=X) #/\ (Y1#=Y-1)       % south
    #\/
    (D#=4) #/\ (X1#=X-1) #/\ (Y1#=Y).      % west
```

Figure 4.9: Defining adjacency with the help of arithmetic constraints.

Theorem 4.12 *Consider a fluent calculus state signature with fluent functions $\mathcal{F}$, and let ϑ^+ and ϑ^- be finite states and $[\![\Phi(z_1)]\!]$ a FLUX state, then for any $[\![\vartheta^+]\!]$ and $[\![\vartheta^-]\!]$, the query*

$$P_{kernel} \cup \{\leftarrow [\![\Phi(z_1)]\!], \mathit{Update}(z_1, [\![\vartheta^+]\!], [\![\vartheta^-]\!], z_2)\}$$

succeeds with answer θ such that

$$\Sigma_{state} \cup \mathit{UNA}[\mathcal{F}] \models \Phi(z_1)\theta \supset (z_2 = z_1 - \vartheta^- + \vartheta^+)\theta$$

Proof: The claim follows from Lemma 4.10 and 4.11 along with the macro definition for update equations. ■

Example 2 (continued) Figure 4.9 shows the encoding of the notion of a location being adjacent to another one in one of the four directions. Consider the following sample query, where the predicate *Consistent*(z) shall be defined as in Figure 4.1, encoding the domain constraints of the cleaning robot world:

```
init(Z0) :- Z0 = [at(X,Y),facing(1) | Z],
            X :: 4..5, Y :: 1..3,
            consistent(Z0).

?- init(Z0),
   holds(at(X,Y),Z0), holds(facing(D),Z0),
   adjacent(X,Y,D,X1,Y1),
   update(Z0,[at(X1,Y1)],[at(X,Y)],Z1).

Z1 = [at(X1,Y1),facing(1) | Z]

Constraints:
X  :: [4..5]
X1 :: [4..5]
Y  :: [1..3]
Y1 :: [2..4]
...
```

Although they are non-ground, the status of both the negative effect $At(x, y)$ and the positive effect $At(x_1, y_1)$ is known wrt. the specification of the initial state: The former is true while the latter is false due to the constraint $NotHoldsAll(At(v, w), z)$, which derives from the domain constraint that there be a unique location.

As an example of an effect with unknown status, consider the state specification $(\exists x, y)\, \neg Holds(Cleaned(x, y), z_1)$ along with the update equation $z_2 = z_1 + Cleaned(1, 1)$. Macro expansion and decomposition plus irreducibility imply that $Holds(Cleaned(x, y), z_2)$ if $x = 1 \wedge y = 1$ and $\neg Holds(Cleaned(x, y), z_2)$ if $x \neq 1 \vee y \neq 1$. Hence, it cannot be decided whether $Cleaned(x, y)$ is affected or not by adding $Cleaned(1, 1)$. Consequently, FLUX gives the following answer:

```
?- not_holds(cleaned(X,Y),Z1), duplicate_free(Z1),
   update(Z1,[cleaned(1,1)],[],Z2).

Z2 = [cleaned(1,1) | Z1]

Constraints:
not_holds(cleaned(1,1),Z1)
duplicate_free(Z1)
```

□

While update in general FLUX is sound, as shown with Theorem 4.12, it is incomplete in the sense that it may not entail all logical consequences of an update equation. Consider, for example, the state specification

$$(\exists x, y)\, (Holds(F(x), z_1) \wedge Holds(F(y), z_1) \wedge x \neq y) \tag{4.10}$$

along with the update equation $z_2 = z_1 - F(7)$. Although it is not entailed that $F(x)$ holds in z_2 nor that $F(y)$ holds (because both $x = 7$ and $y = 7$ are consistent), it does follow that $Holds(F(x), z_2) \vee Holds(F(y), z_2)$ (because $x = y = 7$ is inconsistent with $x \neq y$). FLUX can only infer a weaker state specification:

```
?- Z1 = [f(X),f(Y) | Z], duplicate_free(Z1),
   update(Z1,[],[f(7)],Z2).

Z2 = Z

Constraints:
not_holds(f(7),Z)
duplicate_free(Z)

X#\=Y
```

This incompleteness is not an obstacle towards sound agent programs, because a state specification entails any logical consequence which is derived from a possibly weaker specification.

4.5 Bibliographical Notes

Constraint logic programming has emerged in the early 1980's as a powerful combination of declarative programming with efficient algorithms for particular data structures. The survey article [Jaffar and Maher, 1994] provides a gentle introduction to the theory and gives an overview of a variety of constraint solvers. The theory and practice of arithmetic constraints have been discussed in [Jaffar *et al.*, 1992]. Constraint Handling Rules have been introduced in [Frühwirth, 1998]. The article [Holzbaur and Frühwirth, 2000] provides an overview of applications which use systems of CHRs. The FLUX constraint solver has first been presented and analyzed in [Thielscher, 2002b].

4.6 Exercises

4.1. Formulate the following questions as FLUX queries and answer them using the encoding of the cleaning robot domain.

   (a) If light is perceived in square $(3,3)$ but not in square $(2,4)$, what can be said as to whether light would be sensed in square $(2,3)$?

   (b) Consider a state in which the robot is somewhere in the top row, in which all offices in the fourth row are occupied, and in which the robot perceives no light at its current location. Where is the robot?

4.2. While universally quantified positive state knowledge cannot be directly expressed in FLUX, you can often encode it indirectly by your own Constraint Handling Rules.

   (a) Consider this additional domain constraint for the world of the cleaning robot:

$$Holds(Cleaned(x,y),s) \supset 1 \leq x,y \leq 5$$

   Let $CleanedConsistent(z)$ represent that state z is consistent wrt. the domain constraint just mentioned. Construct suitable CHRs for $CleanedConsistent$, by which, e.g.,

$$y :: [1..5],\ CleanedConsistent([At(1,y),\ Cleaned(1,y-3) \mid z])$$

   is reduced to $y :: [4..5]$, $CleanedConsistent(z)$.

   (b) Recall from Chapter 1 these domain constraints of the mail delivery world:

$$\begin{array}{l} Holds(Empty(b),s) \supset \neg Holds(Carries(b,r),s) \wedge 1 \leq b \leq k \\ Holds(Carries(b,r),s) \supset 1 \leq b \leq k \wedge 1 \leq r \leq n \\ Holds(Carries(b,r_1),s) \wedge Holds(Carries(b,r_2),s) \supset r_1 = r_2 \end{array}$$

   Write CHRs (for $k=3$ and $n=6$) by which incomplete states can be verified against these constraints.

   (c) Let the fluents $Carries(x)$ and $InsideOf(x,y)$ denote, respectively, that a robot carries x and that x is inside of y. Write CHRs to verify that an incomplete state satisfies this domain constraint:

$$\begin{array}{l} Holds(InsideOf(x,y),s) \supset \\ \quad [Holds(Carries(x),s) \equiv Holds(Carries(y),s)] \end{array}$$

   Hint: Any positive instance of $InsideOf(x,y)$ should trigger an auxiliary constraint stipulating that $Carries(x,y)$ is symmetric.

   Prove correctness of the CHRs using their declarative interpretation.

4.3. Extend the FLUX constraint solver by a set of Constraint Handling Rules for the constraint $XorHolds([f_1,\ldots,f_k],z)$ with the intended meaning that one and only one of $f_1,\ldots,f_k$ holds in z $(k \geq 1)$. Prove that the rules are correct under the foundational axioms of fluent calculus.

4.4. Use the *Xor*-constraint of the previous exercise to formalize the Litmus test: Encode the state knowledge that a chemical solution is acidic just in case the Litmus strip is not red. What follows after adding that the strip is red, and what follows by adding that it is not red?

4.5. Extend the FLUX constraint solver by a set of Constraint Handling Rules for the constraint $AllHolds(f, [f_1, \ldots, f_k], z)$ with the intended meaning that all instances of f but $f_1, \ldots, f_k$ hold in z $(k \geq 0)$. Prove that the rules are correct under the foundational axioms of fluent calculus.

4.6. Solve Exercise 3.2 in FLUX: Encode the initial state z_0 of the cleaning robot to be in $(1, 1)$ and to face either north or west. Use FLUX to show that

(a) initially, the robot does not face east;

(b) let $z_1 = z_0 - Facing(d) + Facing(d \bmod 4 + 1)$ such that d satisfies $Holds(Facing(d), z_0)$, then the robot faces north or east in z_1;

(c) it cannot be that in z_1 of Exercise (b) there is no square adjacent to the location of the robot in the direction it faces;

(d) let $z_2 = z_1 - At(x, y) + At(x', y')$ such that $Holds(At(x, y), z_1)$, $Holds(Facing(d), z_1)$, and $Adjacent(x, y, d, x', y')$, and suppose that $Holds(At(1, 2), z_2)$, then the robot faces north in z_2.

4.7. Prove Lemma 4.11.

4.8. This exercise is concerned with a simple computer game known as the "Wumpus World" (see also Figure 4.10): An agent moves in a grid of cells, some of which contain, initially unknown to the agent, bottomless pits or gold, and there is one square which houses the hostile Wumpus. Its sensing capabilities allow the agent to perceive a breeze (a stench, respectively) if it is adjacent to a cell containing a pit (the Wumpus, respectively), and the agent notices a glitter in any cell containing gold. Consider the fluents $At(x, y)$, $Pit(x, y)$, $Wumpus(x, y)$, and $Gold(x, y)$, denoting that in square (x, y) there is the agent, a pit, the Wumpus, and gold, respectively; the fluent $Facing(d)$ as in the cleaning robot world; the fluent *Alive* denoting that the Wumpus is alive; and the fluent $Has(x)$ for $x \in \{Arrow, Gold\}$ to denote that the agent is in possession of an arrow and gold, respectively.

(a) Design suitable domain constraints for the Wumpus World. Encode these constraints in FLUX by a clause which defines the predicate $Consistent(z)$.

(b) Define auxiliary predicates $Breeze(x, y, v, z)$, $Stench(x, y, v, z)$, and $Glitter(x, y, v, z)$ such that $v = True$ if the respective sensation is to be expected at square (x, y) in state z, and $v = False$ otherwise.

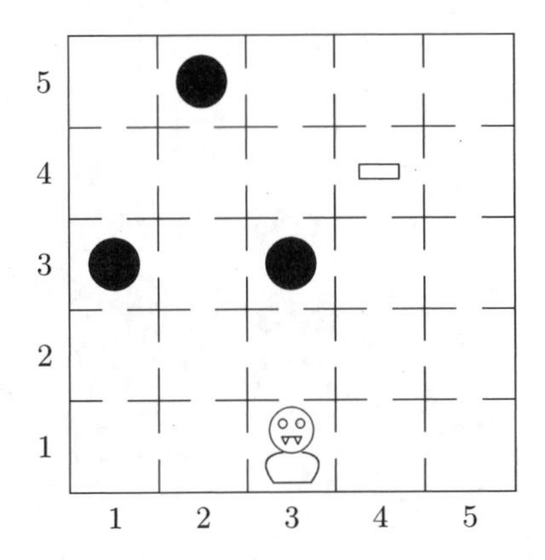

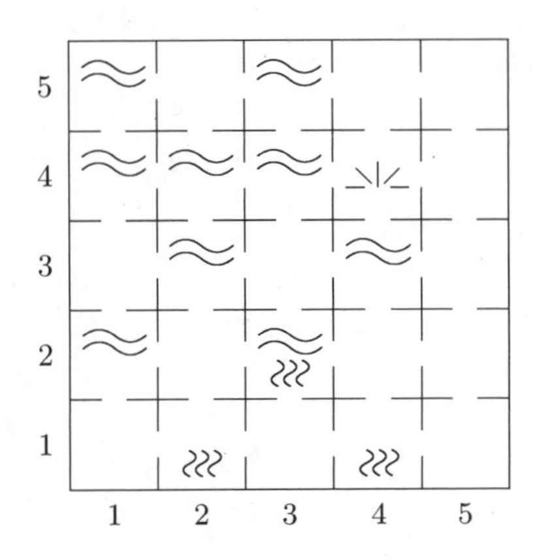

Figure 4.10: A sample scenario in a Wumpus World where the 5×5 cave features three pits, the Wumpus in cell $(3,1)$, and gold in cell $(4,4)$. In the right hand side are depicted the corresponding perceptions (breeze, stench, glitter) for each location.

(c) Use the FLUX constraint solver to prove that if it is only a breeze that is perceived in cell $(1,2)$ and only a stench in cell $(2,1)$ (as depicted in the right hand side of Figure 4.10), then it follows that there is no gold in either of these squares, that there is a pit in cell $(1,3)$, and that the Wumpus hides in cell $(3,1)$.

Chapter 5

Knowledge Programming

Robotic agents should always behave cautiously: No matter whether a certain action would in fact be possible and successful, an agent should not attempt the action unless it *knows* that the action can be performed and has the desired effect. Otherwise the agent program runs the risk to send a command to the physical apparatus which cannot be executed in the current state of affairs, or which has unintended or even dangerous effects. The programmer therefore has to take the subjective perspective of an agent. In this chapter, we show how this perspective can be integrated into fluent calculus and used in agent programs.

The difference between the objective and the subjective view is most obvious when agents use their sensors. Sensing helps the agent with learning more about its environment. This has usually no effect on the external world itself while it does affect the subjective perspective. Hence, the full effect of actions involving sensing is not reflected by a mere state update. For instance, when the office cleaning robot moves to another square, then from the objective perspective it just changes its location. This is precisely what the state update axiom for action Go says. Whether or not the cleanbot senses light at the new place is irrelevant to the physical world, but it enhances the knowledge of the robot about the world. A suitable model of knowledge and sensing will allow us to write control programs for robotic agents which have to make decisions under uncertainty and which can use their sensors to acquire additional information. To this end, we develop

5.1. a definition of what it means to **know** something under incomplete information;

5.2. a method for FLUX agents to infer what they know;

5.3. a method for FLUX agents to **update their knowledge** upon performing actions and receiving sensor input;

5.4. a way of encoding domain descriptions in FLUX which deal with incomplete knowledge;

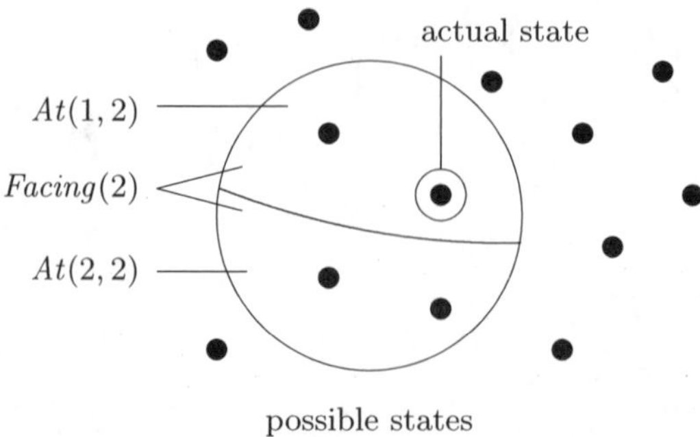

Figure 5.1: A sample knowledge state in the cleaning robot domain: Possible are the states in which *Facing*(2) and either $At(1,2)$ or $At(2,2)$ hold. All other states are considered impossible. Hence, in this situation the robot knows it faces 2 and is either at location $(1,2)$ or at location $(2,2)$.

5.5. an exemplary control program that guides our cleanbot safely through its environment while cleaning as many locations as (cognitively!) possible.

5.1 Representing State Knowledge

Incomplete state knowledge is represented in fluent calculus via the notion of **possible states**. A possible state is any world state in which the agent could be according to what it knows. The diversity of possible states is an indicator of how much information an agent has. In the ideal case of complete knowledge, the only possible state is the actual one. At the other end of the scale, a fully ignorant agent (say, a robot which has just been switched on without being told where it is) considers possible just any state. A typical partial knowledge state, however, is characterized by several possible and impossible states. An example is sketched in Figure 5.1.

For the representation of possible states, the basic signature of fluent calculus is extended by the predicate $KState(s, z)$ (for: knowledge state). The intended meaning is that—as far as the agent knows— z is a possible state in situation s.

Definition 5.1 A tuple $\mathcal{S} \cup \langle KState \rangle$ is a **fluent calculus signature for knowledge** if $\mathcal{S}$ is a fluent calculus signature and

$$KState : \text{SIT} \times \text{STATE}$$

A **knowledge state** for a situation σ is a formula $KState(\sigma, z) \equiv \Phi(z)$ where $\Phi(z)$ is a state formula. □

For example, suppose that initially the cleaning robot only knows that it faces east and is either at $(1,2)$ or at $(2,2)$ and that none of the waste bins has been cleaned yet. This, then, is the knowledge state of the robot in situation S_0:

$$\begin{aligned} KState(S_0, z) \equiv\ & Holds(Facing(2), z) \wedge \\ & (\exists x_0)\,(Holds(At(x_0, 2), z) \wedge [x_0 = 1 \vee x_0 = 2]) \wedge \\ & (\forall u, v)\,\neg Holds(Cleaned(u, v), z) \wedge \\ & Consistent(z) \end{aligned} \tag{5.1}$$

where $Consistent(z)$ stands for the domain constraints in the cleanbot world, that is, axioms (3.2) on page 61, with s being replaced by z. In this way, we assume that the cleanbot is aware of all domain constraints: It considers possible only those states which are physically plausible. Hence, the robot shall always at the very least know that it is at a unique location facing a unique direction, etc.

The new predicate enables us to define formally what it means for agents to know certain aspects of the state of their environment. Informally speaking, a state property is known just in case it is shared by all possible states. Equivalently, we can say that a property is known if the opposite is considered impossible. Conversely, unknown to the agent is everything which is false in at least one possible state. Knowledge state (5.1), for example, implies that the cleaning robot knows that it faces east and that it is not in a column higher than 2. On the other hand, the robot does not know that, say, room $(5, 1)$ is not occupied, because one possible state is $Facing(1) \circ At(1, 2) \circ Occupied(5, 1)$. Likewise, its precise location is unknown to the robot. Nonetheless, with a little logical reasoning the robot should be able to conclude that a Go action is possible in the initial situation: No matter which of $At(1, 2)$ or $At(2, 2)$ holds, there is a square in direction 2 (i.e., east) from the current location. Hence, according to precondition axioms (3.12) on page 67, $Poss(Go, z)$ holds in all possible states in S_0; formally, $(\forall z)\,(KState(S_0, z) \supset Poss(Go, z))$.

The examples illustrate that it is natural to be interested in knowledge of complex matters, such as whether it is known that the cleaning robot faces a direction such that the adjacent square is inside of the boundaries. The following notion of a knowledge expression allows to form such compound properties of states. While they resemble genuine formulas, these expressions acquire meaning only via a macro for knowledge. This macro says that a knowledge expression is known in a situation just in case it holds in all possible states in that situation.

Definition 5.2 A **knowledge expression** is composed of the standard logical connectives and atoms of the form

1. f, where f fluent;
2. $Poss(a)$, where a action;
3. atoms $P(\vec{t})$ without terms of sort STATE or SIT.

Let ϕ be a knowledge expression, then

$$Knows(\phi, s) \stackrel{\text{def}}{=} (\forall z)\,(KState(s, z) \supset HOLDS(\phi, z))$$

where $HOLDS(\phi, z)$ is obtained by replacing in ϕ each fluent f by $Holds(f, z)$ and each $Poss(a)$ by $Poss(a, z)$. Furthermore, for an n-ary fluent function F, let $\vec{x}$ be a sequence of n pairwise different variables ($n \geq 1$) and let $\vec{x}_1$ be a non-empty subset thereof, then

$$KnowsVal(\vec{x}_1, F(\vec{x}), s) \stackrel{\text{def}}{=} (\exists \vec{x}_1)\, Knows((\exists \vec{x}_2)\, F(\vec{x}), s)$$

where $\vec{x}_2$ are the variables in $\vec{x}$ but not in $\vec{x}_1$. □

As an example, $(\exists x, y)\,(At(x, y) \wedge x < 3)$ is a knowledge expression which is known wrt. knowledge state (5.1):

$$\begin{array}{rl} & Knows((\exists x, y)\,(At(x, y) \wedge x < 3), S_0) \\ \stackrel{\text{def}}{=} & KState(S_0, z) \supset HOLDS((\exists x, y)\,(At(x, y) \wedge x < 3), z) \\ \stackrel{\text{def}}{=} & KState(S_0, z) \supset (\exists x, y)\,(Holds(At(x, y), z) \wedge x < 3) \\ \Leftarrow & KState(S_0, z) \supset (\exists x_0)\,(Holds(At(x_0, 2), z) \wedge [x_0 = 1 \vee x_0 = 2]) \end{array}$$

The second macro, $KnowsVal$, is used to express that an agent knows a **value** of a fluent, that is, an instance for the given variables which is true in all possible states. For example, knowledge state (5.1) entails $KnowsVal(d, Facing(d), S_0)$ since there exists d such that $Facing(d)$ holds in all possible states (namely, $d = 2$). On the other hand, (5.1) entails that no value for $At(x, y)$ is known. To prove this, consider the two possible states $z_1 = At(1, 2) \circ Facing(2)$ and $z_2 = At(2, 2) \circ Facing(2)$. From $UNA[At, Facing]$ along with decomposition and irreducibility it follows

$$\neg\,(\exists x, y)\,(Holds(At(x, y), z_1) \wedge Holds(At(x, y), z_2))$$

Hence, since $KState(S_0, z_1) \wedge KState(S_0, z_2)$,

$$\neg\,(\exists x, y)(\forall z)\,(KState(S_0, z) \supset Holds(At(x, y), z)) \tag{5.2}$$

that is, $\neg KnowsVal((x, y), At(x, y), S_0)$. A *partial* value, however, is known for fluent $At(x, y)$, namely, the y-coordinate:

$$\begin{array}{rl} & KnowsVal(y, At(x, y), S_0) \\ \stackrel{\text{def}}{=} & (\exists y)\, Knows((\exists x)\, At(x, y), S_0) \\ \stackrel{\text{def}}{=} & (\exists y)\,(KState(S_0, z) \supset (\exists x)\, Holds(At(x, y), z)) \\ \Leftarrow & KState(S_0, z) \supset (\exists x_0)\,(Holds(At(x_0, 2), z) \wedge [x_0 = 1 \vee x_0 = 2]) \end{array}$$

When verifying knowledge, the choice of the right knowledge expression is crucial. For instance, knowing a value for a fluent is much stronger than merely knowing that the fluent has a value. With regard to knowledge state (5.1), for example, the knowledge expression $(\exists x, y)\, At(x, y)$ is true in all possible states; hence,

$$(\forall z)(\exists x, y)\,(KState(S_0, z) \supset Holds(At(x, y), z)) \tag{5.3}$$

Contrasting this to formula (5.2) illustrates how important it is to choose the quantifier structure which correctly reflects the intended assertion. Formula (5.3) can also be written as $Knows((\exists x, y)\, At(x, y), S_0)$, which represents knowledge **de dicto** (that is, "of a proposition"), as opposed to knowledge **de re** (that is, "of an object") as in $(\exists x, y)\, Knows(At(x, y), S_0)$. It is not difficult to prove that the latter implies the former but not vice versa (see also Exercise 5.2).

A universal property of knowledge is to be true. In fluent calculus, this is formally captured by a simple additional foundational axiom.

Definition 5.3 The **foundational axioms of fluent calculus for knowledge** are $\Sigma_{state} \cup \Sigma_{knows}$ where Σ_{state} is as in Definition 3.1 (page 63) and Σ_{knows} contains the single axiom

$$(\forall s)\, KState(s, State(s))$$

□

The following proposition says that under this axiomatic foundation everything an agent knows of a state does indeed hold in the actual state. Another property of the axiomatization of knowledge is that agents are logically omniscient. That is to say, agents know all deductive consequences of what they know, at least in theory.

Proposition 5.4 *Let ϕ and ψ be knowledge expressions, then foundational axioms $\Sigma_{state} \cup \Sigma_{knows}$ entail*

1. $Knows(\phi, s) \supset HOLDS(\phi, State(s))$.
2. $Knows(\phi, s) \supset Knows(\psi, s)$, *if* $HOLDS(\phi, z) \models HOLDS(\psi, z)$.

Proof:

1. Suppose $Knows(\phi, s)$, then macro expansion implies $HOLDS(\phi, z)$ for all z such that $KState(s, z)$; hence, $HOLDS(\phi, State(s))$ by Σ_{knows}.
2. $HOLDS(\phi, z) \models HOLDS(\psi, z)$ implies $HOLDS(\phi, z) \supset HOLDS(\psi, z)$ for all z such that $KState(s, z)$; hence, by macro expansion it follows that $Knows(\phi, s)$ implies $Knows(\psi, s)$. ■

In practice, logical omniscience is not necessarily a desirable property, because universal inference capabilities are usually computationally infeasible. Due to the incompleteness of update in FLUX (cf. Section 4.4), FLUX agents are in general not logically omniscient.

5.2 Inferring Knowledge in FLUX

The concept of knowing a property of the state is essential for the evaluation of conditions in agent programs under incomplete information. It replaces the simpler notion of a state property to hold, which we have used in special FLUX to condition the behavior of agents.

By definition, a property is known just in case it is true in all possible states. From a computational perspective, it is of course impractical to evaluate a condition by literally checking every possible state, since there is usually quite a number, often even infinitely many of them. Fortunately, there is a feasible alternative. Instead of verifying that all states satisfy a property, we can just as well prove that the *negation* of the property is *unsatisfiable* under a given knowledge state. This approach is justified by the following theorem.

Theorem 5.5 *Let* $KState(\sigma, z) \equiv \Phi(z)$ *be a knowledge state and* ϕ *be a knowledge expression, then*

$$\{KState(\sigma, z) \equiv \Phi(z)\} \models Knows(\phi, \sigma) \text{ iff } \{\Phi(z), HOLDS(\neg\phi, z)\} \models \bot$$

Proof:

$$\begin{array}{ll} & \{KState(\sigma, z) \equiv \Phi(z)\} \models Knows(\phi, \sigma) \\ \text{iff} & \{KState(\sigma, z) \equiv \Phi(z)\} \models (\forall z)\,(KState(\sigma, z) \supset HOLDS(\phi, z)) \\ \text{iff} & \models (\forall z)\,(\Phi(z) \supset HOLDS(\phi, z)) \\ \text{iff} & \models \neg(\exists z)\,(\Phi(z) \wedge HOLDS(\neg\phi, z)) \\ \text{iff} & \{\Phi(z), HOLDS(\neg\phi, z)\} \models \bot \end{array}$$

■

This fundamental result suggests a rather elegant way of encoding knowledge in FLUX. To begin with, if we want to verify knowledge in a situation σ with knowledge state $KState(\sigma, z) \equiv \Phi(z)$, then we can just refer to the (incomplete) state specification $\Phi(z)$. Furthermore, in order to verify that $\{\Phi(z), HOLDS(\neg\phi, z)\}$ is unsatisfiable, negation-as-failure can be employed to prove that $HOLDS(\neg\phi, z)$ cannot be consistently asserted under state specification $\Phi(z)$. Figure 5.2 shows how this is realized in FLUX for the basic knowledge expressions f and $\neg f$, where f is a fluent. More complex knowledge expressions, such as disjunctive knowledge, can be encoded in a similar fashion (see also Exercise 5.7). It is important to observe that both $Knows(f, z)$ and $KnowsNot(f, z)$ should only be used for ground fluents f. The auxiliary predicate $KHolds(f, z)$ can be used instead in order to infer known instances of a fluent expression.

Figure 5.2 includes a FLUX definition of macro $KnowsVal$. To begin with, the auxiliary predicate $KHolds(f, z)$ *matches* fluent expression f against a fluent f_1 that occurs positively in z. Matching requires to test that f_1 is an instance of f before unifying the two terms. For the sake of efficiency, the further encoding is based on the extra-logical Prolog predicates *Assert* and *Retract* used in conjunction with backtracking: The auxiliary, unary $KnowsVal(\vec{x})$ is true just in case the current constraints entail a unique instance for the variables in $\vec{x}$. To this end, $Dom(\vec{x})$ infers a consistent instance for all variables in $\vec{x}$ with the help of the standard FD-predicates $IsDomain(x)$ (which is true if x

```
knows(F, Z) :- \+ not_holds(F, Z).

knows_not(F, Z) :- \+ holds(F, Z).

knows_val(X, F, Z) :- k_holds(F, Z), knows_val(X).

k_holds(F, Z) :- nonvar(Z), Z=[F1|Z1],
                 ( instance(F1, F), F=F1 ; k_holds(F, Z1) ).

:- dynamic known_val/1.

knows_val(X) :- dom(X), ground(X), ambiguous(X) -> false.
knows_val(X) :- retract(known_val(X)).

dom([]).
dom([X|Xs]) :- dom(Xs), ( is_domain(X) -> indomain(X)
                                       ; true ).

ambiguous(X) :- retract(known_val(_)) -> true
                ;
                assert(known_val(X)), false.
```

Figure 5.2: Knowledge in FLUX.

is a constrained variable) and $Indomain(x)$ (which derives a consistent value for x). If x is not a constrained variable, it needs to be ground in order to have a unique value. The auxiliary predicate $Ambiguous(\vec{x})$ fails upon asserting the first inferred instance for $\vec{x}$. The predicate succeeds upon the second call, that is, after successfully backtracking over $Dom(\vec{x})$, which indicates multiple possible values for $\vec{x}$.

Example 2 (continued) Recall from Figure 4.1 (page 78) the FLUX specification for the initial state in the cleaning robot scenario. Suppose, for the sake of argument, it is also given that there is no light perceived in square $(1,2)$ but in square $(1,3)$. Then we can use FLUX to show that the robot knows that room $(1,3)$ is not occupied, while it does not know that office $(1,4)$ is free, nor that it is not so:

```
?- init(Z), light(1,2,false,Z),
            light(1,3,true,Z),
   knows_not(occupied(1,3),Z),
   \+ knows(occupied(1,4),Z),
   \+ knows_not(occupied(1,4),Z).

yes.
```

In terms of fluent calculus, let

$$KState(S_0, z) \equiv \Phi_0(z) \wedge \neg Light(1,2,z) \wedge Light(1,3,z)$$

where $\Phi_0(z)$ is the initial state description for the cleanbot, axiom (3.11) on page 66, with $State(S_0)$ replaced by z. Then $Knows(\neg Occupied(1,3), S_0)$ and $\neg Knows(Occupied(1,4), S_0)$ as well as $\neg Knows(\neg Occupied(1,4), S_0)$.

As an example for the FLUX definition of knowing values for fluents, consider the FLUX encoding of knowledge state (5.1):

```
init(Z0) :- Z0 = [at(X,2),facing(2) | Z],
            X#=1 #\/ X#=2,
            not_holds_all(cleaned(_,_),Z0),
            consistent(Z0).

?- init(Z0),
   knows_val([D],facing(D),Z0),
   \+ knows_val([X,Y],at(X,Y),Z0),
   knows_val([Y],at(_,Y),Z0).

D = 2
Y = 2
```

While functional fluents like the position or orientation of the robot allow at most one value to be known in consistent knowledge states, other fluents may admit several known values for their arguments:

```
init(Z0) :- Z0 = [cleaned(1,2),cleaned(4,Y) | Z], Y :: 2..4,
            consistent(Z0).

?- init(Z0), knows_val([X],cleaned(X,_),Z0).

X = 1  More?

X = 4
```

□

Let the FLUX kernel program P_{kernel} be extended by the clauses of Figure 5.2. Soundness of the definition of knowledge in FLUX follows from the soundness of the underlying constraint system, Theorem 4.7, and Theorem 5.5.

Theorem 5.6 *Consider a FLUX state* $\Phi(z)$*, fluents* φ *and* $F(\vec{x})$*, and a non-empty subset* $\vec{x}_1$ *of* $\vec{x}$*. Let* Σ *be* $\Sigma_{state} \cup \{KState(\sigma, z) \equiv \Phi(z)\}$*.*

1. *If the derivation tree for* $P_{kernel} \cup \{\leftarrow \Phi(z), Knows(\varphi, z)\}$ *is successful, then* $\Sigma \models Knows(\varphi, \sigma)$*.*

2. *If the derivation tree for* $P_{kernel} \cup \{\leftarrow \Phi(z), KnowsNot(\varphi, z)\}$ *is successful, then* $\Sigma \models Knows(\neg\varphi, \sigma)$*.*

3. If θ is computed answer to $P_{kernel} \cup \{\leftarrow \Phi(z), KnowsVal(\vec{x}_1, F(\vec{x}), z)\}$, then

(a) $\Sigma \models KnowsVal(\vec{x}_1, F(\vec{x}), \sigma)$.

(b) $\Sigma \models Knows((\exists \vec{x}_2)\, F(\vec{x})\theta, \sigma)$.

where $\vec{x}_2$ are the variables in $\vec{x}$ besides $\vec{x}_1$.

Proof: We prove each statement separately.

1. Suppose $P_{kernel} \cup \{\leftarrow \Phi(z), Knows(\varphi, z)\}$ is successful, then the query $P_{kernel} \cup \{\leftarrow \Phi(z), NotHolds(\varphi, z)\}$ fails. From Theorem 4.7 it follows that $\Sigma_{state} \cup \{\Phi(z), \neg Holds(\varphi, z)\} \models \bot$. Therefore, $\Sigma \models Knows(\varphi, \sigma)$ according to Theorem 5.5.

2. Suppose $P_{kernel} \cup \{\leftarrow \Phi(z), KnowsNot(\varphi, z)\}$ is successful, then the query $P_{kernel} \cup \{\leftarrow \Phi(z), Holds(\varphi, z)\}$ fails. From Theorem 4.7 it follows that $\Sigma_{state} \cup \{\Phi(z), Holds(\varphi, z)\} \models \bot$. Therefore, $\Sigma \models Knows(\neg\varphi, \sigma)$ according to Theorem 5.5.

3. Suppose θ is an answer to $P_{kernel} \cup \{\leftarrow \Phi(z), Holds(F(\vec{x}), z)\}$, and $\vec{x}_1\theta$ is ground. By Theorem 4.8,

$$\Sigma_{state} \models \Phi(z)\theta \supset Holds(F(\vec{x}), z)\theta \tag{5.4}$$

Let $z = [f_1, \ldots, f_n \mid z']$ be the state equation in FLUX state $\Phi(z)$. As $\vec{x}_1$ is non-empty and θ is a combination of most general unifiers, groundness of $\vec{x}_1\theta$ implies that θ does not affect $[f_1, \ldots, f_n \mid z']$, so $\Phi(z)\theta = \Phi(z)$. Then (5.4) implies

$$\Sigma_{state} \models \Phi(z) \supset Holds(F(\vec{x})\theta, z)$$

Therefore, $\Sigma_{state} \cup \{\Phi(z)\}$ is inconsistent with $\neg Holds(F(\vec{x})\theta, z)$. Hence, since $\vec{x}_1\theta$ is ground, $\Sigma \models Knows((\exists \vec{x}_2)\, F(\vec{x})\theta, \sigma)$, which implies that $\Sigma \models KnowsVal(\vec{x}_1, F(\vec{x}), \sigma)$. ■

5.3 Knowledge Update Axioms

Knowledge of state properties is dynamic. Upon performing an action, agents need to update their knowledge according to the effect of the action and by incorporating acquired sensor information. Much like the update of the actual state is determined by the relation between some $State(s)$ and the successor $State(Do(a, s))$, knowledge update means to infer the knowledge state in the successor situation $Do(a, s)$ relative to the knowledge state in situation s. To this end, the states which are possible after the performance of an action need to be related to the possible states prior to it.

A schematic illustration of knowledge update is given in Figure 5.3. Being uncertain about which state the cleanbot is in, the knowledge state after

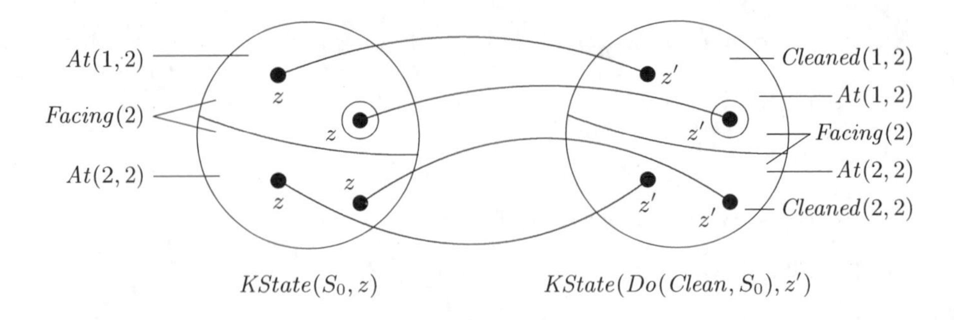

Figure 5.3: Knowledge update: The set of possible states after the action is determined by the effects on the possible states in the original situation.

---

performing the action *Clean* is obtained by updating each one of the initially possible states according to the effects of this action. Because the precise initial location of the robot is not known, we arrive at two kinds of possible states in the successor situation: The robot is either at square $(1, 2)$ and the waste bin there has been cleaned, or it is at square $(2, 2)$ and the waste bin at this location has been cleaned. Other combinations are not possible, that is, the robot cannot be at $(1, 2)$ while having cleaned the bin at location $(2, 2)$ or vice versa. Moreover, unaffected knowledge, like $Facing(2)$, is still known in the resulting situation as it continues to hold in all possible states.

If an action involves sensing, knowledge update has the additional effect that some states are rendered impossible. Thus the set of possible states shrinks according to the acquired knowledge. Figure 5.4 depicts an example of the update caused by a pure information gathering action, where the robot uses a locating sensor. Assuming that the robot is actually at location $(1, 2)$ in situation $S_1 = Do(Clean, S_0)$, only those possible states remain which contain $At(1, 2)$. Incidentally, since all possible states z' satisfy $Holds(Cleaned(1, 2), z')$, the robot also learns which of the waste bins it has cleaned.

In fluent calculus, the effect of an action on the knowledge of the agent is formally specified by a so-called knowledge update axiom, which adds a cognitive effect to the physical effect as described by mere state update axioms. Cognitive and physical effect together determine the knowledge state $KState(Do(\alpha, \sigma), z')$ relative to $KState(\sigma, z)$.

Definition 5.7 Consider a fluent calculus signature with action functions $\mathcal{A}$ and a set Σ_{sua} of state update axioms for $\mathcal{A}$. Consider an action $A \in \mathcal{A}$ along with its state update axiom in Σ_{sua},

$$Poss(A(\vec{x}), s) \supset \Psi[State(Do(A(\vec{x}), s)), State(s)] \tag{5.5}$$

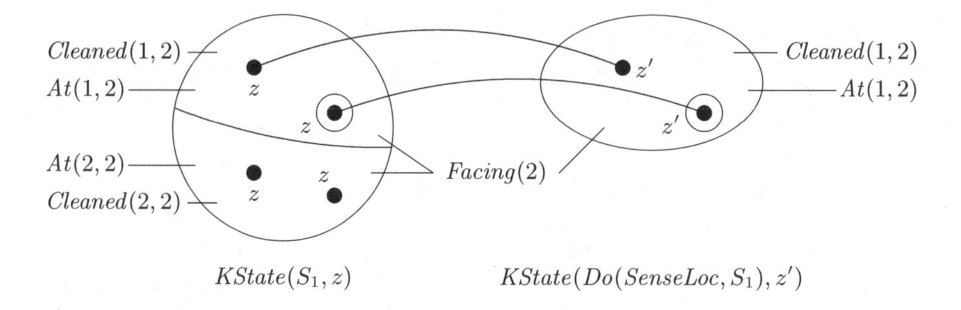

Figure 5.4: Knowledge update with sensing: Continuing with the knowledge state of Figure 5.3, all possible states in the situation after doing *SenseLoc* agree with the actual state on what has been sensed.

A **knowledge update axiom** for A is a formula

$$\begin{array}{l} Poss(A(\vec{x}), s) \supset \\ \quad (\exists \vec{y})(\forall z')\,(\,KState(Do(A(\vec{x}), s), z') \equiv \\ \qquad\qquad (\exists z)\,(KState(s, z) \wedge \Psi[z', z]) \wedge \Pi(z', Do(A(\vec{x}), s))\,) \end{array} \tag{5.6}$$

where

1. the **physical effect** $\Psi[z', z]$ is $\Psi[State(Do(A(\vec{x}), s)), State(s)]$ of (5.5) but with $State(Do(A(\vec{x}), s))$ replaced by z' and $State(s)$ by z, respectively; and

2. the **cognitive effect** $\Pi(z', Do(A(\vec{x}), s))$ is a formula of the form

$$\begin{array}{l} [\,\Pi_1(z') \equiv \Pi_1(Do(A(\vec{x}), s))\,] \wedge \ldots \wedge [\,\Pi_k(z') \equiv \Pi_k(Do(A(\vec{x}), s))\,] \\ \wedge \\ Holds(F_1(\vec{t}_1), z') \wedge \ldots \wedge Holds(F_l(\vec{t}_l), z') \wedge \Pi(\vec{x}, \vec{y}) \end{array}$$

   such that

   - $k, l \geq 0$;
   - for $i = 1, \ldots, k$, $\Pi_i(z')$ is a state formula whose free variables are among $\vec{x}$, and $\Pi_i(s) \stackrel{\text{def}}{=} \Pi_i(State(s))$;
   - for $i = 1, \ldots, l$, $F_i(\vec{t}_i)$ is a fluent with variables among $\vec{x}, \vec{y}$;
   - $\Pi(\vec{x}, \vec{y})$ is an auxiliary formula without states and whose free variables are among $\vec{x}, \vec{y}$. □

Pure non-sensing actions are characterized by the cognitive effect $\top$ (which can simply be omitted). The knowledge update axiom then simply mirrors the state update axiom for this action.

Example 2 (continued) The actions *Turn* and *Clean* of our cleanbot have no cognitive effect. Hence, their knowledge update axioms are determined by the corresponding state update axioms, (3.14) on page 67:

$$
\begin{aligned}
&Poss(Turn, s) \supset \\
&\quad [KState(Do(Turn, s), z') \equiv (\exists z)\,(KState(s, z) \wedge \\
&\qquad (\exists d)\,(\,Holds(Facing(d), z) \wedge \\
&\qquad\qquad z' = z - Facing(d) + Facing(d \bmod 4 + 1)\,)\,)\,]
\end{aligned}
\tag{5.7}
$$

$$
\begin{aligned}
&Poss(Clean, s) \supset \\
&\quad [KState(Do(Clean, s), z') \equiv (\exists z)\,(KState(s, z) \wedge \\
&\qquad (\exists x, y)\,(Holds(At(x, y), z) \wedge z' = z + Cleaned(x, y)\,)\,)\,]
\end{aligned}
$$

Thus z' is possible after turning or cleaning, respectively, just in case z' is the result of turning or cleaning in one of the previously possible states z. □

The second component $\Pi(z', Do(a, s))$ of a knowledge update axiom accounts for sensed information by constraining the resulting possible states z' so as to agree with the actual state in situation $Do(a, s)$ on the sensed properties and the sensed fluent values. Sensors may tell an agent either binary information or information about the value of a certain fluent. This is the reason for sub-formula $\Pi(z', Do(a, s))$ in a knowledge update axiom consisting of two components.

Sensing some state property Π_i is specified by $\Pi_i(z') \equiv \Pi_i(Do(a, s))$. Thus all possible states z' have to agree with $State(Do(a, s))$ on this property. Sensing a fluent value is specified by sub-formulas of the form $Holds(F_i(\vec{t_i}), z')$. Usually, the argument $\vec{t_i}$ contains variables quantified by $(\exists \vec{y})$ in (5.6), which may be further restricted by sub-formula $\Pi(\vec{x}, \vec{y})$. In this way it is stipulated that there be an instance of the sensed fluent which is shared by all possible states. The instance in question must then also be true in the actual state, according to the additional foundational axiom for knowledge.

When specifying the action of sensing a value, it is crucial to observe the order of the quantifiers in (5.6); using $(\forall z')(\exists \vec{y})$ instead would be too weak to imply that the sensed values $\vec{y}$ be the same in all possible states. As a fundamental feature of knowledge update axioms it can be formally shown that sensed properties and values are known in the successor situation (see Exercise 5.4).[1]

Example 2 (continued) Recall the definition of $Light(x, y, z)$ in Chapter 3 by formula (3.3) on page 61. The following knowledge update axiom for *Go* combines the physical effect of going forward with information about whether

[1] The definition of a knowledge update axiom requires that sensed properties and values are specified with reference to the successor situation $Do(a, s)$. This restriction has been imposed just for the sake of clarity; the definition can easily be generalized by extending the range of the cognitive effect to z and s, respectively. For example, a sub-formula $Holds(F(\vec{t}), z) \wedge Holds(F(\vec{t}), s)$ would say that a value for F is obtained *before* the physical effects of the action in question materialize.

light is sensed at the new location:

$$\begin{aligned} Poss(Go, s) \supset [&KState(Do(Go, s), z') \equiv \\ &(\exists z)\,(KState(s, z) \wedge \Psi_{Go}(z', z)) \wedge \\ &[\Pi_{Light}(z') \equiv \Pi_{Light}(Do(Go, s))]] \end{aligned} \tag{5.8}$$

where $\Psi_{Go}(z', z)$ derives from the state update axiom for Go, that is,

$$\begin{aligned} \Psi_{Go}(z', z) \stackrel{\text{def}}{=} (\exists d, x, y, x', y')\,(&Holds(At(x, y), z) \wedge Holds(Facing(d), z) \wedge \\ &Adjacent(x, y, d, x', y') \wedge \\ &z' = z - At(x, y) + At(x', y') \end{aligned}$$

and the sensors reveal whether there is light at the current location of the robot:

$$\Pi_{Light}(z) \stackrel{\text{def}}{=} (\exists x, y)\,(Holds(At(x, y), z) \wedge Light(x, y, z)) \tag{5.9}$$

Put in words, z' is a possible state after going forward if z' is the result of doing this action in some previously possible state and there is light at the current location in z' just in case it is so in $State(Do(Go, s))$.[2] □

As an example of sensing a fluent value rather than a property, consider the specification of a locating sensor. As a pure sensing action, self-location has no physical effect. This is mirrored in the knowledge update axiom by the sub-formula $(\exists z)\,(KState(s, z) \wedge z' = z)$. For the sake of compactness, this sub-formula has been simplified to $KState(s, z')$ in the following axiom:

$$\begin{aligned} &Poss(SenseLoc, s) \supset \\ &\quad (\exists x, y)(\forall z')\,(KState(Do(SenseLoc, s), z') \equiv \\ &\qquad\qquad KState(s, z') \wedge Holds(At(x, y), z')) \end{aligned} \tag{5.10}$$

Although a knowledge update axiom is defined as a piece-wise relation between possible states, it is not at all necessary to update each state separately in order to update a knowledge state. For inferring the effect of an action an inference scheme can be employed which is rather similar to the one used in Chapter 3 for the update of incomplete state specifications. Consider a situation σ with knowledge state

$$KState(\sigma, z) \equiv \Phi(z) \tag{5.11}$$

along with an action α with knowledge update axiom (5.6). Suppose that $Poss(\alpha, s)$, then the update axiom implies

$$\begin{aligned} (\exists \vec{y})(\forall z')\,(&KState(Do(\alpha, \sigma), z') \equiv \\ &(\exists z)\,(KState(\sigma, z) \wedge \Psi[z', z]) \wedge \Pi(z', Do(\alpha, \sigma))\,) \end{aligned}$$

[2] According to (5.9), the sensing outcome (i.e., light or no light) is always attributed to the assumed location. If the robot is uncertain about its location, then the light sensor may thus enable it to indirectly infer its position after going forward, namely, by excluding candidate locations which would necessitate an observation contradicting the one that has been made; see Exercise 5.5.

Substituting $KState(\sigma, z)$ by an equivalent formula according to (5.11) yields

$$(\exists \vec{y})(\forall z')\,(KState(Do(\alpha,\sigma), z') \equiv (\exists z)\,(\Phi(z) \wedge \Psi[z', z]) \wedge \Pi(z', Do(\alpha,\sigma))\,)$$

Thus we obtain a logical characterization of the resulting knowledge state. Because of the outer existential quantification, and due to the occurrence of a situation to the right of the equivalence, the resulting formula itself is not a knowledge state. These two aspects reflect that the successor knowledge state depends on the outcome of sensing.

Example 2 (continued) Recall the initial knowledge state in which the cleanbot is uncertain about its location, axiom (5.1), and suppose we are interested in inferring the knowledge state after performing a *Clean* action. To begin with, the rewriting technique of Chapter 3 can be applied: With the help of the rewrite law, Proposition 3.3 (page 64), and the underlying domain constraints, (3.2) on page 61, knowledge state (5.1) can be equivalently written as

$$\begin{aligned}
&KState(S_0, z) \equiv \\
&\quad (\exists x_0, z'')\,(\, z = At(x_0, 2) \circ Facing(2) \circ z'' \,\wedge\, [x_0 = 1 \vee x_0 = 2] \\
&\qquad \wedge\, (\forall u, v)\,\neg Holds(Cleaned(u, v), z'') \\
&\qquad \wedge\, (\forall u, v)\,\neg Holds(At(u, v), z'') \\
&\qquad \wedge\, (\forall d)\,\neg Holds(Facing(d), z'') \\
&\qquad \wedge\, (\forall x)\,\neg Holds(Occupied(x, 0), z'') \\
&\qquad \wedge\, (\forall x)\,\neg Holds(Occupied(x, 6), z'') \\
&\qquad \wedge\, (\forall y)\,\neg Holds(Occupied(0, y), z'') \\
&\qquad \wedge\, (\forall y)\,\neg Holds(Occupied(6, y), z'')\,)
\end{aligned} \tag{5.12}$$

The precondition axioms imply that $Poss(Clean, S_0)$ since this action is always possible; hence, the knowledge update axiom for *Clean*, (5.7), combined with (5.12) yields

$$\begin{aligned}
&KState(Do(Clean, S_0), z') \equiv \\
&\quad (\exists x_0, z, z'')\,(\, z = At(x_0, 2) \circ Facing(2) \circ z'' \,\wedge\, [x_0 = 1 \vee x_0 = 2] \\
&\qquad \wedge\, (\forall u, v)\,\neg Holds(Cleaned(u, v), z'') \\
&\qquad \wedge\, (\forall u, v)\,\neg Holds(At(u, v), z'') \\
&\qquad \wedge\, \ldots \\
&\qquad \wedge\, (\exists x, y)\,(\, Holds(At(x, y), z) \,\wedge \\
&\qquad\qquad z' = z + Cleaned(x, y))\,)
\end{aligned}$$

Because of $(\forall u, v)\,\neg Holds(At(u, v), z'')$ and $UNA[At, Facing, Cleaned]$, subformula $Holds(At(x, y), z)$ is equivalent to $x = x_0 \wedge y = 2$. Hence,

$$\begin{aligned}
&KState(Do(Clean, S_0), z') \equiv \\
&\quad (\exists x_0, z, z'')\,(\, z = At(x_0, 2) \circ Facing(2) \circ z'' \,\wedge\, [x_0 = 1 \vee x_0 = 2] \\
&\qquad \wedge\, (\forall u, v)\,\neg Holds(Cleaned(u, v), z'') \\
&\qquad \wedge\, (\forall u, v)\,\neg Holds(At(u, v), z'') \\
&\qquad \wedge\, \ldots \\
&\qquad \wedge\, z' = z + Cleaned(x_0, 2)\,)
\end{aligned}$$

Combining the two state equations in this formula yields

$$\begin{aligned}&KState(Do(Clean, S_0), z') \equiv \\ &\quad (\exists x_0, z'')\,(\,[x_0 = 1 \vee x_0 = 2] \\ &\qquad \wedge (\forall u, v)\, \neg Holds(Cleaned(u, v), z'') \\ &\qquad \wedge (\forall u, v)\, \neg Holds(At(u, v), z'') \\ &\qquad \wedge \ldots \\ &\qquad \wedge z' = At(x_0, 2) \circ Facing(2) \circ z'' + Cleaned(x_0, 2)\,)\end{aligned}$$

From $UNA[At, Facing, Cleaned]$ and the rewrite law we then conclude

$$\begin{aligned}&KState(Do(Clean, S_0), z') \equiv \\ &\quad (\exists x_0, z'')\,(\,z' = At(x_0, 2) \circ Facing(2) \circ Cleaned(x_0, 2) \circ z'' \\ &\qquad \wedge [x_0 = 1 \vee x_0 = 2] \\ &\qquad \wedge (\forall u, v)\, \neg Holds(Cleaned(u, v), z'') \\ &\qquad \wedge (\forall u, v)\, \neg Holds(At(u, v), z'') \\ &\qquad \wedge (\forall d)\, \neg Holds(Facing(d), z'') \\ &\qquad \wedge (\forall x)\, \neg Holds(Occupied(x, 0), z'') \\ &\qquad \wedge (\forall x)\, \neg Holds(Occupied(x, 6), z'') \\ &\qquad \wedge (\forall y)\, \neg Holds(Occupied(0, y), z'') \\ &\qquad \wedge (\forall y)\, \neg Holds(Occupied(6, y), z'')\,)\end{aligned} \tag{5.13}$$

Thus we have derived the knowledge state for the successor situation.

As an example for inferring updated knowledge states where sensing is involved, recall knowledge update axiom (5.8) for action Go, and consider the original scenario of the cleaning robot depicted in Figure 3.1 (page 60), where

$$\begin{aligned}Holds(Occupied(x, y), S_0) \equiv\ & x = 1 \wedge y = 4 \ \vee\ x = 3 \wedge y = 1 \vee \\ & x = 3 \wedge y = 3 \ \vee\ x = 5 \wedge y = 3\end{aligned} \tag{5.14}$$

Let the initial knowledge of the cleanbot be axiomatized by this knowledge state:

$$\begin{aligned}&KState(S_0, z) \equiv \\ &\quad (\exists z'')\,(\,z = At(1, 1) \circ Facing(1) \circ z'' \\ &\qquad \wedge \neg Holds(Occupied(1, 1), z'') \\ &\qquad \wedge \ \ldots \ \wedge \neg Holds(Occupied(4, 5), z'') \\ &\qquad \wedge (\forall x, y)\, \neg Holds(Cleaned(x, y), z'') \\ &\qquad \wedge (\forall x, y)\, \neg Holds(At(x, y), z'') \\ &\qquad \wedge (\forall d)\, \neg Holds(Facing(d), z'') \\ &\qquad \wedge (\forall x)\, \neg Holds(Occupied(x, 0), z'') \\ &\qquad \wedge (\forall x)\, \neg Holds(Occupied(x, 6), z'') \\ &\qquad \wedge (\forall y)\, \neg Holds(Occupied(0, y), z'') \\ &\qquad \wedge (\forall y)\, \neg Holds(Occupied(6, y), z'')\,)\end{aligned} \tag{5.15}$$

The left diagram in Figure 5.5 gives a graphical interpretation of this specification. The precondition axiom for Go (cf. axioms (3.12), page 67) implies $Knows(Poss(Go), S_0)$; hence, according to Proposition 5.4, $Poss(Go, S_0)$. Applying knowledge update axiom (5.8) to knowledge state (5.15) yields, after

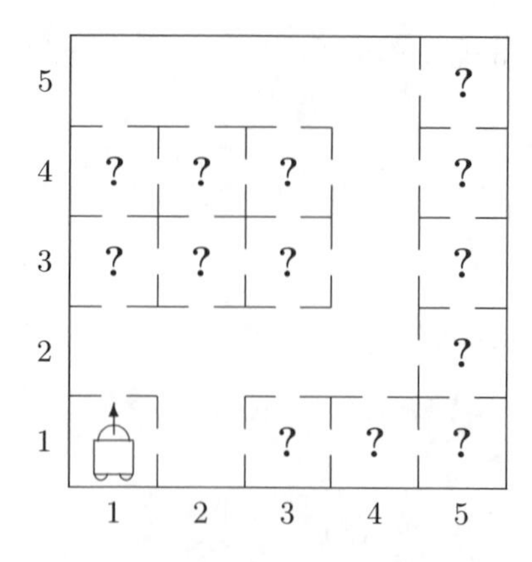

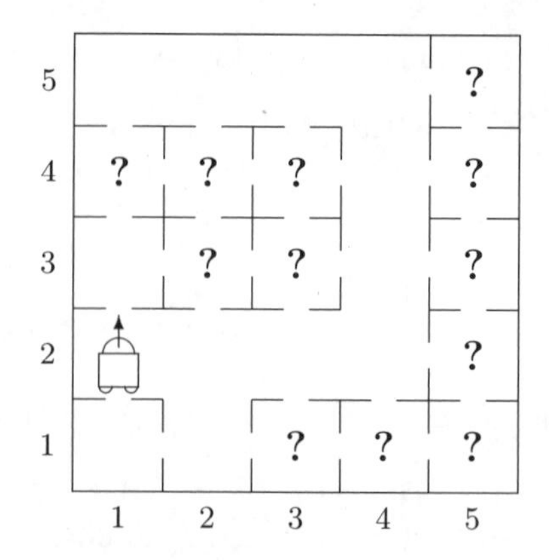

Figure 5.5: Left hand side: Initially, the robot has no information about whether any of the rooms (besides its own) is occupied. Right hand side: After moving to $(1,2)$, by sensing no light there the robot has learned that office $(1,3)$ is not occupied.

---

inferring the physical effect of the action,

$$
\begin{aligned}
& KState(Do(Go, S_0), z') \equiv \\
& \quad (\exists z'')\,(\, z' = At(1,2) \circ Facing(1) \circ z'' \\
& \qquad\qquad \wedge \neg Holds(Occupied(1,1), z'') \\
& \qquad\qquad \wedge \ \ldots \\
& \quad \wedge \ \Pi_{Light}(z') \equiv \Pi_{Light}(Do(Go, S_0))
\end{aligned}
$$

The assumption (5.14), the state update axiom for Go, (3.14) on page 67, and the definition of perceiving light, axiom (3.3) on page 61, imply that $Holds(At(1,2), Do(Go, S_0))$ and $\neg Light(1,2, Do(Go, S_0))$ because neither this nor any of the surrounding squares is occupied. Hence, the possible states z' are restricted by $\neg\Pi_{Light}(z')$, which implies

$$
\begin{aligned}
& KState(Do(Go, S_0), z') \equiv \\
& \quad (\exists z'')\,(\, z' = At(1,2) \circ Facing(1) \circ z'' \\
& \qquad\qquad \wedge \neg Holds(Occupied(1,1), z'') \\
& \qquad\qquad \wedge \ \ldots \\
& \qquad\qquad \wedge \neg Holds(Occupied(1,3), z'')\,)
\end{aligned} \tag{5.16}
$$

This resulting knowledge state is depicted in the right hand side of Figure 5.5. Besides the changed location of the robot, the knowledge state includes the new information that office $(1,3)$ is not occupied.

Suppose the cleanbot continues its way with a second Go action. From knowledge state (5.16) and the precondition axiom for Go it follows that $Knows(Poss(Go), Do(Go, S_0))$; hence, $Poss(Go, Do(Go, S_0))$. The application of knowledge update axiom (5.8) to knowledge state (5.16) yields, after inferring the physical effect,

$$
\begin{aligned}
& KState(Do(Go, Do(Go, S_0)), z') \equiv \\
& \quad (\exists z'')\,(z' = At(1,3) \circ Facing(1) \circ z'' \\
& \qquad \wedge \neg Holds(Occupied(1,1), z'') \\
& \qquad \wedge \ldots \wedge \neg Holds(Occupied(4,5), z'') \\
& \qquad \wedge (\forall x, y)\, \neg Holds(Cleaned(x,y), z'') \\
& \qquad \wedge (\forall x, y)\, \neg Holds(At(x,y), z'') \\
& \qquad \wedge (\forall d)\, \neg Holds(Facing(d), z'') \\
& \qquad \wedge (\forall x)\, \neg Holds(Occupied(x,0), z'') \\
& \qquad \wedge (\forall x)\, \neg Holds(Occupied(x,6), z'') \\
& \qquad \wedge (\forall y)\, \neg Holds(Occupied(0,y), z'') \\
& \qquad \wedge (\forall y)\, \neg Holds(Occupied(6,y), z'') \\
& \qquad \wedge \neg Holds(Occupied(1,3), z'')\,) \\
& \quad \wedge \Pi_{Light}(z') \equiv \Pi_{Light}(Do(Go, Do(Go, S_0)))
\end{aligned}
$$

This, along with Σ_{knows}, implies that $Holds(At(1,3), Do(Go, Do(Go, S_0)))$. From $Holds(Occupied(1,4), S_0)$ according to (5.14), and the state update axiom for Go it follows that $Holds(Occupied(1,4), Do(Go, Do(Go, S_0)))$. Thus there is light at the current location $(1,3)$, and hence the possible states z' are restricted by $\Pi_{Light}(z')$. Decomposition and irreducibility along with the fact that $\neg Holds(Occupied(1,3), z'')$, that $\neg Holds(Occupied(1,2), z'')$, and that $(\forall y)\, \neg Holds(Occupied(0,y), z'')$, imply

$$
\begin{aligned}
& KState(Do(Go, Do(Go, S_0)), z') \equiv \\
& \quad (\exists z'')\,(z' = At(1,3) \circ Facing(1) \circ z'' \\
& \qquad \wedge \neg Holds(Occupied(1,1), z'') \\
& \qquad \wedge \ldots \\
& \qquad \wedge \neg Holds(Occupied(1,3), z'') \\
& \qquad \wedge [Holds(Occupied(1,4), z'') \vee Holds(Occupied(2,3), z'')]\,)
\end{aligned}
$$

Thus the robot knows that either of the locations $(1,4)$ or $(2,3)$ must be occupied, without being able to tell which of the two. This knowledge state is depicted in the left hand side of Figure 5.6. The reader may verify that the cleaning robot reaches the knowledge state depicted in the right hand side of this figure after turning twice, going, turning three times, and going forward again without sensing light at $(2,2)$. $\square$

As an example for inferring the update of knowledge when sensing values, recall the knowledge update axiom for the pure sensing action $SenseLoc$, and suppose that $Poss(SenseLoc, z) \equiv \top$. Consider the sample knowledge state with the uncertain location, formula (5.13). Applying knowledge update axiom (5.10)

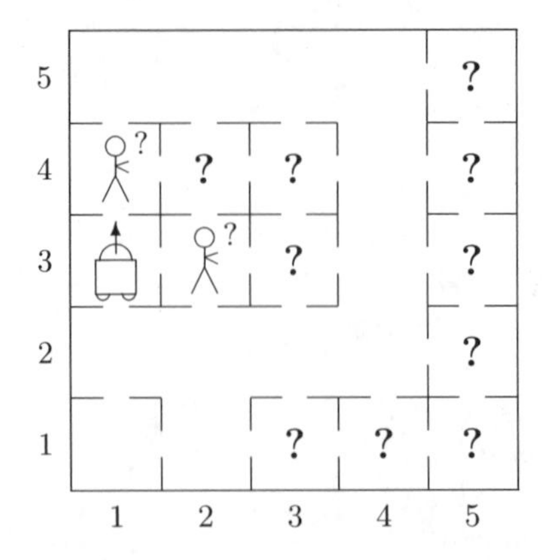

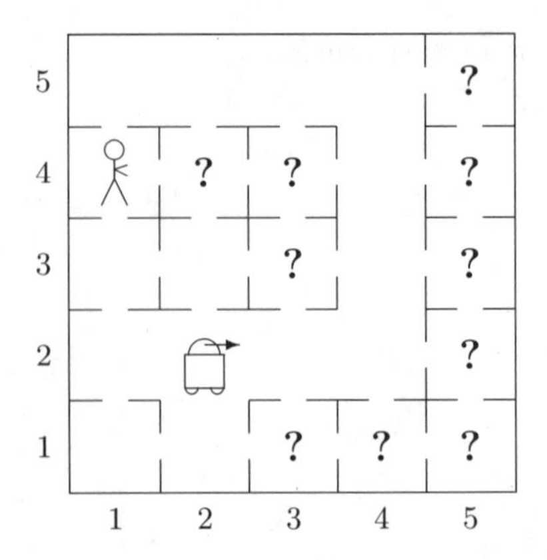

Figure 5.6: Left hand side: Continuing its way to $(1,3)$ and sensing light at this place, the resulting knowledge state indicates that office $(1,4)$ or office $(2,3)$ must be occupied (or both). Right hand side: Knowledge state reached after performing the sequence of actions *Turn*, *Turn*, *Go*, *Turn*, *Turn*, *Turn*, *Go*.

---

yields

$$
\begin{array}{l}
(\exists x, y)(\forall z')\,(\mathit{KState}(\mathit{Do}(\mathit{SenseLoc}, \mathit{Do}(\mathit{Clean}, S_0)), z') \equiv \\
\quad (\exists x_0, z'')\,(\, z' = \mathit{At}(x_0, 2) \circ \mathit{Facing}(2) \circ \mathit{Cleaned}(x_0, 2) \circ z'' \\
\qquad\qquad \wedge\ [x_0 = 1 \vee x_0 = 2] \\
\qquad\qquad \wedge\ (\forall u, v)\,\neg \mathit{Holds}(\mathit{Cleaned}(u, v), z'') \\
\qquad\qquad \wedge\ (\forall u, v)\,\neg \mathit{Holds}(\mathit{At}(u, v), z'') \\
\qquad\qquad \wedge\ \ldots) \\
\quad \wedge\ \mathit{Holds}(\mathit{At}(x, y), z) \\
\quad \wedge\ \mathit{Holds}(\mathit{At}(x, y), \mathit{Do}(\mathit{SenseLoc}, \mathit{Do}(\mathit{Clean}, S_0))))
\end{array}
$$

If we assume that the robot is in fact at location $(1,2)$ in the resulting situation $\mathit{Do}(\mathit{SenseLoc}, \mathit{Do}(\mathit{Go}, S_0))$, then

$$
\begin{array}{l}
\mathit{KState}(\mathit{Do}(\mathit{SenseLoc}, \mathit{Do}(\mathit{Clean}, S_0)), z) \equiv \\
\quad (\exists z'')\,(\, z' = \mathit{At}(1, 2) \circ \mathit{Facing}(2) \circ \mathit{Cleaned}(1, 2) \circ z'' \\
\qquad\qquad \wedge\ (\forall u, v)\,\neg \mathit{Holds}(\mathit{Cleaned}(u, v), z'') \\
\qquad\qquad \wedge\ (\forall u, v)\,\neg \mathit{Holds}(\mathit{At}(u, v), z'') \\
\qquad\qquad \wedge\ (\forall d)\,\neg \mathit{Holds}(\mathit{Facing}(d), z'') \\
\qquad\qquad \wedge\ (\forall x)\,\neg \mathit{Holds}(\mathit{Occupied}(x, 0), z'') \\
\qquad\qquad \wedge\ (\forall x)\,\neg \mathit{Holds}(\mathit{Occupied}(x, 6), z'') \\
\qquad\qquad \wedge\ (\forall y)\,\neg \mathit{Holds}(\mathit{Occupied}(0, y), z'') \\
\qquad\qquad \wedge\ (\forall y)\,\neg \mathit{Holds}(\mathit{Occupied}(6, y), z''))
\end{array}
$$

Thus the robot has learned where it is.

The discussion on knowledge and update is summarized in the following definition of domain axiomatizations in general fluent calculus. Besides requiring a knowledge update axiom for each action rather than a mere state update axiom, some of the ingredients of domain axiomatizations in special fluent calculus need to be generalized. The initial conditions are given as a knowledge state for S_0 rather than a (complete) specification of $State(S_0)$. In a similar fashion, domain constraints in general fluent calculus are defined as restrictions on knowledge states rather than just actual states. Since according to Proposition 5.4, foundational axiom Σ_{knows} implies that everything that is known actually holds, both the initial condition and all domain constraints are necessarily satisfied by actual states.

Definition 5.8 A **domain constraint** (in fluent calculus for knowledge) is a formula of the form $(\forall s)\, Knows(\phi, s)$ where ϕ is a knowledge expression without actions. A knowledge state $KState(\sigma, z) \equiv \Phi(z)$ **satisfies** a set of domain constraints Σ_{dc} wrt. some auxiliary axioms Σ_{aux} iff the set of axioms $\Sigma_{dc} \cup \Sigma_{aux} \cup \Sigma_{state} \cup \{KState(\sigma, z) \equiv \Phi(z)\}$ is consistent.

Consider a fluent calculus signature with action functions $\mathcal{A}$. A **domain axiomatization** (in fluent calculus for knowledge) is a finite set of axioms $\Sigma = \Sigma_{dc} \cup \Sigma_{poss} \cup \Sigma_{kua} \cup \Sigma_{aux} \cup \Sigma_{init} \cup \Sigma_{state} \cup \Sigma_{knows}$ where

- Σ_{dc} set of domain constraints;
- Σ_{poss} set of precondition axioms, one for each $A \in \mathcal{A}$;
- Σ_{kua} set of knowledge update axioms, one for each $A \in \mathcal{A}$;
- Σ_{aux} set of auxiliary axioms, with no occurrence of states except for fluents, no occurrence of situations, and no occurrence of $Poss$;
- $\Sigma_{init} = \{KState(S_0, z) \equiv \Phi(z)\}$ where $\Phi(z)$ state formula;
- $\Sigma_{state} \cup \Sigma_{knows}$ foundational axioms of fluent calculus for knowledge as in Definition 5.3. □

5.4 Specifying a Domain in FLUX

In this section, we will show how to encode in FLUX the various ingredients of domain axiomatizations. This background theory will be needed by agents to initialize and maintain their internal, incomplete model of the world and to test the executability of actions.

Precondition Axioms

Acting cautiously under incomplete information, an agent should trigger the performance of an action only if it knows that the action is possible. Using the standard FLUX predicate $Poss(a, z)$, the precondition axioms should therefore be encoded so as to satisfy the following condition.

Definition 5.9 Consider a domain axiomatization Σ and an action function $A(\vec{x})$. A program P_{poss} is a **sound encoding of the precondition of** $A(\vec{x})$ if the following holds for all state formulas $\Phi(z)$ and ground instances $A(\vec{t})$: If $P_{kernel} \cup P_{poss} \cup \{\leftarrow [\![\Phi(z)]\!], Poss(A(\vec{t}), z)\}$ is successful with computed answer θ, then $\theta = \{\}$ and

$$\Sigma \cup \{KState(\sigma, z) \equiv \Phi(z)\} \models Knows(Poss(A(\vec{t})), \sigma)$$

□

Put in words, a sound encoding allows to derive $Poss(A(\vec{t}), z)$ only if the action is known to be possible in the current knowledge state, encoded by FLUX state $[\![\Phi(z)]\!]$. When an agent checks executability of an action, this should never have any side-effect on its current world model. Hence the additional requirement that the computed answer be empty.

As an example, the following clause encodes the precondition axiom for action *Go* in the cleanbot domain (cf. axioms (3.12), page 67):

```
poss(go,Z) :- \+ ( holds(at(X,Y),Z), holds(facing(D),Z),
                   exists_not_adjacent(X,Y,D) ).

exists_not_adjacent(X,Y,D) :-
   D#=1 #/\ Y#=5 #\/ D#=2 #/\ X#=5 #\/
   D#=3 #/\ Y#=1 #\/ D#=4 #/\ X#=1.
```

Put in words, executability of *Go* is known under an incomplete FLUX state if it cannot be the case that the robot is at a location and faces a direction for which there exists no adjacent square. Thus, for instance,

```
?- Z = [at(X,2),facing(D) | _], D :: [1,3],
   consistent(Z),
   poss(go, Z).

yes.

?- Z = [at(1,Y),facing(D) | _], D :: [2,4],
   consistent(Z),
   poss(go, Z).

no (more) solution.
```

Accordingly, $Knows((\exists x, d)\,(At(x,2) \wedge Facing(d) \wedge [d = 1 \vee d = 3]), s)$ implies $Knows(Poss(Go), s)$ under the axiomatization of the cleaning robot domain, while $Knows((\exists y, d)\,(At(1,y) \wedge Facing(d) \wedge [d = 2 \vee d = 4]), s)$ does not so since the agent may face west while being in one of the westmost locations. Thus the robot acts safely by not attempting an action which is potentially hazardous (e.g., if there are stairways going down at the west end of the floor).

Initial State

As usual, the specification of the initial conditions is encoded in FLUX by a definition of the predicate $Init(z_0)$. Since agent programs use a single call to *Init* in order to initialize their internal model, the clause should not admit backtracking. In particular, disjunctive state knowledge should always be encoded with the help of the *OrHolds*-constraint rather than by standard Prolog disjunction. For the sake of modularity, initial conditions in a particular scenario can be separated from a specification of the general domain constraints as in the example shown in Figure 4.1 (page 78).

The encoding of an initial knowledge state should be sound according to the following condition:

Definition 5.10 Let Σ be a domain axiomatization with initial knowledge state $\Sigma_{init} = \{KState(S_0, z) \equiv \Phi(z)\}$. A program P_{init} is a **sound encoding of the initial knowledge state** if $P_{kernel} \cup P_{init} \cup \{\leftarrow Init(z_0)\}$ has a unique answer θ and $z_0\theta$ is a FLUX state which satisfies

$$\Sigma \models \Phi(z_0) \supset z_0\theta$$

□

Put in words, a sound encoding determines a FLUX state which is implied by the initial knowledge state. The reverse is not required in general. This enables the programmer to encode state specifications which include elements that are not FLUX-expressible. Endowing an agent with a weaker knowledge state is not an obstacle towards sound agent programs. Since FLUX agents are controlled by what they know of the environment, a sound but incomplete knowledge state suffices to ensure that the agent behaves cautiously. This is a consequence of the fact that everything that is known under a weaker knowledge state is also known under the stronger one, as the following result shows.

Proposition 5.11 *Let $\Phi_1(z)$ and $\Phi_2(z)$ be two state formulas which satisfy $\Sigma_{state} \models \Phi_1(z) \supset \Phi_2(z)$ and let σ be a situation, then for any knowledge expression ϕ,*

$$\Sigma_{state} \cup \{KState(\sigma, z) \equiv \Phi_2(z)\} \models Knows(\phi, \sigma)$$

implies

$$\Sigma_{state} \cup \{KState(\sigma, z) \equiv \Phi_1(z)\} \models Knows(\phi, \sigma)$$

Proof: Exercise 5.8 ■

Domain Constraints

A standard way of encoding a set of domain constraints is by defining the predicate $Consistent(z)$. Agent programs use this definition to infer restrictions on their world model which follow from the constraints. It is therefore permitted that query $\leftarrow Consistent(z)$ admits a non-empty computed answer. Consider, for example, the domain constraint which defines fluent $F(x)$ to be functional, that is, to hold for a unique value in every situation. A suitable FLUX encoding of this constraint adds the corresponding restriction to a state:

```
consistent(Z) :-
  holds(f(_),Z,Z1), not_holds_all(f(_),Z1), duplicate_free(Z).

?- consistent(Z)

Z = [f(_) | Z1]

Constraints:
not_holds_all(f(_), Z1)
duplicate_free(Z1)
```

The encoding of a set of domain constraints should be sound according to the following condition:

Definition 5.12 Consider a domain axiomatization Σ. A program P_{dc} is a **sound encoding of the domain constraints** if for every computed answer θ to $P_{kernel} \cup P_{dc} \cup \{\leftarrow Consistent(z)\}$, $z\theta$ is a FLUX state and satisfies

$$\Sigma \models KState(s, z) \supset z\theta$$

□

Put in words, a sound encoding of the domain constraints determines a state specification which is implied by any possible state. Much like in case of initial states, the encoding of a set of domain constraints may be incomplete if some constraints cannot be fully expressed in FLUX. If this is so, then the agent works with a weaker knowledge state, which does not entail every logical consequence of the domain constraints.

By definition, a sound encoding of the domain constraints requires that each answer to the query $\leftarrow Consistent(z)$ is implied by the constraints of a domain. Hence, a disjunctive constraint, for example, should always be encoded with the help of the *OrHolds*-constraint rather than by standard Prolog disjunction—much like in encodings of initial knowledge states. If the predicate $Consistent(z)$ is used as part of the definition of $Init(z)$, then backtracking is to be avoided anyway, that is, the query $\leftarrow Consistent(z)$ should admit a unique computed answer. This needs to be particularly observed when programming domain-dependent CHRs to enforce certain domain constraints.

Knowledge Update Axioms

As in fluent calculus, the update of knowledge in FLUX involves two aspects, the physical and the cognitive effects of an action. In Section 5.2, knowledge states have been identified with (incomplete) FLUX states. Knowledge update according to a given set of physical effects thus amounts to updating a FLUX state specification in the way discussed in Section 4.4.

Having inferred the physical effect of an action, agents need to evaluate possible sensing results to complete the update of their world model. In fluent calculus, sensing is modeled by constraining the set of possible states so as to agree with the actual state on the sensed properties and fluent values. This is realized in FLUX by using the acquired sensing result to constrain the state specification obtained after inferring the physical effects. Recall that sensing is expressed in knowledge update axioms by a sub-formula of the form

$$\bigwedge_{i=1}^{k} [\Pi_i(z) \equiv \Pi_i(Do(a,s))] \wedge \bigwedge_{i=1}^{l} Holds(F_i(\vec{t_i}), z) \wedge \Pi(\vec{x}, \vec{y})$$

Here, the Π_i's are state properties for which the sensors reveal whether or not they actually hold, while the $F_i(\vec{t_i})$'s are fluents for which the sensors return particular values that currently hold. In order to have agents update their knowledge according to acquired sensor information, we need a way to encode sensing results. By the following definition, the result of sensing a state property is either of the constants *True* or *False*, and the result of sensing a value is a ground term of the right sort. Any concrete sensing result determines a state formula that expresses which of the sensed properties are true and for which values the sensed fluents hold. If the sensed information is known in part prior to the action, then certain sensing results are impossible, namely, those which contradict what is already known.

Definition 5.13 Consider a domain axiomatization Σ and a ground action α with knowledge update axiom (5.6) where $\vec{y} = y_1, \ldots, y_m$ $(m \geq 0)$. A **sensing result for** α is a ground list $[\pi_1, \ldots, \pi_k, \pi_{k+1}, \ldots, \pi_{k+m}]$ such that

1. $\pi_i \in \{True, False\}$, if $1 \leq i \leq k$; and

2. $\pi_i \in S_i$ where S_i is the sort of y_i, if $k < i \leq k+m$.

A sensing result $\pi = [\pi_1, \ldots, \pi_k, \vec{\pi}]$ for α **determines** this state formula:

$$\bigwedge_{i=1}^{k} \left\{ \begin{array}{ll} \Pi_i(z) & \text{if } \pi_i = True \\ \neg\Pi_i(z) & \text{if } \pi_i = False \end{array} \right\} \wedge \bigwedge_{i=1}^{l} Holds(F_i(\vec{t_i}), z)\{\vec{y}/\vec{\pi}\} \qquad (5.17)$$

A sensing result π for α in a situation σ is said to be **consistent** with a knowledge state $KState(\sigma, z) \equiv \Phi(z)$ iff

$$\Sigma \cup \{KState(\sigma, z) \equiv \Phi(z)\} \cup \{KState(Do(\alpha, \sigma), z) \supset (5.17)\}$$

is consistent. □

Put in words, a sensing result is consistent if it can be consistently asserted that all successor knowledge states imply the sensing result (or rather the corresponding state formula). For example, the sensing result for knowledge update axiom (5.8) is encoded by $[\pi]$ where $\pi \in \{True, False\}$, depending on whether light is actually sensed at the new location. The sensing result $[False]$, say, determines the formula $\neg\Pi_{Light}(z)$. This assertion is consistent with a knowledge state in which $At(1,2)$ and $Facing(1)$ are known and no information is given as to which offices are occupied, for in this case sensing no light after going forward is a possibility. The sensing result $[False]$ would not be consistent, however, if additionally, say, $Occupied(1,4)$ is known beforehand, since

$$\begin{array}{l} Knows(At(1,2) \wedge Facing(1) \wedge Occupied(1,4), s) \\ \wedge [\, KState(Do(Go, s), z) \supset \neg\Pi_{Light}(z) \,] \end{array}$$

is inconsistent under the axiomatization of the cleanbot domain.

As a second example, recall knowledge update axiom (5.10) for $SenseLoc$, for which the sensing result is encoded by the pair $[\pi_1, \pi_2]$ with $\pi_1, \pi_2 \in \mathbb{N}$, revealing the x- and y-coordinate of the current location of the robot. For instance, the sensing result $[1,2]$ determines the formula $Holds(At(1,2), z)$, which is consistent, say, with prior knowledge $Knows((\exists x)\,(At(x,2) \wedge x < 3), s)$.

Based on the notion of sensing results, knowledge update axioms are encoded in FLUX by clauses for the predicate $StateUpdate(z_1, A(\vec{x}), z_2, y)$ as in special FLUX but where the additional fourth argument conveys the sensing result. A sound encoding of a knowledge update axiom requires that an inferred updated FLUX state is logically entailed given the preceding knowledge state, the update axiom, and the respective sensor result. It suffices to define the update only for those knowledge states which are consistent and under which the action in question is known to be possible. Furthermore, only those sensing results need to be considered which are consistent with the knowledge state to which the update is applied. Formally, the encoding of a knowledge update axiom should be sound according to the following condition:

Definition 5.14 Consider a domain axiomatization Σ and an action function $A(\vec{x})$. A program P_{kua} is a **sound encoding of the knowledge update axiom of** $A(\vec{x})$ if the following holds for all ground instances $A(\vec{t})$ and for all FLUX states $[\![\Phi(z)]\!]$ such that $KState(\sigma, z) \equiv \Phi(z)$ is a consistent knowledge state and such that

$$\Sigma \cup \{KState(\sigma, z) \equiv \Phi(z)\} \models Knows(Poss(A(\vec{t})), \sigma)$$

For all sensing results π for $A(\vec{t})$ consistent with $KState(\sigma, z) \equiv \Phi(z)$, the query $P_{kernel} \cup P_{kua} \cup \{\leftarrow [\![\Phi(z)]\!], StateUpdate(z, A(\vec{t}), z', \pi)\}$ has a unique answer θ and $\Sigma \cup \{KState(\sigma, z) \equiv \Phi(z)\}$ entails

$$\Pi(Do(A(\vec{t}), \sigma)) \wedge KState(Do(A(\vec{t}), \sigma), z') \supset z'\theta$$

where $\Pi(z)$ is the state formula determined by π. □

Put in words, the updated FLUX state must always be implied by the successor knowledge state under the observed sensing result.

```
state_update(Z1, clean, Z2, []) :-
   holds(at(X,Y), Z1),
   update(Z1, [cleaned(X,Y)], [], Z2).

state_update(Z1, turn, Z2, []) :-
   holds(facing(D), Z1),
   (D#<4 #/\ D1#=D+1) #\/ (D#=4 #/\ D1#=1),
   update(Z1, [facing(D1)], [facing(D)], Z2).

state_update(Z1, go, Z2, [Light]) :-
   holds(at(X,Y), Z1),
   holds(facing(D), Z1),
   adjacent(X, Y, D, X1, Y1),
   update(Z1, [at(X1,Y1)], [at(X,Y)], Z2),
   light(X1, Y1, Light, Z2).
```

Figure 5.7: FLUX encoding of the update axioms for the cleanbot. The auxiliary predicates *Light* and *Adjacent* are as defined in Figure 4.4 (page 87) and Figure 4.9 (page 96), respectively.

Example 2 (continued) Figure 5.7 depicts FLUX clauses encoding the knowledge update axioms for the cleaning robot domain. For all three axioms, the update is guaranteed to be unique because the domain constraints entail that in any consistent state there is one and only one instance (possibly with variable arguments) of $At(x, y)$ and of $Facing(d)$, respectively. Action *Go* is the only one which involves sensing. The sensing result is evaluated with the help of the auxiliary predicate *Light*. Recall, for example, the initial FLUX state for the cleaning robot of Figure 4.1 (page 78). Suppose the usual scenario where office $(1, 3)$ is not occupied but office $(1, 4)$ is. Going north twice from room $(1, 1)$, the robot senses no light after the first action but after the second one. With the following query we infer the knowledge update for this sequence of actions and the given sensing results:

```
?- init(Z0), state_update(Z0,go,Z1,[false]),
             state_update(Z1,go,Z2,[true]).

Z0 = [at(1,1),facing(1) | Z]
Z1 = [at(1,2),facing(1) | Z]
Z2 = [at(1,3),facing(1) | Z]

Constraints:
not_holds(occupied(1,3),Z)
or_holds([occupied(2,3),occupied(1,4)],Z)
...
```

Thus the agent has evaluated the acquired sensor data and inferred its resulting position according to the physical effect of *Go*.

As an example for sensing a value of a fluent, consider the following FLUX clause, which encodes the knowledge update axiom for *SenseLoc*:

```
state_update(Z,sense_loc,Z,[X,Y]) :- holds(at(X,Y),Z).
```

That is, no physical effect affects the state but the sensed value is incorporated into the specification. Suppose, for instance, the agent is uncertain as to whether it walked north or east from its initial location $(1,1)$, but the subsequent position tracking reveals that it is at $(1,2)$:

```
init(Z0) :- Z0 = [at(1,1),facing(D) | _],
            D#=1 #\/ D#=2,
            consistent(Z0).

?- init(Z0), state_update(Z0,go,Z1,[false]),
             state_update(Z1,sense_loc,Z2,[1,2]).

Z2 = [at(1,2),facing(1) | Z]

Constraints:
not_holds(occupied(1,3),Z)
...
```

Thus the agent has inferred its actual position and, hence, concluded that it is actually facing north. Incidentally, knowing the location also allows to infer that office $(1,3)$ is not occupied, which follows from the observation that the *Go* action results in no light being sensed. □

Since agent programs use predicate *StateUpdate* to update their internal world model, it is crucial to observe the restriction in Definition 5.14 that knowledge update axioms always admit a unique answer. If an action has conditional effects, the agent programmer needs to define the correct update for the different degrees of knowledge the agent may have concerning the involved fluents. Consider, for example, an action called $Alter(x)$ by which the position of a toggle switch is altered from open to close and vice versa. That is, if x happens to be open (fluent $Open(x)$), then it will be closed afterwards (i.e., not open); else, if it is closed beforehand it will be open after the action. Tacitly assuming that the action is always possible, its conditional effect is specified by the following knowledge update axiom in fluent calculus:

$$\begin{array}{l} KState(Do(Alter(x),s),z') \equiv (\exists z)\,(KState(s,z) \land \\ \quad [\,Holds(Open(x),z) \land z' = z - Open(x) \\ \quad \lor \\ \quad \neg Holds(Open(x),z) \land z' = z + Open(x)\,]\,) \end{array} \tag{5.18}$$

The FLUX encoding of this update axiom requires to distinguish three kinds of knowledge states. In case the current knowledge entails that switch x is open, the resulting knowledge state is obtained through updating by negative effect $Open(x)$. Conversely, in case the current knowledge entails that switch x is not open, the resulting knowledge state is obtained through updating by positive effect $Open(x)$. Finally, if the current knowledge state does not entail the status of the switch, then this uncertainty transfers to the updated knowledge state. Moreover, possible partial (e.g., disjunctive) information regarding the position of the affected switch is no longer valid and, hence, needs to be cancelled:

```
state_update(Z1,alter(X),Z2,[]) :-
   knows(open(X),Z1)     -> update(Z1,[],[open(X)],Z2) ;
   knows_not(open(X),Z1) -> update(Z1,[open(X)],[],Z2) ;
   cancel(open(X),Z1,Z2).
```

For instance,

```
?- not_holds(open(1),Z0),
   or_holds([open(2),open(3)],Z0),
   duplicate_free(Z0),
   state_update(Z0,alter(1),Z1),
   state_update(Z1,alter(2),Z2).

Z2 = [open(1) | Z0]

Constraints:
not_holds(open(1), Z0)
```

That is to say, while switch 1 is known to hold after altering its position, after altering switch 2 it no longer holds that either of the switches 2 or 3 is open.

Actually, the inferred knowledge state in this example is slightly weaker than what is implied by knowledge update axiom (5.18). To see why, suppose that initially switch 2 or 3 is open. Then it follows that after altering the position of switch 2, if this switch is open then so is switch 3! This is so because if switch 2 is open after changing its position, it must have been closed initially, and hence switch 3 was (and still is) open. But although the knowledge expression $Open(2) \supset Open(3)$ could be known after toggling switch 2, it is not entailed by the inferred updated knowledge state.

Generally, because of the incompleteness of update in FLUX (cf. Section 4.4), an updated FLUX state may be weaker than what the knowledge update axiom entails in fluent calculus. This is similar to the characterization of sound but possibly incomplete encodings of initial knowledge states or of the domain constraints.

Auxiliary Axioms

Much like in special FLUX, the universal assumption of uniqueness-of-names is built into the constraint solver of general FLUX. Other auxiliary axioms need to be suitably encoded as logic programming clauses. Examples are the definitions

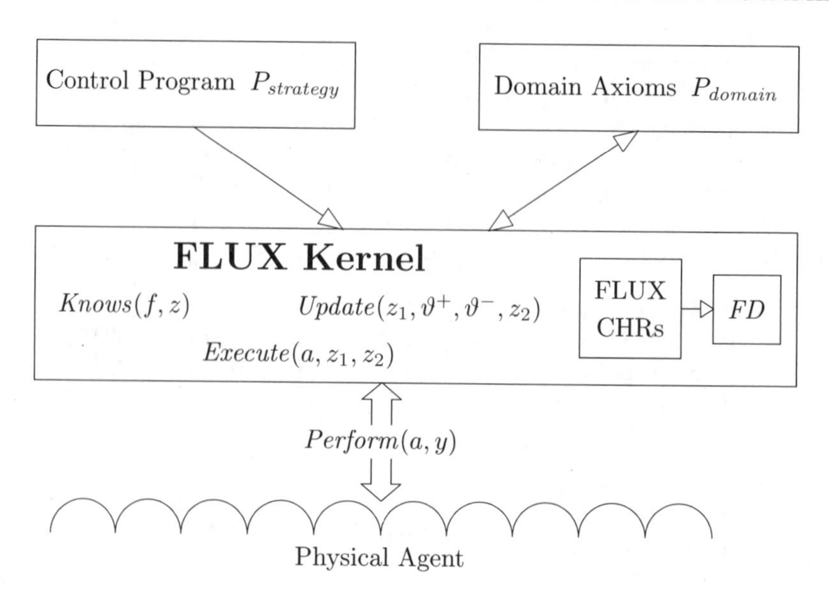

Figure 5.8: Architecture of agent programs written in general FLUX.

of the auxiliary predicates *Light* and *Adjacent* for the cleanbot domain (cf. Figure 4.4 and 4.9 on page 87 and 96, respectively).

5.5 Knowledge Agent Programs

The architecture of general FLUX agents is depicted in Figure 5.8. A submodule of the FLUX kernel is the constraint solver, consisting of the CHRs for the state constraints and a standard constraint solver for finite domains (*FD*). Much like in case of special FLUX with exogenous actions, the interaction of the agent with the outside world is bidirectional as the execution of an action may return sensor information. This is reflected by a modification of the predicate *Perform*, defining the performance of an action. In the presence of sensing, an instance $Perform(a, y)$ means that the agent actually does action a and observes sensing result y.[3] The observed outcome is then used for inferring the updated knowledge state; see Figure 5.9. The second crucial difference to special FLUX is that programs for agents with incomplete information use the basic statement $Knows(f, z)$ (and its derivatives *KnowsNot* and *KnowsVal*) rather than $Holds(f, z)$ when evaluating conditions against the internal world model.

[3] For the sake of clarity, we treat sensing independent from the occurrence of exogenous actions. The combination of these two issues is straightforward, both as far as the theoretical analysis is concerned and the realization in FLUX. The full FLUX system includes the ternary $Perform(a, y, e)$, where y is the sensing result of a and e the exogenous actions that occur during the performance of the action (see Appendix A).

```
execute(A, Z1, Z2) :-
   perform(A, Y), state_update(Z1, A, Z2, Y).
```

Figure 5.9: The execution of an action in general FLUX involves the evaluation of the sensing result when updating the internal world model.

Let P_{kernel} denote the full kernel of general FLUX, consisting of

- the FLUX constraint system of Figure 4.2 and 4.3 plus a standard *FD*-solver;
- the definition of knowledge of Figure 5.2;
- the definition of update of Figure 4.5, 4.7, and 4.8; and
- the definition of execution of Figure 5.9.

Agents in general FLUX are implemented on top of the kernel program. As in special FLUX, the agent programmer provides the background theory P_{domain}, including the update axioms, along with a program $P_{strategy}$ defining the intended behavior.

Example 2 (continued) Figure 5.10 depicts a sample program for our cleaning robot. After initializing the world model and having the physical apparatus clean the waste bin in the home square, the agent program enters the main loop, by which the robot systematically explores and cleans the office floor. To this end, the program employs two parameters containing, respectively, choice points yet to be explored and the current path of the robot, which is used to backtrack from a location once all choices have been considered. A choice point is a list of directions, which are encoded by 1 (for north) to 4 (for west) as usual. The path is represented by the sequence, in reverse order, of the directions the robot took in each step.

In the main loop, the cleanbot selects the first element of the current choices. If the attempt to go into this direction is successful (predicate *GoInDirection*), then the robot empties the waste bin at the new location, and the program creates a new choice point and extends the backtrack path by the direction into which the robot just went. If, on the other hand, the chosen direction cannot be taken, then the main loop is called with a reduced list of current choices. In case no more choices are left, the cleanbot backtracks.

The auxiliary predicate $GoInDirection(d, z_1, z_2)$ succeeds if the cleanbot can safely go into direction d from its current location in state z_1, ending up in state z_2. A direction is only explored if the adjacent square is inside the boundaries, if this location has not been visited already (that is, it is not known to be cleaned), and—most importantly—if the adjacent location is *known* not to be occupied. The auxiliary predicate *Backtrack* causes the robot to take back one step on its current path by reversing the direction. The program terminates

```
main :- init(Z0),
        execute(clean, Z0, Z1),
        Choicepoints = [[1,2,3,4]], Backtrack = [],
        main_loop(Choicepoints, Backtrack, Z1).

main_loop([Choices|Choicepoints], Backtrack, Z) :-
   Choices = [Direction|Directions] ->
   ( go_in_direction(Direction, Z, Z1)
     -> execute(clean, Z1, Z2),
        Choicepoints1 = [[1,2,3,4], Directions | Choicepoints],
        Backtrack1 = [Direction | Backtrack],
        main_loop(Choicepoints1, Backtrack1, Z2)
     ;
     main_loop([Directions|Choicepoints], Backtrack, Z) )
   ;
   backtrack(Choicepoints, Backtrack, Z).

go_in_direction(D, Z1, Z2) :-
   knows_val([X,Y], at(X,Y), Z1),
   adjacent(X, Y, D, X1, Y1),
   \+ knows(cleaned(X1,Y1), Z1),
   knows_not(occupied(X1,Y1), Z1),
   turn_to_go(D, Z1, Z2).

backtrack(_, [], _).
backtrack(Choicepoints, [Direction|Backtrack], Z) :-
   Reverse is (Direction+1) mod 4 + 1,
   turn_to_go(Reverse, Z, Z1),
   main_loop(Choicepoints, Backtrack, Z1).

turn_to_go(D, Z1, Z2) :-
   knows(facing(D), Z1) -> execute(go, Z1, Z2) ;
   execute(turn, Z1, Z), turn_to_go(D, Z, Z2).
```

Figure 5.10: A cleanbot agent in FLUX.

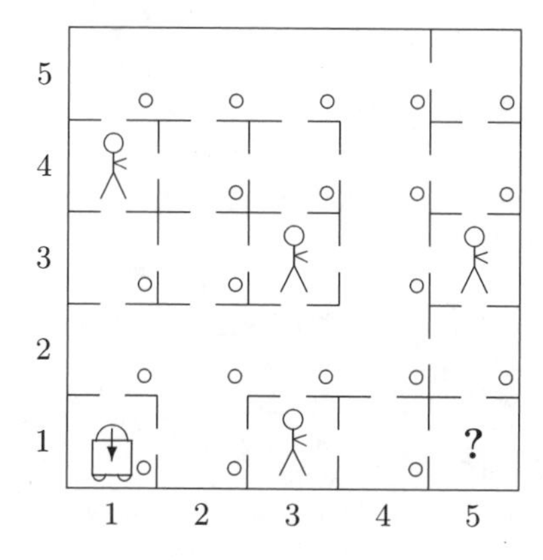

Figure 5.11: The final knowledge state in the cleaning robot scenario. The small circles indicate the cleaned locations.

---

once this path is empty, which implies that the robot has returned to its home after having visited and cleaned as many locations as possible. The two auxiliary predicates *GoInDirection* and *Backtrack* in turn call the predicate *TurnToGo*, by which the robot makes turns until it faces the intended direction, and then goes forward.

The following table illustrates what happens in the first nine calls to the main loop when running the program with the initial state of Figure 4.1 (page 78) and the scenario depicted in Figure 3.1 (page 60).

At	Choicepoints	Backtrack	Actions
$(1,1)$	`[[1,2,3,4]]`	`[]`	*GC*
$(1,2)$	`[[1,2,3,4],[2,3,4]]`	`[1]`	*GC*
$(1,3)$	`[[1,2,3,4],[2,3,4],[2,3,4]]`	`[1,1]`	–
$(1,3)$	`[[2,3,4],[2,3,4],[2,3,4]]`	`[1,1]`	–
$(1,3)$	`[[3,4],[2,3,4],[2,3,4]]`	`[1,1]`	–
$(1,3)$	`[[4],[2,3,4],[2,3,4]]`	`[1,1]`	–
$(1,3)$	`[[],[2,3,4],[2,3,4]]`	`[1,1]`	*TTG*
$(1,2)$	`[[2,3,4],[2,3,4]]`	`[1]`	*TTTGC*
$(2,2)$	`[[1,2,3,4],[3,4],[2,3,4]]`	`[2,1]`	*TTTGC*

The letters G, C, T are abbreviations for the actions *Go*, *Clean*, and *Turn*, respectively. After going north twice to office $(1,3)$, the cleanbot cannot continue in direction 1 or 2 because both offices $(1,4)$ and $(2,3)$ may be occupied according to its current knowledge. Direction 3 is not explored since location $(1,2)$ has already been cleaned, and direction 4 is ruled out as $(0,3)$ is outside of the boundaries. Hence, the cleanbot backtracks to $(1,2)$ and continues with the next choice there, direction 2, which brings it to location $(2,2)$. From there it can go north, and so on. Figure 5.11 depicts the knowledge state at the time the program terminates. Back home, the cleanbot has acquired knowl-

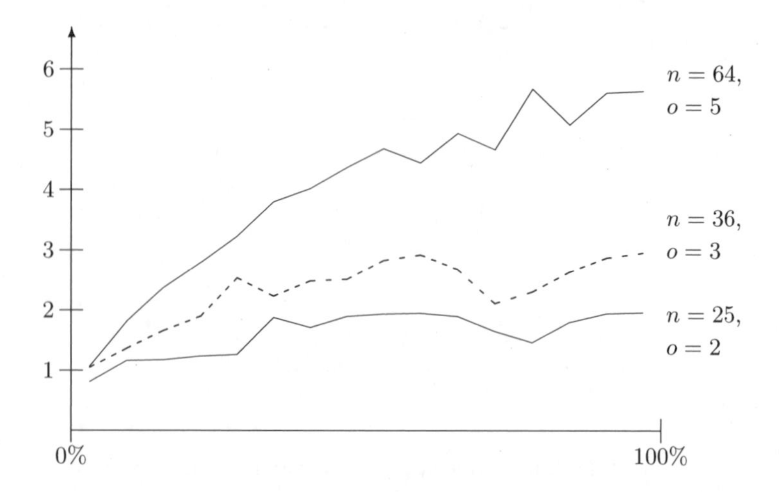

Figure 5.12: The average computational behavior of the program for the office cleaning robot in the course of its execution. The horizontal axis depicts the degree to which the run is completed while the vertical scale is in seconds CPU time per 100 actions.

---

edge of all four occupied offices. Moreover, it has emptied all waste bins but the ones in these four offices and the bin in office $(5, 1)$. The latter has not been cleaned because the robot cannot know that this office is not occupied—the light sensors have been activated at both surrounding locations, $(4, 1)$ and $(5, 2)$!

The computational behavior of the program can be analyzed by running it with different scenarios and considering the average computation time. Figure 5.12 depicts the results of a series of experiments with square office floors of different size (and without corridor, for the sake of simplicity) in which offices are randomly chosen to be occupied. The curves depict the average of 10 runs each with 25 rooms (and two being occupied), 36 rooms (three), and 64 rooms (five), respectively. The graphs indicate two execution phases: In the first phase, the robot acquires more and more knowledge of the environment while acting. Here, the curves show a linear increase of the average computation cost for action selection, update computation, and evaluation of sensor information. In the second phase, where the agent continues to act under the acquired but still incomplete knowledge, the average time remains basically constant. □

Soundness is of course a fundamental property of agent programs under incomplete information, too. In general FLUX, sensing results play a role similar to exogenous actions: They may have significant influence on what is actually computed by an agent program. Hence, a program is sound just in case, informally speaking, the domain axiomatization *plus the acquired sensing results* entail that all executed actions are possible. To this end, we introduce the notion of an observation as a sequence of sensing results.

Definition 5.15 Let Σ be a domain axiomatization. A sequence of sensing results $\vec{\pi} = \pi_1, \pi_2, \ldots$ for a sequence of actions $\vec{\alpha} = \alpha_1, \alpha_2, \ldots$ is an **observation**. Let $\Pi_1(z), \Pi_2(z), \ldots$ be the state formulas determined by $\vec{\pi}$, then the observation is **consistent** wrt. Σ and $\vec{\alpha}$ iff for any $n > 0$,

$$\Sigma \cup \{\Pi_1(\sigma_1), \ldots, \Pi_n(\sigma_n)\} \not\models \bot$$

where σ_i abbreviates the situation $Do([\alpha_1, \ldots, \alpha_i], S_0)$ $(1 \leq i \leq n)$. □

Recall, for example, the particular scenario in which we ran the cleanbot program. The actions of the robot were

$$\mathit{Clean},\ \mathit{Go},\ \mathit{Clean},\ \mathit{Go},\ \mathit{Clean},\ \mathit{Turn},\ \mathit{Turn},\ \mathit{Go},\ \ldots \tag{5.19}$$

and the observation was

$$[\,],\ [\mathit{False}],\ [\,],\ [\mathit{True}],\ [\,],\ [\,],\ [\,],\ [\mathit{False}],\ \ldots \tag{5.20}$$

This observation is consistent with the domain axiomatization and the actions of (5.19), whereas the following observation is not so:

$$[\,],\ [\mathit{False}],\ [\,],\ [\mathit{False}],\ [\,],\ [\,],\ [\,],\ [\mathit{True}],\ \ldots$$

Here, the last observation (that is, light in cell $(2,1)$) contradicts the second one (no light in $(2,1)$). When considering soundness of programs, we will restrict attention to consistent observations.[4]

As in case of exogenous actions, it is necessary to characterize the different computation trees of a program by the input which is being processed. In case of general FLUX, the input is determined by the observation, that is, the sequence of substitutions for y in each resolved $\mathit{Perform}(a, y)$ instance.

Definition 5.16 Let T be a computation tree for an agent program and a query. Let $\mathit{Execute}(\alpha_1, \_, \_), \mathit{Execute}(\alpha_2, \_, \_), \ldots$ be the ordered sequence of execution nodes in T. Tree T is said to **process** the observation $\pi_1, \pi_2, \ldots$ if $\mathit{Perform}(\alpha_1, y_1), \mathit{Perform}(\alpha_2, y_2), \ldots$ is the ordered sequence of child nodes of the execution nodes in T and $\{y_1/\pi_1\}, \{y_2/\pi_2\}, \ldots$ are the substitutions used to resolve these child nodes. Tree T is said to **generate** the sequence of actions $\alpha_1, \alpha_2, \ldots$. □

As an example, Figure 5.13 depicts skeletons of two different computation trees for the cleanbot program. The tree sketched on the left hand side processes observation (5.20) and generates actions (5.19) mentioned before, whereas the tree in the right hand side processes the observation

$$[\,],\ [\mathit{True}],\ [\,],\ [\,],\ [\mathit{True}],\ [\,],\ [\mathit{False}],\ [\,],\ \ldots$$

---

[4] Thereby the assumptions are made that the environment does not change dynamically and that the agent is never mistaken in what it senses and what it knows about the actual state. Dynamic changes in the environment are modeled in Chapter 7 while uncertain sensors and effectors are dealt with in Chapter 8, and Chapter 10 is concerned with how agents can revise mistaken beliefs about the environment.

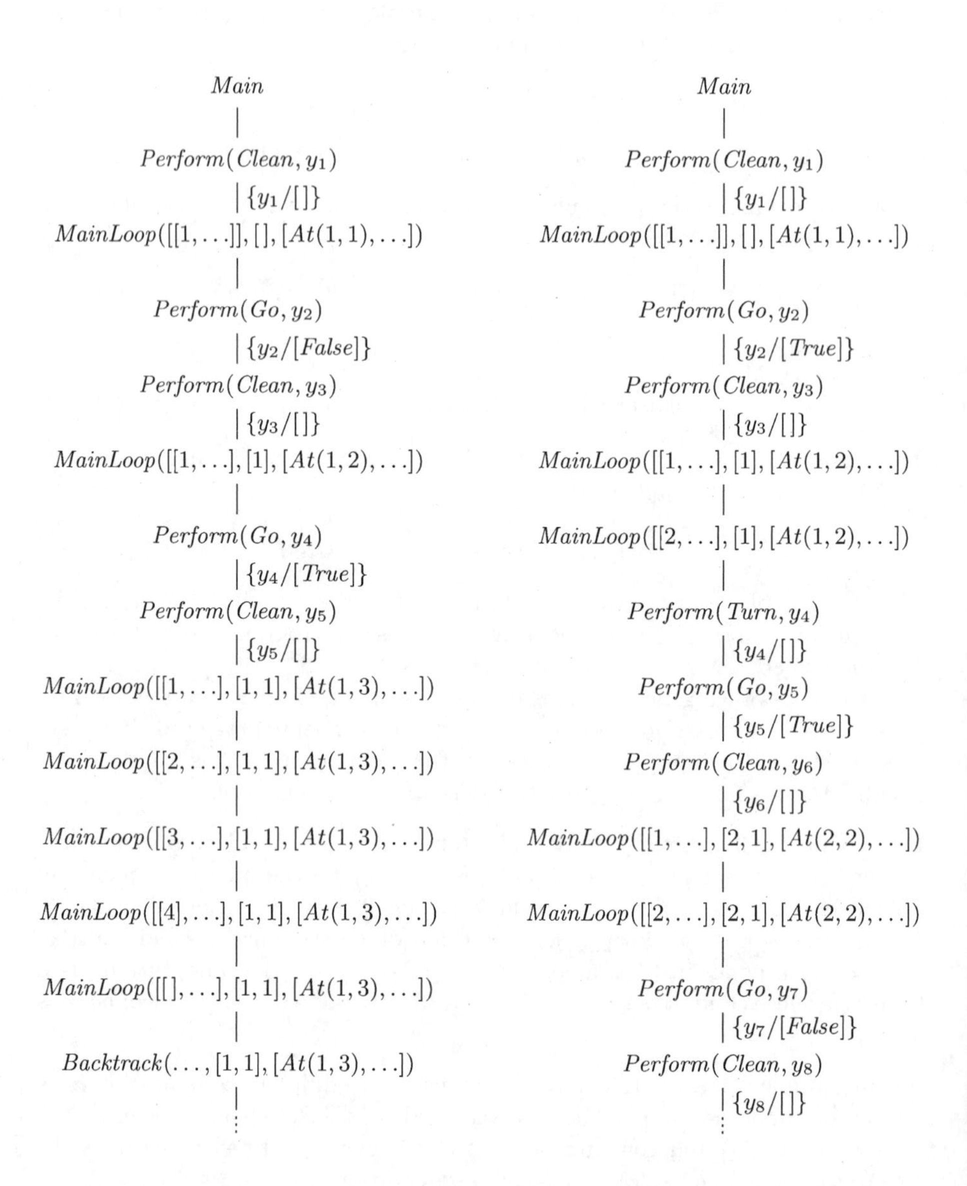

Figure 5.13: Two different computation trees for the cleanbot program.

while generating the actions

$$Clean,\ Go,\ Clean,\ Turn,\ Go,\ Clean,\ Go,\ Clean,\ \ldots$$

We are now in a position to define soundness of agent programs in general FLUX in the presence of sensing.

Definition 5.17 A program P and a query Q_0 are **sound** wrt. a domain axiomatization Σ if for every observation $\vec{\pi}$, the computation tree T for $P \cup \{\leftarrow Q_0\}$ which processes $\vec{\pi}$ satisfies the following: Let $\vec{\alpha} = \alpha_1, \alpha_2, \ldots$ be the actions generated by T. If $\vec{\pi}$ is consistent wrt. Σ and $\vec{\alpha}$, then for each execution node Q starting with atom $Execute(\alpha, \_, \_)$, α is ground and

$$\Sigma \cup \{\Pi_1(\sigma_1), \ldots \Pi_n(\sigma_n)\} \models Poss(\alpha, \sigma_n)$$

where $\Pi_1(z), \Pi_2(z), \ldots$ are the state formulas determined by π, and where $\sigma_n = Do([\alpha_1, \ldots, \alpha_n], S_0)$ is the situation associated with Q. □

The following result generalizes Theorem 2.9 (page 43) by providing a strategy for writing sound programs in general FLUX. In particular, if actions are executed only if they are known to be executable according to the current knowledge state, then soundness of the program follows by soundness of the knowledge states.

Theorem 5.18 *Let Σ be a domain axiomatization, P an agent program including a sound encoding of Σ, and Q_0 a query. Suppose that for every observation $\vec{\pi}$, the computation tree T for $P \cup \{\leftarrow Q_0\}$ which processes $\vec{\pi}$ satisfies the following: If $\vec{\pi}$ is consistent wrt. Σ and the action sequence generated by T, then for every execution node Q with leading atom $Execute(\alpha, \tau_1, \tau_2)$ and associated situation σ,*

1. *Q is in the main tree and there is no branch to the right of Q;*
2. *α is ground and τ_2 is a variable;*
3. *if $\sigma \neq S_0$, then Q is linked to its preceding execution node;*
4. *$\Sigma \models \Phi(z) \supset Poss(\alpha, z)$, where $\Phi(z)$ is the state to which τ_1 is bound.*

Then P and Q_0 are sound wrt. Σ.

Proof: Consider an execution node Q starting with $Execute(\alpha, \tau_1, \tau_2)$ and whose associated situation is $\sigma_n = Do([\alpha_1, \ldots, \alpha_n], S_0)$. According to condition 1, all execution nodes lie on one branch; hence, unless $\sigma = S_0$, Q is linked to an execution node with associated situation $\sigma_{n-1} = Do([\alpha_1, \ldots, \alpha_{n-1}], S_0)$. Let $\Phi(z)$ be the FLUX state which τ_1 is bound to, and let $\Pi_1(z), \Pi_2(z), \ldots$ be the state formulas determined by $\vec{\pi}$. By induction on n, we prove that

$$\Sigma \cup \{\Pi_1(\sigma_1), \ldots, \Pi_n(\sigma_n)\} \models KState(\sigma_n, z) \supset \Phi(z)$$

which implies the claim according to condition 4 and Proposition 5.11. The base case $n = 0$ follows by Definition 5.10. In case $n > 0$, let $Execute(\alpha_n, \tau_1', \tau_2')$ be the execution node to which Q is linked, and let $\Phi_{n-1}(z)$ be the FLUX state to which τ_1' is bound. Condition 4 and the induction hypothesis imply that $\Sigma \cup \{KState(\sigma_{n-1}, z)\} \equiv \Phi_{n-1}(z)$ entails $Knows(Poss(\alpha_n), \sigma_{n-1})$. The claim follows by Definition 5.14. ■

Example 2 (continued) With the help of the theorem, our agent program for the cleanbot of Figure 5.10 can be proved sound wrt. the axiomatization of this domain.

1. There is no negation in the program; hence, all execution nodes are in the main tree. Moreover, since none of the clauses admits backtracking over execution nodes, there are no branches to the right of any execution node.

2. All actions of the cleanbot are constants and therefore ground. The third argument in every occurrence of $Execute(a, z_1, z_2)$, on the other hand, remains variable until the atom is resolved.

3. The first execution node in the body of the clause for *Main* takes as second argument the FLUX state resulting from the call to $Init(z)$. The construction of the main loop and of the clauses for *Backtrack* and *TurnToGo* imply that all further execution nodes are linked to their predecessors.

4. Actions *Clean* and *Turn* are always possible. Action *Go* is executed according to *TurnToGo* only if the robot knows it faces a direction in which there is an adjacent square, following the definitions of the two auxiliary predicates: The body of the clause for *GoInDirection* includes atom *Adjacent* while *Backtrack* is defined in such a way as to send back the robot on its path. □

5.6 Bibliographical Notes

Knowledge and sensing in the action formalism of situation calculus has been investigated in [Moore, 1985] using modal logic. In [Scherl and Levesque, 2003] this approach has been combined with the solution to the frame problem provided by successor state axioms [Reiter, 1991]. An alternative way of modeling knowledge and sensing in situation calculus is by introducing special knowledge fluents [Demolombe and del Parra, 2000]. These two approaches have been compared in [Petrick and Levesque, 2002]. A modal-like approach has also been proposed in [Lobo *et al.*, 1997] as an extension of the action description language [Gelfond and Lifschitz, 1993]. The axiomatization of state knowledge in fluent calculus has been developed in [Thielscher, 2000b], and the foundations for the FLUX encoding of knowledge and update have been laid in [Thielscher, 2001a]. The situation calculus-based programming language GOLOG [Levesque *et al.*, 1997] has been extended to knowledge programming in [Reiter, 2001b]. In [Golden and Weld, 1996] a different implementation has been described for which a semantics is given based on situation calculus but where the implementation is based on the notion of an incomplete state as a triple of true, false, and unknown propositional fluents. The same representation has been used in the logic programming systems [Baral and Son, 1997; Lobo, 1998], which have both been given semantics by a three-valued variant [Baral and Son, 1997] of the action description language [Gelfond and Lifschitz, 1993]. Yet another approach uses description logic to represent knowledge [Giacomo *et al.*, 1997]. The approach [Herzig *et al.*, 2000], which is based on a special epistemic propositional logic, can be used for the verification of plans on the basis of incomplete state knowledge. Several systems for planning with sensing actions extend the early STRIPS approach [Fikes and Nilsson, 1971], for example, [Etzoni *et al.*, 1997; Weld *et al.*, 1998].

5.7 Exercises

5.1. Formalize as a knowledge state for situation S_0 that the mailbot of Chapter 1 knows that it is at room 3, for which it carries a package in one of its bags. Prove that the robot knows that some delivery action is possible in S_0, but that there is no delivery action of which the robot knows that it is possible in S_0.

5.2. For each of the following statements, either prove that it is entailed by foundational axioms $\Sigma_{state} \cup \Sigma_{knows}$, or find a counter-example.

(a) $Knows(\neg\phi, s) \supset \neg Knows(\phi, s)$.

(b) $\neg Knows(\phi, s) \supset Knows(\neg\phi, s)$.

(c) $Knows(\phi \wedge \psi, s) \supset Knows(\phi, s) \wedge Knows(\psi, s)$.

(d) $Knows(\phi, s) \wedge Knows(\psi, s) \supset Knows(\phi \wedge \psi, s)$.

(e) $Knows(\phi \vee \psi, s) \supset Knows(\phi, s) \vee Knows(\psi, s)$.

(f) $Knows(\phi, s) \vee Knows(\psi, s) \supset Knows(\phi \vee \psi, s)$.

(g) $(\exists x)\, Knows(\phi, s) \supset Knows((\exists x)\, \phi, s)$.

(h) $Knows((\exists x)\, \phi, s) \supset (\exists x)\, Knows(\phi, s)$.

(i) $(\forall x)\, Knows(\phi, s) \supset Knows((\forall x)\, \phi, s)$.

(j) $Knows((\forall x)\, \phi, s) \supset (\forall x)\, Knows(\phi, s)$.

5.3. (a) Formalize the knowledge state of knowing that $F(n)$ holds for all odd numbers n. Prove that in this knowledge state it is not known whether $F(n)$ holds for some even number n.

(b) Formalize the knowledge state of knowing that $F(n)$ holds for all instances $n \in \mathbb{N}$ but one. Prove that in this knowledge state no value for F is known.

5.4. Show that the formalization of sensing in knowledge update axioms implies that sensed properties and values are known in the successor situation. Give, to this end, suitable knowledge update axioms for the action $Sense_\phi$ of sensing whether knowledge expression ϕ holds, and of the action $Sense_F$ of sensing a value for fluent F. Prove the following:

(a) Let ϕ be a knowledge expression, then

$$\begin{aligned}&KState(Do(Sense_\phi, s), z') \equiv \\ &\quad KState(s, z') \wedge [HOLDS(\phi, z') \equiv HOLDS(\phi, Do(Sense_\phi, s))]\end{aligned}$$

implies $Knows(\phi, Do(Sense_\phi, s)) \vee Knows(\neg\phi, Do(Sense_\phi, s))$.

(b) Let F be a fluent function, then

$$\begin{aligned}(\exists \vec{y})(\forall z')\,(&KState(Do(Sense_F, s), z') \equiv \\ &KState(s, z') \wedge \\ &Holds(F(\vec{y}), z') \wedge Holds(F(\vec{y}), Do(Sense_F, s))\,)\end{aligned}$$

implies $KnowsVal(\vec{y}, F(\vec{y}), Do(Sense_F, s))$.

5.5. Formalize as a knowledge state for situation S_0 that the cleaning robot knows it is at square $(4,5)$ facing either east or south, and it also knows all domain constraints and that none of the squares in the corridor is occupied. Suppose the robot performs a *Go* action, after which it senses light, followed by a *Turn* action and another *Go* action, after which it does not sense light. Use fluent calculus to prove that in the resulting situation the robot knows which direction it faces, and that it knows of one occupied office.

5.6. Solve the previous exercise in FLUX.

5.7. (a) Extend FLUX by a clause for $KnowsOr([f_1, \ldots, f_k], z)$, defining that the knowledge expression $Holds(f_1) \vee \ldots \vee Holds(f_k)$ is known wrt. the specification of z. Prove correctness of the encoding along the line of Theorem 5.6.

(b) Find a state specification $\Phi(z)$ and fluents f_1, f_2 such that the query $\Phi(z)$, $KnowsOr([f_1, f_2], z)$ succeeds while

$$\Phi(z),\ (Knows(f_1, z)\ ;\ Knows(f_2, z))$$

fails.

5.8. Prove Proposition 5.11.

5.9. Answer the following questions for the cleanbot program of Figure 5.10:

(a) Consider a computation tree which processes an observation that starts with

$$[\,],\ [False],\ [\,],\ [False]$$

What are the first 16 actions generated by this tree?

(b) Consider a computation tree which generates an action sequence that starts with

Clean, *Go*, *Clean*, *Turn*, *Go*, *Clean*, *Turn*

What are the first nine sensing results processed by this tree?

5.10. The actions of the agent in the Wumpus World (cf. Exercise 4.8) are to enter and exit the cave at cell $(1,1)$, turning clockwise by $90°$, going forward one square in the direction it faces, grabbing gold, and shooting an arrow in the direction it faces. When the agent enters the cave, it first faces north. The effect of shooting the arrow is to kill the Wumpus if the latter happens to be somewhere in the direction taken by the arrow. The agent must not enter a cell with a pit or which houses the Wumpus if the latter is alive. Upon entering the cave or moving to another square, the agent gathers sensor information as defined in Exercise 4.8. In addition, immediately after shooting the arrow the agent hears a scream if the Wumpus gets killed.

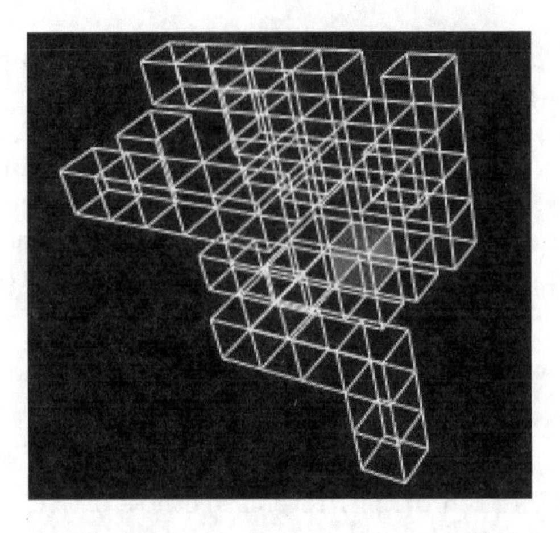

Figure 5.14: A regular 3-dimensional maze.

(a) Formalize suitable state and knowledge update axioms for the actions *Enter*, *Exit*, *Turn*, *Go*, *Grab*, and *Shoot*, respectively.

(b) Axiomatize the initial knowledge that the agent is not inside of the cave, that it has an arrow but not gold, and that the Wumpus is still alive. Suppose that when the agent enters the cave it senses a stench but no breeze nor a glitter. Suppose further that the next action of the agent is to shoot its arrow. Use fluent calculus to prove that afterwards the agent knows it can go forward.

(c) Specify the Wumpus World in FLUX and write a program for the agent to systematically explore the cave with the goal of exiting with gold.

5.11. Write a program for a robot to find its way out of an unknown, 3-dimensional maze. Assume that the maze is structured as a grid where all tunnels lead straight into one of six directions (north, east, south, west, upwards, downwards); see Figure 5.14. The two actions of the robot are to turn and to go forward in the direction it faces, provided it does not face a wall. When turning or stepping forward, the sensors tell the robot whether or not it faces a wall afterwards and whether or not it has reached a location outside of the maze. Initially, the robot just knows that it is inside of the maze facing a direction in which there is no wall. The robot has no prior knowledge of size or shape of the maze.

Chapter 6

Planning

Personal software assistants are programs which carry out a variety of organizational tasks for and on behalf of their users. Among the many things these agents can do is to manage their user's daily schedule. Imagine yourself in the morning requesting your personal software assistant to arrange several appointments during the day. This may include, say, meeting your GP and having lunch with a business partner. The first task for your agent then is to proactively contact everyone involved in order to collect information about possible times and places for the intended meetings. Ideally, your colleagues and business partners have their own autonomous software assistants with which your agent can make arrangements, and even your GP may provide an interactive web page through which agents can inquire about consultation hours and make appointments.

After your software assistant has collected information about free time slots and possible places for all requested meetings, it needs to compose a consistent schedule. In other words, the agent must make a **plan** prior to actually fixing the appointments. Devising and comparing plans is a powerful ability of reasoning agents. Planning means to entertain the effects of various possible action sequences before actually executing one. With a specific goal in mind, agents are thus able to solve certain problems on their own. This increases their autonomy and makes programming much easier in cases where it would require considerable effort to find and implement a strategy which is good for all situations that the agent may encounter.

In this chapter, we develop

6.1. a notion of planning problems and plan search;

6.2. a method for comparative evaluation of different plans;

6.3. a way to enhance the performance by planning with user-defined actions;

6.4. a method to generate conditional plans under uncertainty.

6.1 Planning Problems

Generally speaking, a planning problem is to find a way which leads from the current state to a state that satisfies certain desired properties. These are commonly called the **goal** of the planning task. Putting aside, for the moment, the question of quality, a solution to a planning problem is given by any executable sequence of actions for which the underlying domain axiomatization entails that it achieves the goal. In terms of fluent calculus, this is formalized as follows:

Definition 6.1 Consider a domain axiomatization Σ. A **planning problem** is given by an initial knowledge state $KState(\sigma, z) \equiv \Phi(z)$ and a goal state formula $\Gamma(z)$. A **plan** is a sequence $[\alpha_1, \ldots, \alpha_n]$ of ground actions ($n \geq 0$). The plan is a **solution** iff $\Sigma \cup \{KState(\sigma, z) \equiv \Phi(z)\}$ entails

$$Poss([\alpha_1, \ldots, \alpha_n], \sigma) \wedge \Gamma\{z/State(Do([\alpha_1, \ldots, \alpha_n], \sigma))\}$$

□

This definition requires agents to know that a plan achieves the goal; it does not suffice if a sequence of actions just happens to lead to the goal without the agent being aware of it.

Example 3 For the problem of planning a personal meeting schedule, our software assistant uses three fluents, for which we assume given a sort for the various parties to be met (a single person or a group), for the possible meeting places, and for the time of day:

$$\begin{aligned} Request &: \text{PARTY} \times \text{TIME} \mapsto \text{FLUENT} \\ Available &: \text{PARTY} \times \text{PLACE} \times \text{TIME} \mapsto \text{FLUENT} \\ At &: \text{PLACE} \times \text{TIME} \mapsto \text{FLUENT} \end{aligned}$$

Fluent $Request(p, d)$ means that the user has requested a meeting with p for a duration d. Fluent $Available(p, x, t)$ is used to encode the proposal made by party p to meet at location x at time t. Fluent $At(x, t)$ is used by the software agent to plan the day for the user: It represents the fact the user has to be at location x at time t. As an example, suppose that the agent has been given three requests and that it has already exchanged information with the agents of the involved partners about possible time slots and places for the meetings:

$$\begin{aligned} & KState(S_0, z) \equiv \\ & \quad z = Request(Staff, 0{:}30) \circ Request(GP, 0{:}45) \circ Request(Bob, 1{:}00) \\ & \qquad \circ Available(Staff, Office, 10{:}00) \circ Available(Staff, CityA, 12{:}00) \\ & \qquad \circ Available(GP, CityB, 10{:}45) \circ Available(GP, CityB, 12{:}45) \\ & \qquad \circ Available(Bob, CityA, 10{:}30) \circ Available(Bob, CityB, 12{:}45) \\ & \qquad \circ At(Office, 9{:}30) \end{aligned} \tag{6.1}$$

In order to solve the problem of finding a consistent schedule by planning, we introduce the following two actions:

$$\begin{aligned} Go &: \text{PLACE} \times \text{TIME} \mapsto \text{ACTION} && Go(x, t) \mathrel{\hat{=}} \text{go to } x \text{ at time } t \\ Meet &: \text{PARTY} \times \text{TIME} \mapsto \text{ACTION} && Meet(p, t) \mathrel{\hat{=}} \text{meet } p \text{ at time } t \end{aligned}$$

The preconditions are as follows:

$$\begin{array}{l} Poss(Go(x,t),z) \equiv \\ \quad (\exists x_1,t_1)\,(Holds(At(x_1,t_1),z) \wedge x_1 \neq x \wedge t_1 \leq t) \\ \\ Poss(Meet(p,t),z) \equiv \\ \quad (\exists x,t_1)\,(Holds(Available(p,x,t),z) \wedge Holds(At(x,t_1),z) \wedge t_1 \leq t) \end{array} \tag{6.2}$$

Put in words, going from one place to another is possible at any later time. Meeting a party requires to choose an available appointment and to be at the right place and time. The effects of the first action shall be formalized by this knowledge update axiom:

$$\begin{array}{l} Poss(Go(x,t),s) \supset \\ \quad [KState(Do(Go(x,t),s),z') \equiv (\exists z)\,(KState(s,z) \wedge \\ \quad\quad (\exists x_1,t_1)\,(\,Holds(At(x_1,t_1),z) \wedge \\ \quad\quad\quad\quad z' = z - At(x_1,t_1) + At(x,t + Distance(x,x'))\,)\,)\,] \end{array} \tag{6.3}$$

That is to say, going to x at time t has the effect to be at x after having traveled the distance between the current location x_1 and x. For our example, consider the following values for auxiliary function *Distance*, which indicates the time it takes to go from one place to the other:

$$\begin{array}{l} Distance(Office, CityA) = Distance(CityA, Office) = 0{:}15 \\ Distance(Office, CityB) = Distance(CityB, Office) = 0{:}20 \\ Distance(CityA, CityB) = Distance(CityA, CityB) = 0{:}05 \end{array}$$

The effect of the second action can be axiomatized as follows:

$$\begin{array}{l} Poss(Meet(p,t),s) \supset \\ \quad [KState(Do(Meet(p,t),s),z') \equiv (\exists z)\,(KState(s,z) \wedge \\ \quad\quad (\exists d,x,t_1)\,(\,Holds(Request(p,d),z) \wedge \\ \quad\quad\quad\quad Holds(At(x,t_1),z) \wedge \\ \quad\quad\quad\quad z' = z - At(x,t_1) \circ Request(p,d) + At(x,t+d)\,)\,)\,] \end{array} \tag{6.4}$$

That is to say, the result of meeting party p is to fulfill the request $Request(p,d)$. The meeting ends after the requested duration d. Based on this domain axiomatization, we can formulate the planning problem of finding a meeting schedule by a goal formula $\Gamma(z)$ which says that all requests should be satisfied:

$$\neg(\exists p,d)\,Holds(Request(p,d),z) \tag{6.5}$$

Given our example initial knowledge state (6.1), the planning problem admits, among others, these two plans as solutions:

$$\begin{array}{l} [Go(CityB, 10{:}25), Meet(GP, 10{:}45), Go(CityA, 11{:}55), \\ \quad Meet(Staff, 12{:}00), Go(CityB, 12{:}40), Meet(Bob, 12{:}45)] \\ [Go(CityA, 10{:}15), Meet(Bob, 10{:}30), \\ \quad Meet(Staff, 12{:}00), Go(CityB, 12{:}40), Meet(GP, 12{:}45)] \end{array} \tag{6.6}$$

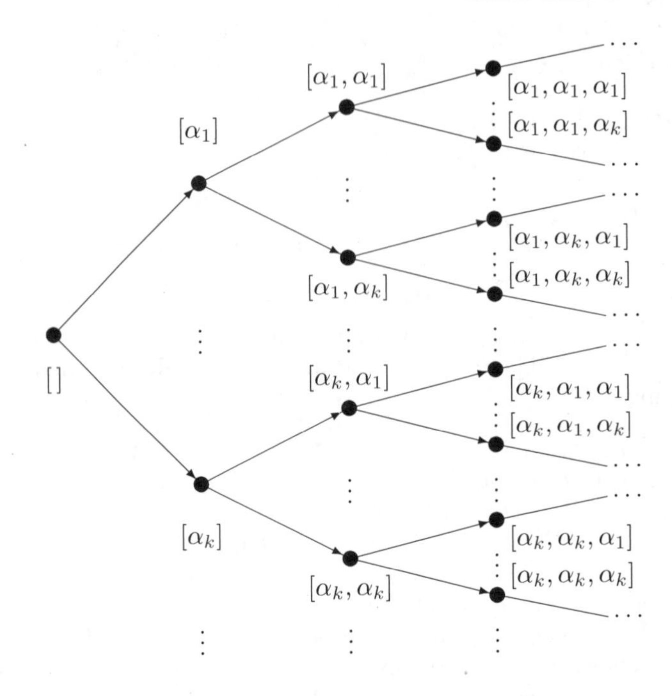

Figure 6.1: A tree of situations spanning the search space for a planning problem starting in situation $S = Do([], S)$.

The following two action sequences, on the other hand, do not solve the problem:

$$[Meet(Staff, 10{:}00), Go(CityB, 12{:}25), Meet(GP, 12{:}45)]$$
$$[Meet(Staff, 10{:}00), Go(CityB, 10{:}30), Meet(GP, 10{:}45), Meet(Bob, 12{:}45)]$$

Here, the first sequence is not a solution because the request to meet Bob remains while the second sequence includes the impossible action of meeting the GP at 10:45 because it takes 20 minutes to go from the office to the practice in *CityB*. □

When an agent faces a planning problem, it has to search for a solution. Since plans are sequences of actions leading away from the current state, the search space corresponds to a situation tree which is rooted in the current situation; see Figure 6.1. In order to find a plan, the agent needs to determine a branch in this tree which contains an action sequence that leads to the goal at hand. For this purpose, agents can employ their background theory to infer executability and effects of potential action sequences. Planning can thus be viewed as the "virtual" execution of actions.

Since planning means to look into the future, sensing actions constitute a particular challenge. An agent cannot foretell the outcome of sensing a yet unknown property. For this reason, we defer discussing the problem of planning with sensing until Section 6.4 and concentrate on plans in which actions have merely physical effects.

Planning in FLUX is based on inferring the expected effects of possible actions with the help of the corresponding state update axioms but without actually executing actions. In this way, the agent uses its internal model to decide whether each action in the sequence would be executable and whether the sequence would yield a state which satisfies the planning goal. The following clauses constitute a **sound and complete** encoding of a planning problem named P whose goal is specified by $Goal(z)$:

$$P(z, p, z) \leftarrow Goal(z),\ p = [\,]$$
$$P(z, [a|p], z_n) \leftarrow Poss(a, z),\ StateUpdate(z, a, z_1, [\,]),\ P(z_1, p, z_n)$$

According to this definition, $P(z, p, z_n)$ is true if in initial state z plan p is executable and results in state z_n satisfying the planning goal. Hence, if $KState(\sigma, z) \equiv \Phi(z)$ is the initial knowledge state of the planning problem, then any computed answer θ to the query

$$\leftarrow [\![\Phi(z)]\!], P(z, p, z_n)$$

determines a sequence of actions $p\theta = [\alpha_1, \ldots, \alpha_n]$ which constitutes a solution to the planning problem. In this way the two clauses induce a systematic search through all executable sequences of actions. In principle, we can thus tell the agent to look for and execute the first plan it finds, or to generate all plans and select the one which is best according to a suitable quality measure.

However, searching through all possible sequences of actions causes a number of computational difficulties. To begin with, the branches in a situation tree are usually infinite.[1] Without additional measurements, a Prolog system therefore easily runs into a loop due to its depth-first computation rule. To avoid infinite branches, a (possibly dynamic) limit needs to be put on the length of the plans that are being searched. Yet if infinitely many ground instances of actions are possible in a state, then there is also no limit to the breadth of the situation tree. The action $Go(x, t)$ in our schedule domain is an example: If time is modeled with infinite granularity, then there are infinitely many instances of this action. In this case, search does not terminate even if the depth of the search tree is restricted. Moreover, it is in general undecidable whether a planning problem is solvable at all. Hence, it may happen that an agent iteratively extends the search space without ever finding a solution. While this can be avoided by eventually breaking off the search, in so doing the agent may not be able to find a plan for a solvable problem.

Putting the fundamental problem of termination aside, finite search can still be computationally infeasible due to the exponential blowup of situation trees. Usually, most actions of an agent are irrelevant for a particular planning

[1] Otherwise there must be states in which no action at all is possible, which is rarely the case except in highly artificial domains.

goal. Moreover, even if the agent considers only the relevant actions, there may be many redundant actions or action sequences, so that the search through exponentially many permutations makes planning practically impossible, too.

In FLUX it is possible to cope with the complexity of planning by allowing for sound but incomplete encodings of planning problems. These encodings define restrictions on the possible plans beyond the basic requirement to achieve the goal. This enables the programmer to provide the agent with a problem-dependent definition of what kind of plans to look for. Specifically, the encoding of a planning problem is a program which defines the predicate $P(z, p, z_n)$ as p being a solution *of a certain form* to planning problem P and such that p leads from the initial state z to state z_n.

Definition 6.2 **A heuristic encoding of a planning problem** is a program P_{plan} defining a predicate $P(z, p, z_n)$. If Σ is a domain axiomatization with FLUX encoding P_Σ, and P_{plan} is the encoding of a planning problem with initial knowledge state $\{KState(\sigma, z) \equiv \Phi(z)\}$ and goal $\Gamma(z)$, then the encoding is **sound** iff the following holds: For every computed answer θ to

$$P_{kernel} \cup P_\Sigma \cup P_{plan} \cup \{\leftarrow [\![\Phi(z)]\!], P(z, p, z_n)\}$$

$p\theta$ is a solution to the planning problem and $z_n\theta$ is a FLUX state which satisfies

$$\Sigma \cup \{KState(\sigma, z) \equiv \Phi(z)\} \models KState(Do(p\theta, \sigma), z_n) \supset z_n\theta$$

□

Put in words, every computed answer must both constitute a solution to the planning problem and determine a sound resulting FLUX state $z_n\theta$.

Example 3 (continued) Figure 6.2 shows a particular heuristic encoding of the planning problem for the personal software assistant. The definition of the predicate $SchedulePlan(z, p, z_n)$ spans a search tree containing executable action sequences consisting of all possible $Meet(p, x)$ actions (predicate *SchedulePlan*1), possibly preceded by a $Go(x, t)$ action (predicate *SchedulePlan*2). In order to avoid searching through too many instances for the parameter t, by the auxiliary predicate *SelectGo* only those timepoints are considered which brings the user to a potential meeting place just in time for an available appointment. This search space is always finite since the initial state cannot contain infinitely many instances of the fluent *Request*. The base case for *SchedulePlan* defines the empty plan to be a solution if a goal state has been reached according to specification (6.5).

Figure 6.2 includes the background theory for the personal software assistant. A time point like 10:45 is encoded by a natural number, in this case 1045. The conversion between the last two digits and minutes is part of the auxiliary predicates *TimeAdd* and *TimeDiff*. In the encoding of our example initial state (6.1), the staff, the GP, and Bob are represented, respectively, by 1, 2, and 3, as are the three places *Office*, *CityA*, and *CityB*.

```
schedule_plan(Z,[],Z) :- not_holds_all(request(_,_),Z).
schedule_plan(Z,P,Zn) :- schedule_plan1(Z,P,Zn)
                         ; schedule_plan2(Z,P,Zn).

schedule_plan1(Z,[A|P],Zn) :-
   A = meet(_,_), poss(A,Z), state_update(Z,A,Z1,[]),
   schedule_plan(Z1,P,Zn).
schedule_plan2(Z,[A|P],Zn) :-
   A = go(_,_), select_go(A,Z), state_update(Z,A,Z1,[]),
   schedule_plan1(Z1,P,Zn).

select_go(go(X,T),Z) :-
   knows_val([X1,T1],at(X1,T1),Z), knows_val([P],request(P,_),Z),
   knows_val([X,T2],available(P,X,T2),Z),
   distance(X1,X,D), time_diff(T2,D,T), poss(go(X,T),Z).

time_diff(T1,T2,T) :- H1 is T1 // 100, M1 is T1 mod 100,
                      H2 is T2 // 100, M2 is T2 mod 100,
                      M is (H1-H2)*60 + M1-M2,
                      T is (M//60)*100 + M mod 60.
time_add(T1,T2,T) :-  H1 is T1 // 100, M1 is T1 mod 100,
                      H2 is T2 // 100, M2 is T2 mod 100,
                      H is H1+H2+(M1+M2)//60,
                      T is H*100+(M1+M2) mod 60.

dist(1,2,15). dist(1,3,20). dist(2,3,5).
distance(X,Y,D) :- dist(X,Y,D) ; dist(Y,X,D).

poss(go(X,T),Z)    :- knows_val([X1,T1],at(X1,T1),Z), X1 \= X, T1 =< T.
poss(meet(P,T),Z) :- knows_val([P,X,T],available(P,X,T),Z),
                     knows_val([T1],at(X,T1),Z), T1 =< T.

state_update(Z1,go(X,T),Z2,[]) :-
   holds(at(X1,T1),Z1), distance(X1,X,D), time_add(T,D,T2),
   update(Z1,[at(X,T2)],[at(X1,T1)],Z2).

state_update(Z1,meet(P,T),Z2,[]) :-
   holds(request(P,D),Z1), holds(at(X,T1),Z1), time_add(T,D,T2),
   update(Z1,[at(X,T2)],[at(X,T1),request(P,D)],Z2).

init(Z0) :- Z0 = [request(1,30),request(2,45),request(3,60),
                  available(1,1,1000),available(1,2,1200),
                  available(2,3,1045),available(2,3,1245),
                  available(3,2,1030),available(3,3,1245),
                  at(1,930)].
```

Figure 6.2: Heuristic encoding of the schedule planning problem.

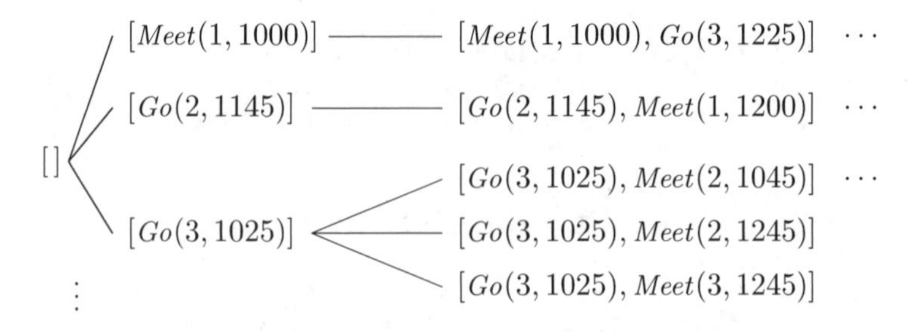

Figure 6.3: The initial fragment of a tree of plans in the software assistant domain. The fourth and fifth branch cannot be continued because no further action is possible according to the heuristic encoding.

Given the initial state at the bottom of Figure 6.2, Figure 6.3 shows the initial segment of the search tree. Of all potential plans in this tree, three branches end in a goal state, the first of which corresponds to the first solution given above, (6.6):

```
[go(3,1025), meet(2,1045), go(2,1155),
   meet(1,1200), go(3,1240), meet(3,1245)]
```

The heuristic encoding is truly incomplete because many solutions are not generated, e.g.,

```
[go(3,1017), meet(2,1045), go(2,1155),
   meet(1,1200), go(3,1240), meet(3,1245)]
```

(where the only difference to the solution above is the starting time). Also not included are solutions which contain redundant movements between the meeting places, e.g.,

```
[go(2,1020), go(3,1035), meet(2,1045), go(2,1155),
   meet(1,1200), go(3,1240), meet(3,1245)]
```

The definition of $SchedulePlan(z, p, z_n)$ can be proved sound wrt. the axioms for the software assistant domain and the planning goal of satisfying all meeting requests:

1. The clauses for $Poss(Meet(p, x), z)$ and $SelectGo(Go(x, t), z)$ ground the arguments of the actions; hence, all actions in the plan are ground.

2. Both actions $Meet(p, x)$ and $Go(x, t)$ are verified to be possible prior to adding them to the plan.

3. The base case for $SchedulePlan(z, p, z_n)$ is defined to be true just in case z is known to satisfy goal formula (6.5). □

```
:- dynamic plan_search_best/2.

plan(Problem,Z,P) :-
   assert(plan_search_best(_,-1)), plan_search(Problem,Z),
   retract(plan_search_best(P,C)), C =\= -1.

plan_search(Problem,Z) :-
   PlanningProblem =.. [Problem,Z,P,Zn],
   call(PlanningProblem), plan_cost(Problem,P,Zn,C),
   plan_search_best(_,C1),
   ( C1 =< C, C1 =\= -1 -> fail
     ; retract(plan_search_best(_,C1)),
       assert(plan_search_best(P,C)), fail )
   ; true.
```

Figure 6.4: FLUX definition for selecting optimal plans. The auxiliary, dynamic predicate *PlanSearchBest*(p, c) holds for the currently best plan p and its cost c. Initially, $c = -1$. The clause for *PlanSearch* continually calls the definition of the planning problem and updates the dynamic predicate upon finding a plan whose cost c is lower than that of the currently best plan.

6.2 Plan Evaluation

Sometimes a solution to a planning problem is just as good as any. In this case it suffices to have the agent execute the first plan it finds without further consideration. Putting an end to the search and starting with the actual execution as soon as the first plan has been found is also suitable in case of planning problems which are known to admit a unique solution. Often, however, it is desirable to compare different plans and to decide in favor of the one that is best according to a specific criterion. The user of the personal software assistant, for example, may prefer to be done with all meetings as early as possible. Another criterion may be to minimize the travel time between meeting places.

Under the provision that the underlying plan tree is finite, agents can compare all solutions to a planning problem and select the best one based on a problem-dependent quality measure. The general FLUX program for plan selection depicted in Figure 6.4 assumes that the programmer has specified this measure as *PlanCost*(P, p, z, c), defining, for planning problem P, the **cost** c wrt. plan p and resulting state z. It is assumed that the cost is represented by a natural number. FLUX predicate *Plan*(P, z, p) is then defined as p being a solution with minimal cost to planning problem P with initial state z. In case there is more than one plan with minimal cost, the first one in the plan tree is chosen. If, on the other hand, the planning problem is unsolvable, then *Plan*(P, z, p) is false for any p.

Example 3 (continued) Figure 6.5 depicts the definition of the cost of a plan for the personal software assistant. The cost is defined as the time when the

```
plan_cost(schedule_plan,P,Z,C) :-
   knows_val([C1],at(_,C1),Z), travel_time(P,C2), C is C1+C2.

travel_time([],0).
travel_time([meet(_,_)|P],C)              :- travel_time(P,C).
travel_time([go(_,T1),meet(_,T2)|P],C) :- time_diff(T2,T1,C1),
                                          travel_time(P,C2),
                                          C is C1+C2.
```

Figure 6.5: Cost of a plan for the planning problem in the software assistant domain.

---

last meeting is over plus the total time spent on traveling between the different meeting places. The latter is inferred by the auxiliary predicate *TravelTime*, which exploits the fact that in all generated plans the *Go* actions are scheduled just in time for the next meeting. Applied to the sample scenario of the previous section, the agent infers the unique least-cost plan

```
[go(2,1015), meet(3,1030),
    meet(1,1200), go(3,1240), meet(2,1245)]
```

with cost 1350, resulting from the time when the last meeting ends, 1330, and the 'travel cost' 20. In contrast, the alternative plan

```
[go(3,1025), meet(2,1045), go(2,1155),
   meet(1,1200), go(3,1240), meet(3,1245)]
```

has total cost $1375 = 1345 + 30$. □

6.3 Planning with Complex Actions

Plans, as considered thus far, are sequences of elementary actions. This results in plans which can be directly executed but requires to plan ahead each single step. This often makes planning unnecessarily complicated, and in many cases the search can be sped up considerably if certain details are ignored and filled in after a plan has been found.

The concept of a **complex action** allows agents to plan on more abstract a level than that of the elementary actions. A complex action is a specific behavior. It is defined in terms of the actions which need to be done in order to realize the behavior. A simple case are complex actions which correspond to a particular sequence of elementary actions. Much more interesting are complex actions whose realization depends on the state in which they are performed.

Example 3 (continued) With the aim to put the focus on how to schedule the various appointments, we introduce the complex action $GoMeet(p, x, t)$ with

```
complex_action(go_meet(P,X,T),Z1,Z2) :-
   knows_val([X1],at(X1,_),Z1),
   ( X1=X -> execute(meet(P,T),Z1,Z2)
     ;
     distance(X1,X,D), time_diff(T,D,T1),
     execute(go(X,T1),Z1,Z), execute(go_meet(P,X,T),Z,Z2).
```

Figure 6.6: FLUX specification of a complex action for the software assistant domain.

---

the intended meaning to meet party p at place x and time t. The following axiom specifies this action in terms of the elementary actions *Go* and *Meet*:

$$\begin{array}{l} Holds(At(x_1,t_1),s) \supset \\ \quad x_1 = x \wedge Do(GoMeet(p,x,t),s) = Do(Meet(p,t),s) \\ \quad \vee \\ \quad x_1 \neq x \wedge Do(GoMeet(p,x,t),s) = \\ \qquad Do(GoMeet(p,x,t), Do(Go(x,t-Distance(x_1,x),s))) \end{array} \tag{6.7}$$

Put in words, a meeting at the current place can be held directly, otherwise a just-in-time *Go* action is required beforehand. Figure 6.6 shows the FLUX encoding of this specification, which tells the agent how to execute the new complex action in a state z_1, leading to state z_2. □

When an agent considers a complex action at planning time, it must verify that the action is executable in the state for which it is planned. The agent also needs to infer the effect of the complex action in order to see where the plan leads to. Given a formal specification of a complex action along the line of axiom (6.7), preconditions and effects of the action can be derived from the respective precondition and update axioms for the underlying elementary actions. However, the whole advantage of planning on an abstract level would be spoiled if the agent had to draw these inferences by simulating a complex action every time the action is considered during planning. It is necessary, therefore, that the programmer provides specifications of the precondition and effects of the complex actions that are being used for planning. These non-elementary actions can then be virtually executed just like ordinary actions in the encoding of a planning problem.

After solving a planning problem, an agent needs to actually execute the plan it has opted for. This requires an extension of the standard FLUX predicate *Execute* to cover both action sequences and complex actions; see Figure 6.7. Regarding the latter, it is assumed that their realization is specified by the predicate *ComplexAction*(a, z_1, z_2) along the line of Figure 6.6.

Example 3 (continued) Recall the precondition and update axioms for *Go*(x,t) and *Meet*(p,t) (cf. (6.2)–(6.4)). From the specification of the action *GoMeet*(p,x,t), axiom (6.7), it follows that a corresponding sequence of

```
execute(A,Z1,Z2) :-
   perform(A,Y) -> state_update(Z1,A,Z2,Y) ;
   A = [A1|A2]  -> execute(A1,Z1,Z), execute(A2,Z,Z2) ;
   A = []       -> Z1=Z2 ;
   complex_action(A,Z1,Z2).
```

Figure 6.7: FLUX definition of execution including plans and complex actions.

---

elementary actions is known to be possible in a situation s if and only if a meeting with party p is requested, the party is available at location x and time t, and there is sufficient time to travel to x if necessary:[2]

$$\begin{aligned}&Poss(GoMeet(p,x,t),z) \equiv\\ &\quad(\exists x_1,t_1)\,(\,Holds(Available(p,x,t),z) \wedge\\ &\qquad\qquad [x_1 = x \wedge t_1 \leq t \vee x_1 \neq x \wedge t_1 + Distance(x_1,x) \leq t])\end{aligned}$$

Furthermore, the overall effect of $GoMeet(p,x,t)$ is to have satisfied the corresponding request and to be at x at the end of the appointment:

$$\begin{aligned}&Poss(GoMeet(p,x,t),s) \supset\\ &\quad[KState(Do(GoMeet(p,x,t),s),z') \equiv (\exists z)\,(KState(s,z) \wedge\\ &\qquad(\exists d,x_1,t_1)\,(\,Holds(Request(p,d),z) \wedge Holds(At(x_1,t_1),z) \wedge\\ &\qquad\qquad z' = z - At(x_1,t_1) \circ Request(p,d) + At(x,t+d))\,)\,]\end{aligned}$$

Under these axioms for the new action, the following sequences, too, are solutions to the example planning problem of Section 6.1:

$$\begin{aligned}&[GoMeet(GP,CityB,10{:}45),\\ &\quad GoMeet(Staff,CityA,12{:}00),GoMeet(Bob,CityB,12{:}45)]\\ &[GoMeet(Bob,CityA,10{:}30),\\ &\quad GoMeet(Staff,CityA,12{:}00),GoMeet(GP,CityB,12{:}45)]\end{aligned}$$

while these sequences fail to solve the problem:

$$\begin{aligned}&[GoMeet(Staff,Office,10{:}00),GoMeet(GP,CityB,12{:}45]\\ &[GoMeet(Staff,Office,10{:}00),\\ &\quad GoMeet(GP,CityB,10{:}45,GoMeet(Bob,CityB,12{:}45]\end{aligned}$$

Figure 6.8 depicts the FLUX encoding of the precondition and effect axiom for the complex action $GoMeet$, along with a new heuristic encoding of the planning problem by which systematically all possible meeting schedules are generated. It is easy to see that the encoding is sound, too. Applied to our usual scenario, the following optimal plan is computed.

---

[2] To be formally precise, the following axiom would be logically entailed if we define $Poss(GoMeet(p,x,t),State(s))$ as true just in case all elementary actions in the situation $Do(GoMeet(p,x,t),s)$ are possible.

```
schedule_plan(Z,[],Z) :- not_holds_all(request(_,_),Z).
schedule_plan(Z,[A|P],Zn) :-
   A = go_meet(_,_,_), poss(A,Z), state_update(Z,A,Z1,[]),
   schedule_plan(Z1,P,Zn).

poss(go_meet(P,X,T),Z) :-
   knows_val([P,X,T],available(P,X,T),Z),
   knows_val([X1,T1],at(X1,T1),Z),
   ( X1 = X -> T1 =< T ;
     distance(X1,X,D), time_add(T1,D,T2), T2 =< T ).

state_update(Z1,go_meet(P,X,T),Z2,[]) :-
   holds(request(P,D),Z1), holds(at(X1,T1),Z1), time_add(T,D,T2),
   update(Z1,[at(X,T2)],[at(X1,T1),request(P,D)],Z2).

plan_cost(schedule_plan,P,Z,C) :-
   knows_val([C1],at(_,C1),Z), travel_time(P,1,C2), C is C1+C2.

travel_time([],_,0).
travel_time([go_meet(_,X,_)|P],X1,C) :-
   ( X1=X -> C1=0 ; distance(X1,X,C1) ),
   travel_time(P,X,C2), C is C1+C2.
```

Figure 6.8: Planning with a single complex action for the software assistant.

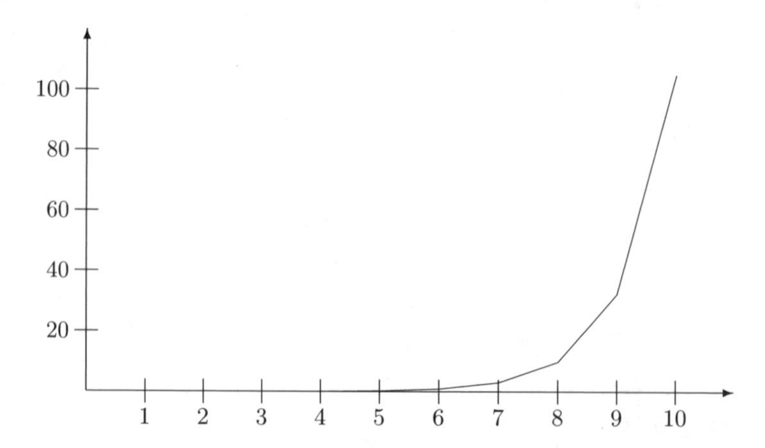

Figure 6.9: Planning time for the software assistant. The horizontal axis shows the number of requested meetings while the vertical scale is in seconds CPU time.

```
[go_meet(3,2,1030), go_meet(1,2,1200), go_meet(2,3,1245)]
```

As before the cost is 1350. This plan is computed much faster since the entire search tree consists of just 16 virtual execution nodes, as opposed to the search tree of the previous section containing 44 of these nodes. □

Although the concept of heuristic encodings of planning problems and the introduction of complex actions are ways to enhance efficiency, planning is an inherently difficult problem. Even if the number of choice points in the definition of a planning problem increases just linearly with the problem size, the search tree itself grows exponentially. For illustration, Figure 6.9 shows the runtime behavior of planning for our software assistant using the virtual action *GoMeet*. The curve shows the overall time it takes to solve the planning problem in depending on the number of requested meetings. The runtime was measured on problems where each party is available at two different meeting places. The results show that the runtime grows exponentially. As a consequence, the planning algorithm would not scale up to problems of much larger size. A partial solution is to add stronger heuristics, giving a very restricted definition of possible plans such that the search tree is cut down further to a large extent.

The fact that planning in general is infeasible requires agents to very carefully employ their ability to plan, lest they spend an unreasonable amount of time trying to compute optimal behaviors. A restrictive encoding of a planning problem, including strong heuristics, can help improving the computational behavior drastically. Still, unless the plan tree is defined in such a way that the number of choice points grows less than linearly with the size of the problem, planning problems in agent programs should always be restricted in size.

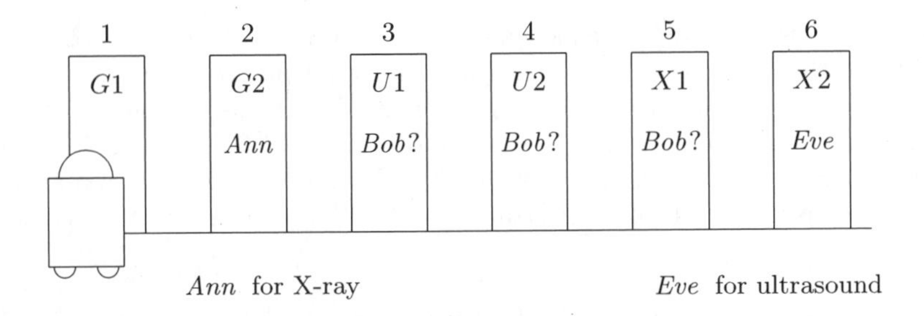

Figure 6.10: The initial state of a sample nursebot planning problem.

6.4 Conditional Planning

Thus far we have confined ourselves to planning with non-sensing actions. In order to devise plans, agents must take a look into the future, predicting the outcome of possible action sequences. Actions which involve sensing are therefore difficult to plan, unless the agent has sufficient knowledge to foresee the result. If several outcomes of a planned action are conceivable, then the plan can be considered a solution to the problem at hand only if it is valid under any possible sensing result. However, a plan being a unique sequence of actions is too strict a concept to this end, because it does not allow agents to plan a sensing action followed by different reactions to different outcomes.

Example 4 Imagine a robot in a hospital floor whose task is to bring patients from one examination room to another. Since there are several rooms which can be used for the same examination method, the "nursebot" is sometimes allowed to decide on its own where exactly to take a patient. The robot may have incomplete information as to which rooms are currently occupied. There is, however, a red light above each door which indicates whether the room is currently in use. Figure 6.10 illustrates an example layout of a hospital floor with six rooms, two of which are used for general examination $(G1, G2)$, two for ultrasound $(U1, U2)$, and two for X-ray $(X1, X2)$. In the depicted scenario, the nursebot does not know whether patient Bob is currently ultrasound examined or X-rayed. The robot is requested to bring patient Ann to an X-ray and patient Eve to an ultrasound examination. This task requires the robot to devise a plan. Suppose, to this end, that the nursebot has at its disposal the two elementary actions $CheckLight(r)$, going to room r and checking whether it is in use, and $Bring(r_1, r_2)$, bringing the patient who is currently in room r_1 to room r_2. The latter action is possible if there is a patient in r_1 and if r_2 is free. If *Bring* were the only action the robot can perform, then the depicted planning problem would be unsolvable: Neither $[Bring(2,5), Bring(6,4)]$ nor $[Bring(6,3), Bring(2,6)]$ or any other sequence of instantiated *Bring* actions is provably executable and achieves the goal under the given incomplete knowledge state. But even with the additional sensing action, there is no straight sequence of ground actions which solves the problem: Since the outcome of the sensing action cannot be predicted, neither, say, $[CheckLight(3), Bring(2,5), Bring(6,4)]$ is a valid plan (as it would work only in case Bob is in room 3), nor $[CheckLight(3), Bring(6,3), Bring(2,6)]$ (which would work in case Bob is not in room 3). Solving this problem requires the nursebot to plan ahead different sequences of actions for different outcomes of sensing. □

The concept of **conditional planning** enables agents to plan sensing actions accompanied by different continuations, of which the appropriate one is selected at execution time. In so doing, an agent can base its decisions on information which it does not have at planning time but will acquire when it actually performs its planned actions. Plans of this kind are formalized with the help of an extended notion of actions in fluent calculus: Two actions a_1 and a_2 can be composed to the conditional action $If(f, a_1, a_2)$, which denotes the action of doing a_1 if fluent f holds and doing a_2 otherwise. Furthermore, actions can be sequentially composed to a new action, written $a_1; a_2$. The constant ε, finally, represents the "empty" action.

Definition 6.3 A tuple $\mathcal{S} \cup \langle \varepsilon, ;, If \rangle$ is a **fluent calculus signature for conditional planning** if $\mathcal{S}$ is a fluent calculus signature for knowledge and

- ε : ACTION
- ; : ACTION × ACTION ↦ ACTION
- *If* : FLUENT × ACTION × ACTION ↦ ACTION. □

For example,

$$CheckLight(3)\,;\ If(Occupied(3),\ (Bring(2,5); Bring(6,4)),\ (Bring(6,3); Bring(2,6))) \tag{6.8}$$

is a ground action using the fluent $Occupied(r)$. Compound actions can be depicted as directed graphs where the nodes are either elementary actions or decision points. Figure 6.11 gives this graphical interpretation for the action just mentioned.

The new function symbols for constructing compound actions acquire meaning through additional foundational axioms. These define preconditions and effects of compound actions in relation to the axioms for the elementary actions. In particular, a conditional $If(f, a_1, a_2)$ is executable in a situation if and only if it is known whether condition f is true or false, and if f is known to be true, then a_1 must be possible, else a_2. The other axioms are straightforward given the intuitive meaning of compound actions.

Definition 6.4 The **foundational axioms of fluent calculus for conditional planning** are $\Sigma_{state} \cup \Sigma_{knows} \cup \Sigma_{plan}$ where $\Sigma_{state} \cup \Sigma_{knows}$ is as in Definition 5.3 (page 107) and Σ_{plan} consists of the following.

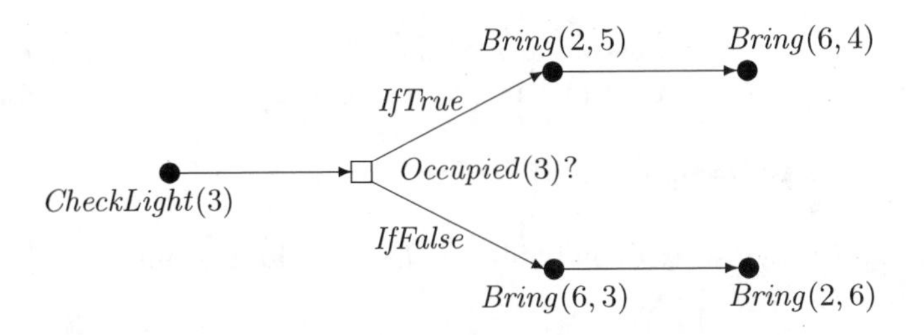

Figure 6.11: A compound action depicted as graph.

---

1. Axioms for preconditions,

$$\begin{aligned} Poss(\varepsilon, s) &\equiv \top \\ Poss(a_1; a_2, s) &\equiv Poss(a_1, s) \wedge Poss(a_2, Do(a_1, s)) \\ Poss(If(f, a_1, a_2), s) &\equiv [Knows(f, s) \wedge Poss(a_1, s)] \vee \\ &\quad\ [Knows(\neg f, s) \wedge Poss(a_2, s)] \end{aligned}$$

2. Axioms for knowledge update,

$$KState(Do(\varepsilon, s), z) \equiv KState(s, z)$$

$$\begin{aligned} &Poss(a_1; a_2, s) \supset \\ &\quad [KState(Do(a_1; a_2, s), z) \equiv KState(Do(a_2, Do(a_1, s)), z)] \end{aligned}$$

$$\begin{aligned} &Poss(If(f, a_1, a_2), s) \supset \\ &\quad Knows(f, s) \wedge \\ &\qquad [KState(Do(If(f, a_1, a_2), s), z) \equiv KState(Do(a_1, s), z)] \vee \\ &\quad Knows(\neg f, s) \wedge \\ &\qquad [KState(Do(If(f, a_1, a_2), s), z) \equiv KState(Do(a_2, s), z)] \end{aligned}$$

□

Preconditions of compound actions are axiomatized wrt. situations, instead of states, because a conditional action is possible only in case the condition is *known* to be true or false, which cannot be expressed based on a state argument.

Example 4 (continued) Let the nursebot domain be formalized by these three fluents, which are defined using the domain sort PATIENT:

$$\begin{array}{rll} InRoom: & \text{PATIENT} \times \mathbb{N} \mapsto \text{FLUENT} & InRoom(p, r) \mathrel{\hat{=}} \text{patient } p \text{ is in } r \\ Occupied: & \mathbb{N} \mapsto \text{FLUENT} & Occupied(r) \mathrel{\hat{=}} \text{room } r \text{ is occupied} \\ Request: & \mathbb{N} \times \{G, U, X\} \mapsto \text{FLUENT} & Request(r, x) \mathrel{\hat{=}} \text{patient in room } r \\ & & \quad \text{sent to an } x\text{-examination room} \end{array}$$

We assume the usual unique-name axiom. The following domain constraint states that a room is occupied just in case there is a patient in that room:

$$Holds(Occupied(r), s) \equiv (\exists p)\, Holds(InRoom(p, r), s) \tag{6.9}$$

The initial knowledge state depicted in Figure 6.10, for instance, can be formalized as

$$\begin{aligned} KState(S_0, z_0) \equiv (\exists r, z)\,(\, z_0 = {} & InRoom(Ann, 2) \circ Occupied(2) \\ & \circ InRoom(Eve, 6) \circ Occupied(6) \\ & \circ InRoom(Bob, r) \circ Occupied(r) \\ & \circ Request(2, X) \circ Request(6, U) \circ z \\ \wedge\ & 3 \leq r \leq 5 \\ \wedge\ & (\forall p, r')\, \neg Holds(InRoom(p, r'), z) \\ \wedge\ & (\forall r')\, \neg Holds(Occupied(r'), z) \\ \wedge\ & (\forall r', x)\, \neg Holds(Request(r', x), z)\,) \end{aligned}$$

The formal specification of the two elementary actions of the nursebot are as follows:

$$\begin{aligned} Bring &: \mathbb{N} \times \mathbb{N} \mapsto \text{ACTION} \quad && Bring(r_1, r_2) \mathrel{\hat{=}} \text{bring patient in } r_1 \text{ to } r_2 \\ CheckLight &: \mathbb{N} \mapsto \text{ACTION} && CheckLight(r) \mathrel{\hat{=}} \text{check occupancy of } r \end{aligned}$$

The precondition axioms for these actions are straightforward:

$$\begin{aligned} Poss(Bring(r_1, r_2), z) \equiv\ & (\exists x)\,(\, Holds(Request(r_1, x), z) \wedge \\ & \qquad ExaminationRoom(x, r_2)\,) \\ & \wedge \neg Holds(Occupied(r_2), z) \\ Poss(CheckLight(r), z) \equiv\ & \top \end{aligned} \tag{6.10}$$

where

$$\begin{aligned} ExaminationRoom(x, r) \stackrel{\text{def}}{=}\ & x = G \wedge [r = 1 \vee r = 2] \vee \\ & x = U \wedge [r = 3 \vee r = 4] \vee \\ & x = X \wedge [r = 5 \vee r = 6] \end{aligned}$$

From this, we can infer the (im-)possibility of compound actions in this domain with the help of the foundational axioms. For example, with regard to the initial knowledge state mentioned before, the domain axiomatization entails

$$\neg Poss(If(Occupied(3), Bring(6, 4), Bring(6, 3)), S_0)$$
$$\neg Poss(If(Occupied(2), Bring(6, 1), Bring(6, 2)), S_0)$$

The first action is impossible because $\neg Knows(Occupied(3), S_0)$ as well as $\neg Knows(\neg Occupied(3), S_0)$ follow from the initial knowledge state; the second conditional action is impossible because of $\neg Poss(Bring(6, 1), S_0)$ (as there is no request asking the patient in room 6 to be brought to 1). The effects of the two elementary nursebot actions can be axiomatized by the following knowledge update axioms:

$$
\begin{array}{l}
Poss(Bring(r_1, r_2), s) \supset \\
\quad [KState(Do(Bring(r_1, r_2), s), z') \equiv (\exists z)\,(KState(s, z) \wedge \\
\qquad (\exists p, x)\,(\,Holds(InRoom(p, r_1), z) \wedge Holds(Request(r_1, x), z) \wedge \\
\qquad\qquad z' = z - InRoom(p, r_1) \circ Occupied(r_1) \circ Request(r_1, x) \\
\qquad\qquad\quad + InRoom(p, r_2) \circ Occupied(r_2)\,)\,)\,]
\end{array}
$$

$$
\begin{array}{l}
Poss(CheckLight(r), s) \supset \\
\quad [KState(Do(CheckLight(r), s), z) \equiv KState(s, z) \wedge \\
\qquad [Holds(Occupied(r), z) \equiv Holds(Occupied(r), Do(CheckLight, s))]\,]
\end{array}
$$

Recall, for instance, conditional plan (6.8). This compound action is provably possible in our initial knowledge state and leads to a situation in which both initial requests have been satisfied.

1. To begin with, the domain axioms entail $Poss(CheckLight(3), S_0)$. Let $S_1 = Do(CheckLight(3), S_0)$, then the knowledge update axiom for this action implies

$$
\begin{array}{l}
(\forall z)\,(KState(S_1, z) \equiv KState(S_0, z) \wedge Holds(Occupied(3), z)) \\
\vee \\
(\forall z)\,(KState(S_1, z) \equiv KState(S_0, z) \wedge \neg Holds(Occupied(3), z))
\end{array}
\tag{6.11}
$$

   Hence, $Knows(Occupied(3), S_1) \vee Knows(\neg Occupied(3), S_1)$.

2. Suppose $Knows(Occupied(3), S_1)$, then $Knows(\neg Occupied(4), S_1)$ as well as $Knows(\neg Occupied(5), S_1)$ according to the the unique-name axiom for *Occupied*, the initial knowledge state, and the knowledge update axiom for *CheckLight*. Hence, $Poss(Bring(2, 5), S_1)$. Furthermore, the knowledge update axiom for *Bring* implies $Poss(Bring(6, 4), Do(Bring(2, 5), S_1))$. Axioms Σ_{plan} thus entail

$$Holds(Occupied(3), S_1) \supset Poss(Bring(2, 5); Bring(6, 4), S_1) \tag{6.12}$$

   Moreover, under the provision that $Holds(Occupied(3), S_1)$, the initial knowledge state along with the knowledge update axioms and the foundational axioms imply

$$\neg(\exists r, x)\, Holds(Request(r, x), Do(Bring(2, 5); Bring(6, 4), S_1))$$

3. On the other hand, suppose $Knows(\neg Occupied(3), S_1)$, then by the initial knowledge state and the knowledge update axiom for *CheckLight* it follows that $Poss(Bring(6, 3), S_1)$. The update axiom for *Bring* implies that $Knows(\neg Occupied(6), Do(Bring(6, 3), S_1))$, which in turn implies $Poss(Bring(2, 6), Do(Bring(6, 3), S_1))$. Axioms Σ_{plan} thus entail

$$\neg Holds(Occupied(3), S_1) \supset Poss(Bring(6, 3); Bring(2, 6), S_1) \tag{6.13}$$

   As before, we also have that if $\neg Holds(Occupied(3), S_1)$ then

$$\neg(\exists r, x)\, Holds(Request(r, x), Do(Bring(6, 3); Bring(2, 6), S_1))$$

4. Putting together (6.12) and (6.13), foundational axioms Σ_{plan} entail

$$Poss(If(Occupied(3), (Bring(2,5); Bring(6,4)), (Bring(6,3); Bring(2,6))), S_1)$$

Thus, the full action (6.8) is possible in S_0. Moreover, no request remains in the resulting situation. □

The projection of a sensing action with unknown outcome results in several possible future knowledge states, as we have just seen in the example with disjunctive formula (6.11). These different knowledge states cannot be combined into a single knowledge state without loosing important information. For it is the very effect of sensing to split a knowledge state: Formula (6.11) is subtly but crucially different from

$$(\forall z)\,(KState(S_1, z) \equiv KState(S_0, z) \wedge [Holds(Occupied(3), z) \vee \neg Holds(Occupied(3), z)]\,)$$

(which is equivalent to $KState(S_1, z) \equiv KState(S_0, z)$), because here it is not entailed that it will be known in S_1 whether or not $Occupied(3)$ holds. This information, however, is crucial for planning a conditional action which depends on the truth-value of $Occupied(3)$.

The existence of multiple projection results makes it necessary to extend the FLUX definition of knowledge. A fluent calculus axiomatization entails a formula $Knows(\phi, s)$ just in case this formula is true in all models for $KState(s, z)$. Hence, in order to verify that a property will be known in a future situation, all possible projections need to be verified to entail the property. To this end, we extend the FLUX kernel by the predicate $Res(\sigma, z_0, z)$, whose intended meaning is that FLUX state z encodes a possible resulting knowledge state of performing actions $\sigma = Do(\alpha_n, \ldots, Do(\alpha_1, [\,])\ldots)$ in FLUX state z_0 $(n \geq 0)$. The FLUX definition is shown in the first part of Figure 6.12. It includes the special functions

$$IfTrue : \text{FLUENT} \mapsto \text{ACTION}$$
$$IfFalse : \text{FLUENT} \mapsto \text{ACTION}$$

These functions are needed—in encodings of planning problems—to identify the different knowledge states in the two branches that follow a conditional action. The effect of $IfTrue(f)$ is that f is known to hold, while $IfFalse(f)$ implies that f is known not to hold.

On this basis, a property ϕ is entailed to be known after performing the actions in σ just in case there is no resulting FLUX state z in which ϕ is not known. To express this formally, FLUX uses a ternary predicate $Knows(f, s, z_0)$ defining fluent f to be known after performing the actions in s starting with the knowledge state encoded by z_0. Figure 6.12 depicts the FLUX definition of this predicate as well as for $KnowsNot(f, s, z_0)$ and $KnowsVal(\vec{x}, f, s, z_0)$. The latter is defined as $\vec{x}$ being a value for f which is known in all possible

```
res([],Z0,Z0).
res(do(A,S),Z0,Z) :- A=if_true(F)  -> res(S,Z0,Z), holds(F,Z) ;
                     A=if_false(F) -> res(S,Z0,Z), not_holds(F,Z) ;
                     res(S,Z0,Z1), state_update(Z1,A,Z,_).

knows(F,S,Z0) :- \+ ( res(S,Z0,Z), not_holds(F,Z) ).

knows_not(F,S,Z0) :- \+ ( res(S,Z0,Z), holds(F,Z) ).

knows_val(X,F,S,Z0) :-
   res(S,Z0,Z) -> findall(X,knows_val(X,F,Z),T),
                  assert(known_val(T)),
                  false.
knows_val(X,F,S,Z0) :-
   known_val(T), retract(known_val(T)), member(X,T),
   \+ ( res(S,Z0,Z), not_holds_all(F,Z) ).

execute(A,Z1,Z2) :-
   perform(A,Y)          -> state_update(Z1,A,Z2,Y) ;
   A = [A1|A2]           -> execute(A1,Z1,Z), execute(A2,Z,Z2) ;
   A = if(F,A1,A2)       -> (holds(F,Z1) -> execute(A1,Z1,Z2)
                                         ; execute(A2,Z1,Z2)) ;
   A = []                -> Z1=Z2 ;
   complex_action(A,Z1,Z2).
```

Figure 6.12: Kernel definitions for planning with sensing actions.

resulting knowledge states.[3] The ultimate clause in Figure 6.12 extends the definition of execution to conditional plans, where sequentially compound actions are encoded as lists.

Example 4 (continued) Figure 6.13 depicts the encoding of the update axioms for the actions of our nursebot along with the initial state as depicted in Figure 6.10. Patients Ann, Bob, and Eve carry, respectively, the numbers 1, 2, and 3; likewise, the three types of examinations G, U, X are encoded by 1–3. The CHR for the auxiliary constraint *RoomOccupied* encodes domain constraint (6.9), which relates the fluents *Occupied*(r) and *InRoom*(p, r). Suppose, for instance, the robot considers checking room 3, then there are two possible resulting knowledge states:

```
?- init(Z0), res(do(check_light(3),[]),Z0,Z1).
```

[3] The FLUX definition of *KnowsVal* is actually stronger than just requiring that in every resulting knowledge state a value is known. To see why, let Σ be the formula $(\forall z)KState(s, F(1) \circ z) \vee (\forall z)KState(s, F(2) \circ z)$, say, then $\Sigma \models KnowsVal(x, F(x), s)$ but there is no $n \in \mathbb{N}$ such that $\Sigma \models Knows(F(n), s)$! The stronger definition is much more useful to planning agents as it allows them to plan with concrete values.

```
state_update(Z1, bring(R1,R2), Z2, []) :-
   holds(in_room(P,R1), Z1), holds(request(R1,X), Z1),
   update(Z1, [in_room(P,R2),occupied(R2)],
              [in_room(P,R1),occupied(R1),request(R1,X)], Z2).

state_update(Z, check_light(R), Z, [Light]) :-
   Light = true,  holds(occupied(R), Z) ;
   Light = false, not_holds(occupied(R), Z).

examination_room(1,1). examination_room(1,2).
examination_room(2,3). examination_room(2,4).
examination_room(3,5). examination_room(3,6).

init(Z0) :- Z0 = [in_room(1,2),in_room(2,R),in_room(3,6),
                  request(2,3), request(6,2) | Z],
            R :: 3..5,
            not_holds_all(in_room(_,_), Z),
            not_holds_all(request(_,_), Z),
            consistent(Z0).

consistent(Z) :- room_occupied(Z), duplicate_free(Z).

room_occupied([F|Z]) <=>
   F=in_room(_,R) -> holds(occupied(R), Z,Z1), room_occupied(Z1) ;
   F=occupied(R)  -> holds(in_room(_,R),Z,Z1), room_occupied(Z1) ;
   room_occupied(Z).
```

Figure 6.13: FLUX encoding of the domain axiomatization for the nursebot.

```
Z1 = [in_room(1,2),in_room(2,3),in_room(3,6),...]     More?

Z1 = [in_room(1,2),in_room(2,R),in_room(3,6),...]
R :: 4..5                                              More?

no (more) solution.
```

Therefore, it does not follow that room 3 is occupied, nor that it is not so:

```
?- init(Z0), S1 = do(check_light(3),[]),
   ( knows(occupied(3),S1,Z0) ;
     knows_not(occupied(3),S1,Z0) ).

no (more) solution.
```

The occupation of rooms 2 and 6, however, would still be known, as would be the fact that room 1 is free:

```
?- init(Z0), S1 = do(check_light(3),[]),
   knows_val([R],occupied(R),S1,Z0),
   knows_not(occupied(1),S1,Z0).

R = 2   More?

R = 6   More?

no (more) solution.
```

Suppose that after sensing the status of room 3, the nursebot intended to branch upon the truth-value of *Occupied*(3). In case this fluent is true, the robot knows the whereabouts of patient 2 (Bob):

```
?- init(Z0),
   S2 = do(if_true(occupied(3)),do(check_light(3),[])),
   knows_val([R],in_room(2,R),S2,Z0).

R = 3
```

Considering the opposite, that is, *Occupied*(3) being false, the nursebot knows that Bob cannot be in this room, but it still does not know where exactly he is:

```
?- init(Z0),
   S2 = do(if_false(occupied(3)),do(check_light(3),[])),
   knows_not(in_room(2,3),S2,Z0),
   \+ knows_val([R],in_room(2,R),S2,Z0).

yes.
```

□

Conditional plans are represented in FLUX by nested lists of actions, where $[\alpha_1, \ldots, \alpha_n]$ denotes the sequential composition $\alpha_1; \ldots; \alpha_n$ and $[\,]$ stands for the empty action ε. Plan (6.8), say, corresponds to

$$[\mathit{CheckLight}(3), \mathit{If}(\mathit{Occupied}(3), [\mathit{Bring}(2,5), \mathit{Bring}(6,4)], [\mathit{Bring}(6,3), \mathit{Bring}(2,6)])]$$

Planning problems with sensing are encoded by defining a binary predicate $P(z,p)$ such that p solves the problem at hand with initial state z. In contrast to Definition 6.2 for planning without sensing, predicate P does not carry an argument for the resulting state, owing to the fact that a conditional plan cannot in general be projected onto a single knowledge state.

Definition 6.5 A **heuristic encoding of a planning problem for conditional planning** is a program P_{plan} defining a predicate $P(z,p)$. If Σ is a domain axiomatization with FLUX encoding P_Σ, and P_{plan} is the encoding of a planning problem with initial knowledge state $\{\mathit{KState}(S,z) \equiv \Phi(z)\}$ and goal $\Gamma(z)$, then the encoding is **sound** iff the following holds: For every computed answer θ to

$$P_{kernel} \cup P_\Sigma \cup P_{plan} \cup \{\leftarrow [\![\Phi(z)]\!], P(z,p)\}$$

$p\theta$ is a solution to the planning problem. □

As usual, a sound encoding of a planning problem requires each planned action to be executable. In particular, according to the foundational axioms on complex actions, Σ_{plan}, a conditional $\mathit{If}(f, p_1, p_2)$ is possible just in case the agent knows the truth-value of f in the respective situation and the two sub-plans p_1 and p_2 are possible under the respective value of f. The latter can be expressed with the help of the two functions $\mathit{IfTrue}(f)$ and $\mathit{IfFalse}(f)$, which bring the agent in a situation in which f is known to be true and false, respectively.

Example 4 (continued) Figure 6.14 depicts a heuristic encoding for the planning problem of the nursebot. Beginning with the empty plan, in each step either a *Bring* action is added or a $\mathit{CheckLight}(r)$ action such that r is a room with as yet unknown status. After each such sensing action, a conditional action is added and the two alternative continuations are planned with the help of $\mathit{IfTrue}(\mathit{Occupied}(r))$ and $\mathit{IfFalse}(\mathit{Occupied}(r))$, respectively. This is the first solution that is being computed:

```
[check_light(5),
 if(occupied(5),
    [bring(6,3),bring(2,6)],
    [bring(2,5),check_light(3),if(occupied(3),[bring(6,4)],
                                               [bring(6,3)])]
   )
]
```

```
request_plan(Z,P) :- request_plan(Z,[],P).

request_plan(S, Z, []) :- knows_not(request(_,_), S, Z).
request_plan(S, Z, P) :-
   knows_val([R1,X], request(R1,X), S, Z),
   examination_room(X, R2), knows_not(occupied(R2), S, Z),
   request_plan(do(bring(R1,R2),S), Z, P1),
   P=[bring(R1,R2)|P1].
request_plan(S, Z, P) :-
   knows_val([X], request(_,X), S, Z),
   examination_room(X, R),
   \+ knows(occupied(R), S, Z), \+ knows_not(occupied(R), S, Z),
   request_plan(do(if_true(occupied(R)),
                   do(check_light(R),S)), Z, P1),
   request_plan(do(if_false(occupied(R)),
                   do(check_light(R),S)), Z, P2),
   P=[check_light(R),if(occupied(R),P1,P2)].
```

Figure 6.14: Planning in the nursebot domain.

According to this plan, the nursebot will first check whether room 5 is occupied. If so, room 3 must be free (as Bob is in 5) and the robot plans to bring Eve to room 3 and, then, Ann to room 6. Otherwise, that is, if 5 is not occupied, then the robot first brings Ann to 5 and thereafter checks whether room 3 is occupied in order to see whether Eve should be taken to 3 or 4. Among the solutions that are included in the plan tree is also the shorter plan from above, viz.

```
[check_light(3),
 if(occupied(3),[bring(2,5),bring(6,4)],
                [bring(6,3),bring(2,6)])]
```

The definition of $RequestPlan(z, p)$ can be proved sound wrt. the axioms for the nursebot domain and the planning goal of solving all requests:

1. An instance of $Bring(r, r_2)$ is virtually executed only if $Request(r, x)$ is known for some x, if this x satisfies $ExaminationRoom(x, r_2)$, and if $\neg Occupied(r_2)$ is known. These conditions imply that the action is possible according to precondition axioms (6.10). According to these axioms, action $CheckLight(r)$ is always possible. The conditioning on $Occupied(r)$, finally, is planned only after a $CheckLight(r)$ action, whose knowledge update axiom implies that the truth-value of $Occupied(r)$ will be known in the successor situation.

2. By the first clause for $RequestPlan(z_0, s, p)$, the planning goal is defined as knowing that no request holds. □

```
:- dynamic plan_search_best/2.

plan(Problem,Z,P) :-
   assert(plan_search_best(_,-1)), plan_search(Problem,Z),
   retract(plan_search_best(P,C)), C =\= -1.

plan_search(Problem,Z) :-
   is_predicate(Problem/2) ->
      ( PlanningProblem =.. [Problem,Z,P],
        call(PlanningProblem), plan_cost(Problem,P,C),
        plan_search_best(_,C1),
        ( C1 =< C, C1 =\= -1 -> fail
          ; retract(plan_search_best(_,C1)),
            assert(plan_search_best(P,C)), fail )
        ; true ) ;
   PlanningProblem =.. [Problem,Z,P,Zn],
   call(PlanningProblem), plan_cost(Problem,P,Zn,C),
   plan_search_best(_,C1),
   ( C1 =< C, C1 =\= -1 -> fail
     ; retract(plan_search_best(_,C1)),
       assert(plan_search_best(P,C)), fail )
   ; true.
```

Figure 6.15: FLUX definition for selecting optimal plans. The first part of the clause for *PlanSearch* is for conditional planning problems, which are characterized by a binary *PlanningProblem* and ternary *PlanCost*.

Figure 6.15 depicts an extended kernel definition to compare different conditional plans and to select an optimal one. Since conditional plans do not, in general, determine a unique resulting knowledge state, planning cost need to be determined solely by the plan itself. Hence, the quality measure is defined by a ternary variant $PlanCost(P, p, c)$, defining, for planning problem P, the cost c wrt. plan p. This is in contrast to planning without sensing, where the cost may be defined by directly referring to the resulting state. Exercise 6.4 is concerned with applying the extended FLUX kernel by giving a suitable definition of plan cost and having the nursebot find optimal plans according to various criteria.

6.5 Bibliographical Notes

The automation of the ability to devise plans has been a major motivation behind the development of theories of actions [McCarthy, 1958]. This might have been the reason for grounding the earliest formalism on the notion of a situation [McCarthy, 1963], which allows to define planning problems as deduction problems [Shanahan, 1989; Reiter, 2001a]. The complexity of planning has been thoroughly analyzed for a variety of problem classes; we just mention [Bäckström and Klein, 1991; Bäckström and Nebel, 1993; Bylander, 1994]. The notion of a heuristic encoding is rooted in the concept of nondeterministic GOLOG programs [Levesque *et al.*, 1997], defining a search tree over situations. The main difference is that GOLOG uses a specific language for these programs whereas heuristic encodings are standard logic programs. Variants of GOLOG [Giacomo and Levesque, 1999; Giacomo *et al.*, 2002] have combined "off-line" and "online" execution of nondeterministic programs, which resembles the embedding of planning problems into a FLUX agent program. A different way of defining the search space for a planning problem in situation calculus is to use temporal logic to define constraints on the future states that are being searched [Bacchus and Kabanza, 2000]. Incorporating domain knowledge has led to very efficient planning systems such as [Kvarnström and Doherty, 2000; Kvarnström, 2002]. An alternative to tackle large-scale propositional planning problems is by general local search algorithms for NP-hard problems [Kautz and Selman, 1996; Ernst *et al.*, 1997]. A variety of special planning algorithms have also been developed, among which [Blum and Furst, 1997] is a classic.

Among the early conditional planners is the system described in [Golden and Weld, 1996] for which a semantics has been given based on situation calculus and the solution to the frame problem for knowledge of [Scherl and Levesque, 1993]. The implementation itself has used the notion of an incomplete state consisting of true, false, and unknown propositional fluents. The same representation has been used in the logic programming systems [Baral and Son, 1997; Lobo, 1998]. This restricted notion of incomplete states does not allow for planning with disjunctive information. This limitation applies also to the approach of [Giacomo *et al.*, 1997], where description logic has been used, and to the program developed in [Lakemeyer, 1999], whose semantics has been given in terms of a general approach to conditional planning in the situation calculus.

6.6 Exercises

6.1. Extend the planning problem for the personal software assistant so as to allow for specifying alternative meeting requests with the goal to schedule exactly one of them. Modify the definition of the cost of a plan with the aim to minimize the overall time spent in meetings.

6.2. Extend the control program for the mailbot of Chapter 2 by a planning problem which allows the agent to look ahead a fixed number of actions with the aim to shorten the length of the overall action sequence. Compare the results of your program with the figures given on page 47.

6.3. The goal of this exercise is to have the agent in the Wumpus World of Exercise 5.10 plan the shortest way out of the cave immediately after having claimed gold.

(a) Give a fluent calculus definition of a complex action called $Go(d)$ with the intended meaning that the agent goes one step into direction d. Derive a precondition and a knowledge update axiom for this complex action from the axiomatization of the elementary actions *Turn* and *Go*.

(b) Give a FLUX encoding of the specification of the complex action $Go(d)$. For the sake of simplicity, assume that this action does not involve sensing.

(c) Program a heuristic encoding for the problem to find a shortest but safe path to square $(1, 1)$ from the current location. Integrate this plan tree into the agent program by having the agent find and execute such a plan after it has claimed gold.

(d) Try to improve the encoding by some heuristics which allow to cut down the search space.

6.4. For each of the following notions of the cost of a plan in the nursebot domain, give a formal specification in FLUX and use it to generate an optimal plan for the scenario of Figure 6.10.

(a) The cost is the number of elementary actions the robot must perform in the worst case.

(b) The cost is the total distance the robot has to walk in the worst case, starting at room 1 and assuming equidistance between consecutive examination rooms.

(c) The cost is the total distance the patients have to walk in the worst case.

6.5. Be a hero and rescue the princess: You find yourself in the outer room (*Cave0*) of the cave system depicted in Figure 6.16. The princess is in one of the side caves. The other two caves may house dragons, which you

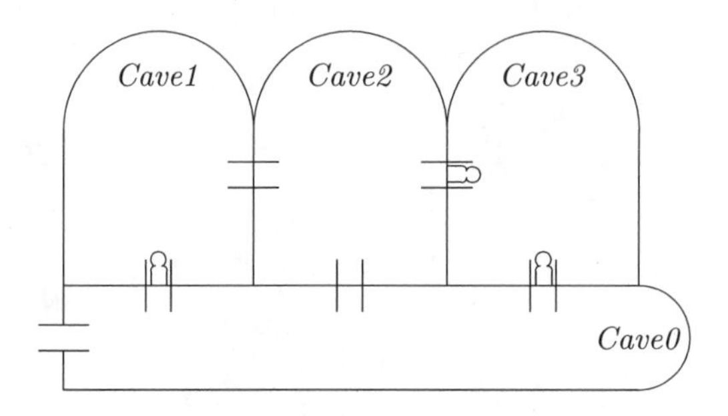

Figure 6.16: A system of four caves with five connecting doors, three of which have a keyhole through which the hero can peek.

should avoid to face by all means. You can go through each door. You can peek through each depicted keyhole (from both sides) to check whether there is a dragon in the adjacent room. Use FLUX to find a conditional plan with the goal to be in the same cave as the princess.

Chapter 7

Nondeterminism

An action is deterministic if its effect is always the same when applied under the same conditions. This allows agents to predict the result of a deterministic action, provided they have sufficient knowledge of the status of the involved fluents. In reality, however, the effects of actions are often uncertain since they depend on circumstances that go beyond the perceptive faculty or control of an agent. Actions are called **nondeterministic** if their result cannot be fully predicted even if the current state of all relevant fluents is known. Reasoning and planning requires to deal with this uncertainty by making sure that the agent functions correctly under any outcome of a nondeterministic action.

Actually, fluent calculus and FLUX as presented thus far are sufficiently general to accommodate actions with uncertain outcome. From the perspective of an agent, reasoning about a nondeterministic action is very similar to reasoning about a deterministic one under incomplete information. If, for example, the cleaning robot of Chapter 5 is uncertain about its current location or the direction it faces, then the result of going forward one step is uncertain although this action is, in principle, deterministic. Hence, the same techniques apply to both acting under incomplete knowledge and reasoning about nondeterministic actions. As will be shown later, nondeterminism can also be used to model dynamic state properties, that is, fluents which may at any time undergo changes unnoticed by the robot. Hence, this chapter is divided into the discussion of

7.1. uncertain effects;

7.2. dynamic changes

7.1 Uncertain Effects

The main purpose of using nondeterminism in fluent calculus is to model actions whose effects are uncertain. Following Definition 1.16 (page 19), a nondeterministic action can be characterized by an update axiom which does not imply a unique successor state for every state in which the action is possible. Specifically, an update specification of the general form,

$$\begin{array}{l}(\exists \vec{y}_1)\,(\Delta_1(z) \wedge z' = z - \vartheta_1^- + \vartheta_1^+)\\ \vee \ldots \vee\\ (\exists \vec{y}_n)\,(\Delta_n(z) \wedge z' = z - \vartheta_n^- + \vartheta_n^+)\end{array}$$

can be nondeterministic for two reasons: First, there may be two different conditions Δ_i and Δ_j which both are satisfied in the current state. If in this case the effects $\vartheta_i^+, \vartheta_i^-$ differ from $\vartheta_j^+, \vartheta_j^-$, then the update axiom does not imply a unique update equation. Second, there may be more than one instance for the existentially quantified variables $\vec{y}_i$ for which a condition Δ_i is satisfied by the current state. If these variables also occur in the effects $\vartheta_i^+, \vartheta_i^-$, then again the update axiom does not imply a unique update equation. In the following, we illustrate both sources of nondeterminism with an elaboration of the mail delivery problem of Chapter 1 and 2: In a realistic setting, the mailbot may sometimes unsuccessfully try to deliver a package, say, if the recipient is absent or refuses to accept the item. Likewise, when picking up a package the robot may not be able to influence which one of possibly several requests a person wants to be handled first.

Example 1 (continued) The mailbot should bring back to the sender a package which cannot be delivered. In order to be prepared for this, the robot needs to record the origin of a package with the help of the ternary fluent

$$Carries\colon\ \mathbb{N} \times \mathbb{N} \times \mathbb{N} \mapsto \text{FLUENT}$$

where $Carries(b, r_1, r_2)$ shall indicate that bag b carries a package sent from room r_1 to r_2. Using this extension, the following nondeterministic knowledge update axiom for action $Deliver(b)$ models the uncertain outcome of the attempt to deliver the contents of mail bag b:

$$\begin{array}{l}Poss(Deliver(b), s) \supset\\ \quad [KState(Do(Deliver(b), s), z') \equiv (\exists z)\,(KState(s, z) \wedge\\ \quad\quad [\,(\exists r_1, r_2)\,(\,Holds(Carries(b, r_1, r_2), z) \wedge\\ \quad\quad\quad\quad z' = z - Carries(b, r_1, r_2) + Empty(b)\,)\\ \quad\quad \vee\\ \quad\quad (\exists r_1, r_2)\,(\,Holds(Carries(b, r_1, r_2), z) \wedge\\ \quad\quad\quad\quad z' = z - Carries(b, r_1, r_2) + Carries(b, r_2, r_1)\,)\,]\,)\,]\end{array} \tag{7.1}$$

Put in words, the package contained in bag b is either taken (first alternative) or remains in the mail bag and is automatically sent back (second alternative). Being identical, the conditions of the two possible updates always apply together, and hence the successor state can never be predicted even if the state in situation s is known completely. This makes the new update nondeterministic. Consider, e.g., the ground state

$$KState(S_0, z) \equiv z = At(2) \circ Carries(1, 1, 2) \circ Empty(2) \circ Empty(3)$$

Let the precondition axiom for $Deliver(b)$ be adapted to the ternary variant of $Carries$ thus:

$$Poss(Deliver(b), z) \equiv (\exists r_1, r_2)\,(Holds(At(r_2), z) \wedge Holds(Carries(b, r_1, r_2), z))$$

Then $Poss(Deliver(1), S_0)$. Given this, knowledge update axiom (7.1) implies

$$KState(S_1, z') \equiv \begin{array}{l} z' = At(2) \circ Empty(1) \circ Empty(2) \circ Empty(3) \\ \vee \\ z' = At(2) \circ Carries(1,2,1) \circ Empty(2) \circ Empty(3) \end{array} \tag{7.2}$$

where $S_1 = Do(Deliver(1), S_0)$. No stronger conclusion can be drawn about the resulting state. This shows that the new update axiom fails to satisfy the defining property of deterministic domains.

The mailbot must of course be able to somehow figure out whether a delivery attempt has succeeded. Suppose, to this end, the robot can at any time check whether a particular mail bag is empty. Formally, we introduce the elementary action

$$CheckBag : \mathbb{N} \mapsto \text{ACTION} \quad CheckBag(b) \mathrel{\hat{=}} \text{check whether } b \text{ is empty}$$

along with the following precondition and knowledge update axioms:

$$Poss(CheckBag(b), z) \equiv \top$$

$$Poss(CheckBag(b), s) \supset [KState(Do(CheckBag(b), s), z') \equiv KState(s, z') \wedge [Holds(Empty(b), z') \equiv Holds(Empty(b), Do(CheckBag(b), s))]]$$

Recall, for example, knowledge state (7.2). If the inspection reveals that bag 1 is not empty, say, then it contains a package to be brought back to room 1: With $\neg Holds(Empty(1), Do(CheckBag(1), S_1))$, knowledge state (7.2) along with the knowledge update axiom for $CheckBag$ entail

$$KState(Do(CheckBag(1), S_1), z) \equiv z = At(2) \circ Carries(1,2,1) \circ Empty(2) \circ Empty(3)$$

If the mailbot has no influence on which package it receives upon collecting one, then this action, too, is nondeterministic. For the effect can be to carry any of the packages that are requested from the current room. To this end, the argument for the destination is omitted in the action function

$$Pickup : \mathbb{N} \mapsto \text{ACTION} \quad Pickup(b) \mathrel{\hat{=}} \text{put some package into bag } b$$

In order to be able to deliver to the right destination, the mailbot must of course be told which packages it receives. This shall be part of the action of picking up. The following knowledge update axiom models the nondeterminism of the physical effect and includes the cognitive effect that the robot learns the addressee of the package in question:

$$
\begin{aligned}
& Poss(Pickup(b), s) \supset \\
& \quad (\exists r_2)(\forall z')\,[\,KState(Do(Pickup(b), s), z') \equiv (\exists z)\,(KState(s, z) \wedge \\
& \qquad\qquad (\exists r_1)\,(\,Holds(At(r_1), z) \wedge Holds(Request(r_1, r_2), z) \wedge \\
& \qquad\qquad\qquad z' = z - Empty(b) \circ Request(r_1, r_2) \\
& \qquad\qquad\qquad\qquad + Carries(b, r_1, r_2)))\,]
\end{aligned}
\tag{7.3}
$$

Put in words, any of the requested packages in room r may end up in the mail bag. Unless there is just a single request in the current room r_1, the condition for the update equation is satisfied by several instances for variable r_2. However, there always exists a destination r_2 such that all possible states agree on the fact that it is the package for r_2 which has been introduced into mail bag b. Since the update differs for different instances, the successor state cannot always be predicted even if the state in situation s is known completely. Take, e.g., the ground state

$$
\begin{aligned}
KState(S_0, z) \equiv z = {} & At(1) \circ Request(1,2) \circ Request(1,3) \\
& \circ Empty(1) \circ Empty(2) \circ Empty(3)
\end{aligned}
\tag{7.4}
$$

Let the precondition axiom be adapted to the unary *Pickup* thus:

$$
\begin{aligned}
& Poss(Pickup(b), z) \equiv \\
& \quad Holds(Empty(b), z) \wedge (\exists r_1, r_2)\,(\,Holds(At(r_1), z) \wedge \\
& \qquad\qquad Holds(Request(r_1, r_2), z))
\end{aligned}
$$

Then $Poss(Pickup(1), S_0)$. Given this, the update axiom implies that the robot reaches either of these two knowledge states:

$$
\begin{aligned}
& (\forall z')\,(KState(S_1, z') \equiv \\
& \quad z' = At(1) \circ Carries(1,1,2) \circ Request(1,3) \circ Empty(2) \circ Empty(3)) \\
& \vee \\
& (\forall z')\,(KState(S_1, z') \equiv \\
& \quad z' = At(1) \circ Carries(1,1,3) \circ Request(1,2) \circ Empty(2) \circ Empty(3))
\end{aligned}
$$

where $S_1 = Do(Pickup(1), S_0)$. Which of the two knowledge states result cannot be predicated. Hence, the new state update axiom for *Pickup*, too, fails to satisfy the defining property of deterministic domains. In both cases, however, the package is known which has been collected.

Figure 7.1 shows a FLUX specification of the background theory which accounts for the uncertainty of picking up and delivering packages in the mail delivery world. In the nondeterministic setting, controlling the mailbot requires to employ general FLUX, because complete states turn into incomplete ones upon performing an action with uncertain effect. The precondition axioms have been adapted to incomplete states, taking into account the modified signature with the unary *Pickup* and the ternary *Carries*. The clauses have been designed so as to allow for being called with variable argument b, which

```
poss(pickup(B),Z) :- knows_val([B1],empty(B1),Z), B=B1,
                     knows_val([R],request(R,_),Z),
                     knows(at(R),Z).

poss(deliver(B),Z) :- knows_val([R],at(R),Z),
                      knows_val([B1],carries(B1,_,R),Z), B=B1.

state_update(Z1,pickup(B),Z2,[R2]) :-
   holds(at(R1),Z1), holds(request(R1,R2),Z1),
   update(Z1,[carries(B,R1,R2)],[empty(B),request(R1,R2)],Z2).

state_update(Z1,deliver(B),Z2,[]) :-
   holds(carries(B,R1,R2),Z1),
   update(Z1,[],[carries(B,R1,R2)],Z),
   cancel(empty(B),Z,Z3), cancel(carries(B,R2,R1),Z3,Z2),
   or_holds([empty(B),carries(B,R2,R1)],Z2).

state_update(Z,check_bag(B),Z,[Empty]) :-
   Empty = true,  holds(empty(B),Z) ;
   Empty = false, not_holds(empty(B),Z).

state_update(Z1,go(D),Z2,[]) :-
   holds(at(R),Z1), ( D = up -> R1 #= R+1 ; R1 #= R-1 ),
   update(Z1,[at(R1)],[at(R)],Z2).

consistent(Z) :-
   holds(at(R),Z,Z1) -> R::[1..6], not_holds_all(at(_),Z1),
   empty_carries(Z),
   request_consistent(Z),
   duplicate_free(Z).

empty_carries([F|Z]) <=>
   ( F=empty(B)          -> B::1..3,
                            not_holds_all(carries(B,_,_),Z) ;
     F=carries(B,R1,R2) -> B::1..3, [R1,R2]::1..6,
                            not_holds(empty(B),Z),
                            not_holds_all(carries(B,_,_),Z) ;
     true ),
   empty_carries(Z).

request_consistent([F|Z]) <=>
   ( F=request(R1,R2) -> [R1,R2]::1..6 ; true ),
   request_consistent(Z).
```

Figure 7.1: FLUX encoding of the new precondition and nondeterministic update axioms for the mailbot.

gets suitably instantiated. Nondeterministic knowledge update axiom (7.3) is straightforwardly encoded by incorporating the sensor information telling the robot which package is actually being picked up. Update axiom (7.1) for *Deliver*(b) is encoded using the definition of cancellation: Aside from the certain effect that *Carries*(b, r_1, r_2) becomes false, both fluents *Empty*(b) and *Carries*(b, r_2, r_1) may be affected. Hence, possible information of these fluents is cancelled and replaced by their disjunction. The encoding can be proved sound wrt. the underlying update axiom (see Exercise 7.3). The encoding of the new *CheckBag*(b) action follows the standard scheme for sensing the truth-value of a fluent. The update axiom for *Go*(d) is the same as in Chapter 2 but with the 4-ary *StateUpdate* of general FLUX. Finally, the auxiliary clause for consistency of states uses the domain-dependent constraints *EmptyCarries* and *RequestConsistent*, whose CHRs model range restrictions and the domain constraints relating emptiness of a mail bag to carrying a package:

$$Holds(Empty(b), s) \equiv (\forall r_1, r_2)\, \neg Holds(Carries(b, r_1, r_2), s)$$
$$Holds(Carries(b, r_1, r_2), s) \wedge Holds(Carries(b, r_1', r_2'), s) \supset r_1 = r_1' \wedge r_2 = r_2'$$

As an example, consider the situation depicted in Figure 7.2. The robot starts with picking up and putting into mail bag 1 a package at its current location, room 1. The result is that it carries the item sent to room 3. After picking up the second package waiting in 1, which is for room 2, the robot goes up and delivers the contents of bag 2. At this stage, it is uncertain whether the package has been accepted or sent back:

```
?- Z0 = [at(1),empty(1),empty(2),empty(3),
         request(1,2),request(1,3),request(2,3) | _],
   consistent(Z0),
   state_update(Z0,pickup(1),Z1,[3]),
   state_update(Z1,pickup(2),Z2,[2]),
   state_update(Z2,go(up),Z3,[]),
   state_update(Z3,deliver(2),Z4,[]).

Z4 = [at(2),carries(1,1,3),empty(3),request(2,3) | Z]

Constraints:
or_holds([empty(2),carries(2,2,1)], Z)
empty_carries(Z)
...
```

If a subsequent inspection *CheckBag*(2) reveals that bag 2 is actually not empty, then the disjunction *OrHolds*$([Empty(2), Carries(2, 2, 1)], z)$ is resolved to *Holds*$(Carries(2, 2, 1), z)$. On the other hand, if checking the bag shows that it is empty, then after effecting *Holds*$(Empty(2), z)$ the disjunction is resolved to *True*, and the auxiliary constraint *EmptyCarries*(z) introduces the constraint *NotHoldsAll*$(Carries(2, \_, \_), z)$.

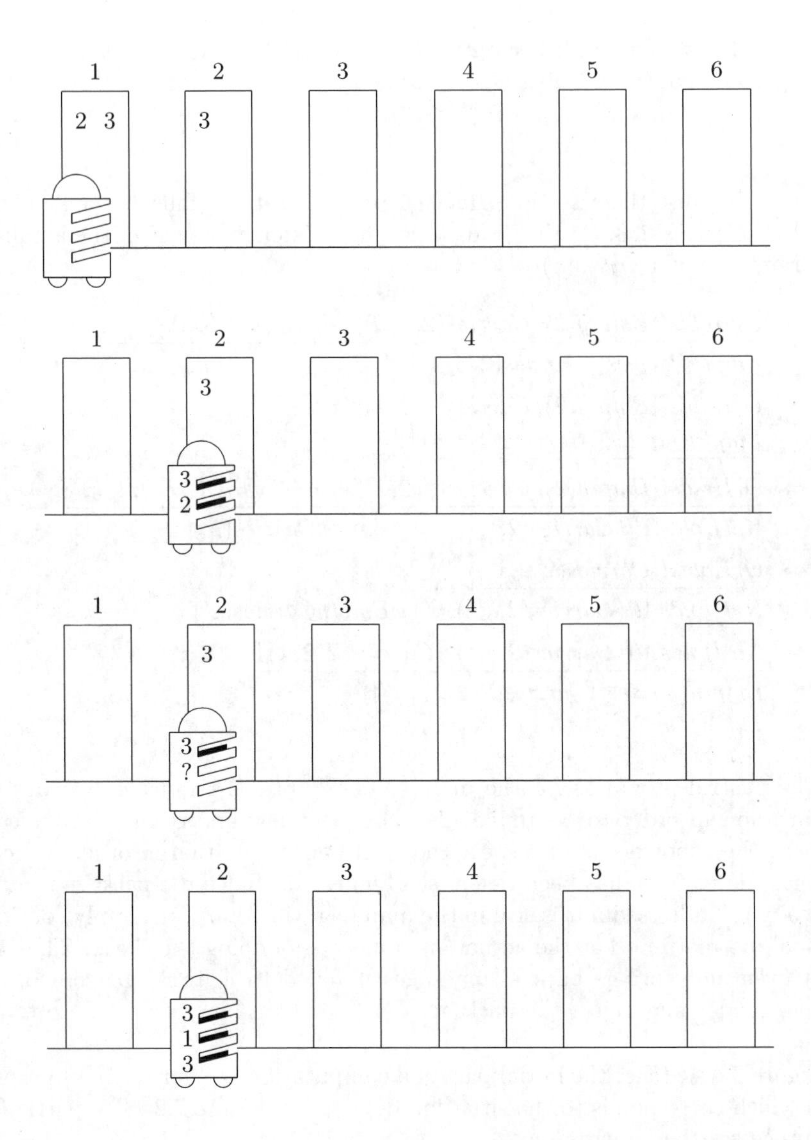

Figure 7.2: Example run where the effect of both picking up and delivering packages is uncertain.

It is worth mentioning that despite the uncertainty whether $Empty(2)$ or $Carries(2,2,1)$ holds in the intermediate state, it still follows that no other package can possibly be contained in mail bag 2; e.g.,

```
?- or_holds([empty(2),carries(2,2,1)], Z), consistent(Z),
   knows_not(carries(2,3,4), Z).

yes
```

This is so because the assertion $Holds(Carries(2,3,4), z)$ fails as the substitution $\{z \setminus [Carries(2,3,4) \mid z']\}$ leads to an inconsistency by way of the auxiliary constraint $EmptyCarries(z)$:

$$
\begin{aligned}
& \underline{OrHolds([Empty(2), Carries(2,2,1)], [Carries(2,3,4) \mid z']),} \\
& EmptyCarries([Carries(2,3,4) \mid z']) \\
\Leftrightarrow\ & OrHolds([Empty(2), Carries(2,2,1)], z'), \\
& \underline{EmptyCarries([Carries(2,3,4) \mid z'])} \\
\Leftrightarrow\ & \underline{OrHolds([Empty(2), Carries(2,2,1)], z')},\ \underline{NotHolds(Empty(2), z')}, \\
& NotHoldsAll(Carries(2,\_,\_), z'),\ EmptyCarries(z') \\
\Leftrightarrow\ & \underline{OrHolds([Carries(2,2,1)], z'),} \\
& NotHoldsAll(Carries(2,\_,\_), z'),\ EmptyCarries(z') \\
\Leftrightarrow\ & \underline{NotHoldsAll(Carries(2,\_,\_), [Carries(2,2,1) \mid z'']),} \\
& EmptyCarries([Carries(2,2,1) \mid z'']) \\
\Leftrightarrow\ & False
\end{aligned}
$$

The control programs for the mailbot of Chapter 2 require only marginal modification in order to be applicable under the new uncertainty. After each delivery, the robot needs to inspect the mail bag in question in order to know whether the package has been taken; see Figure 7.3. Since the package is automatically re-addressed if it is still in the mail bag, the robot can simply continue as usual, no matter what the actual outcome was of doing the check. Thus the robot eventually brings back all packages it failed to deliver. Of course, the original sender may reject the package, too, thereby asking the robot to try to deliver it again.

Figure 7.4 sketches the initial part of a computation tree for the mailbot program which corresponds to the situation depicted in Figure 7.2. The particular tree processes the observation

$$[3],\ [2],\ [\,],\ [\,],\ [False],\ [3],\ \ldots$$

That is to say, in room 1 the package to room 3 is introduced first, followed by the one for room 2. After the attempt to deliver the latter package, the robot senses that the mail bag in question is not empty. Hence, the packages has been re-sent to room 1. Subsequently, the mailbot picks up the only package waiting

```
main :- init(Z), main_loop(Z).

main_loop(Z) :-
   poss(deliver(B), Z)      -> execute(deliver(B), Z, Z1),
                               execute(check_bag(B), Z1, Z2),
                               main_loop(Z2) ;
   poss(pickup(B), Z)       -> execute(pickup(B), Z, Z1),
                               main_loop(Z1) ;
   continue_delivery(D, Z) -> execute(go(D), Z, Z1),
                               main_loop(Z1) ;
   true.

continue_delivery(D, Z) :-
   ( knows_val([B],empty(B),Z), knows_val([R1],request(R1,_),Z)
     ;
     knows_val([R1], carries(_,_,R1), Z) ),
   knows_val([R], at(R), Z),
   ( R < R1 -> D = up
            ; D = down ).
```

Figure 7.3: Modified FLUX program for the mailbot.

in room 2, which goes to room 3. Altogether, the computation tree generates the actions

$$Pickup(1),\ Pickup(2),\ Go(Up),\ Deliver(2),\ CheckBag(2),\ Pickup(3),\ \ldots$$

□

7.2 Dynamic Fluents

Nondeterminism is also a useful concept for modeling **dynamic** state properties. These are fluents which are not under the sole control of the agent and which may change at any time without the agent being aware of it. In this respect, a dynamic fluent changing its value is different from the occurrence of an exogenous action (cf. Section 2.4), of which the agent is informed. Agents need to respect the fact that a fluent is dynamic, lest they erroneously project knowledge of this fluent into the future. In a more realistic setting for the office cleaning robot of Chapter 3, for example, people may leave their offices for the day during the cleaning procedure. This makes the fluent $Occupied(x, y)$ a dynamic one as some of the instances may change from true to false at any time and not immediately noticeable by the cleanbot.

Fluents that possibly alter their value while the agent performs its actions can be accounted for by nondeterministic update axioms which express this uncertainty. In principle, each action of a domain would have to be specified as

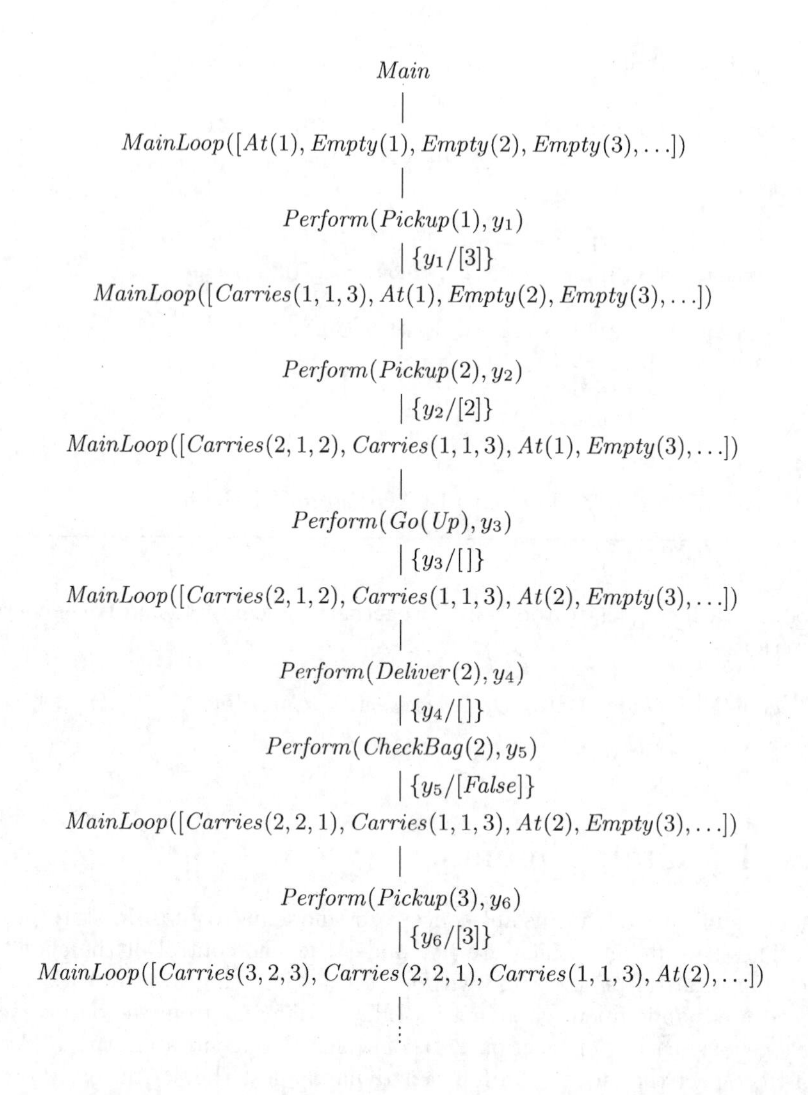

Figure 7.4: The initial segment of a computation tree for the mailbot program with uncertainty.

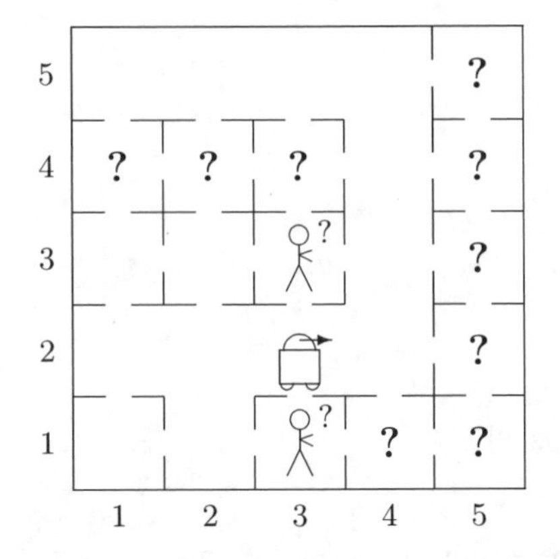

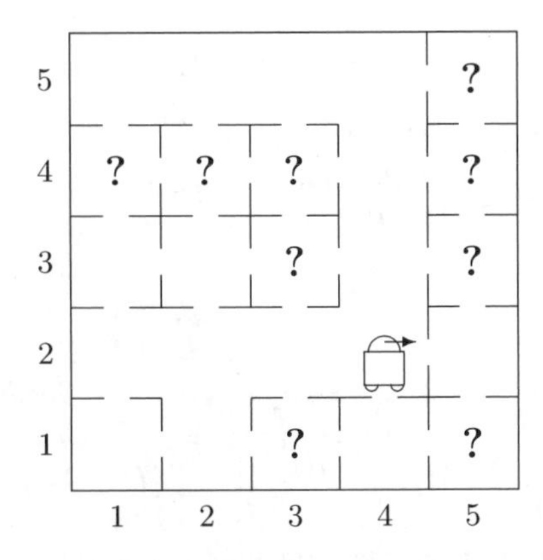

Figure 7.5: Left hand side: Sensing light at $(3,2)$ implies that at least one of the offices in the neighborhood is occupied. Right hand side: After continuing its way to $(4,2)$, the two offices are out of sight and, hence, the partial knowledge of them being in occupation is lost.

---

nondeterministic with regard to all dynamic fluents, unless these are explicitly manipulated by the action. Yet as long as the value of a fluent is unknown to the agent anyway, this uncertainty need not be confirmed by every update. A nondeterministic update is necessary only after the agent looses sight or control of a dynamic fluent of which it has previously acquired information.

Example 2 (continued) Whenever our cleanbot enters a square, it acquires (partial) knowledge of some instances of the dynamic fluent $Occupied(x, y)$. These instances concern the four locations in the immediate vicinity. As soon as the robot goes on to one of these locations, it looses sight of the other ones. Hence, if the cleanbot has learned that some of them may be occupied, then it has to forget this information upon moving on if occupation of offices is not static. As an example, consider the situation depicted in the left hand side of Figure 7.5. The cleanbot has just sensed light at location $(3,2)$, which indicates that at least one of the offices $(3,1)$ or $(3,3)$ is currently occupied. Upon proceeding to square $(4,2)$, the robot has to account for the fact that these two offices may be abandoned now or later. This will be achieved by an update which is nondeterministic regarding the fluents $Occupied(3,1)$ and $Occupied(3,3)$; see the right hand side of Figure 7.5. The cleanbot may then sense no light at both $(4,1)$ and $(4,3)$ later, thereby learning that neither of the two offices mentioned before is occupied any longer. This observation would have been inconsistent wrt. the static model of the office environment employed earlier.

The knowledge update axiom for the *Go* action of the cleanbot can be modified as follows to account for the nondeterministic effect on *Occupied* regarding the offices surrounding the starting point:[1]

$$
\begin{array}{l}
Poss(Go, s) \supset \\
\quad [\, KState(Do(Go, s), z') \equiv (\exists z)\,(KState(s, z) \wedge \\
\qquad (\exists)(Holds(At(x, y), z) \wedge Holds(Facing(d), z) \wedge \\
\qquad\quad Adjacent(x, y, d, x', y') \wedge Surrounding(x, y, d, x_1, y_1, x_2, y_2, x_3, y_3) \wedge \\
\qquad\quad z_1 = z - At(x, y) + At(x', y') \wedge \\
\qquad\quad [z_2 = z_1 \vee z_2 = z_1 - Occupied(x_1, y_1)] \wedge \\
\qquad\quad [z_3 = z_2 \vee z_3 = z_2 - Occupied(x_2, y_2)] \wedge \\
\qquad\quad [z' = z_3 \vee z' = z_3 - Occupied(x_3, y_3)]\,)\,) \wedge \\
\qquad \Pi_{Light}(z') \equiv \Pi_{Light}(Do(Go, s))\,]
\end{array}
$$

Put in words, considered are the three surrounding locations of square (x, y) which are not in the direction d the robot took (the definition of predicate *Surrounding* is given below). Any of these locations may no longer be occupied in the resulting state. If one or more of them are not occupied in the first place, then the removal of the fluent *Occupied* has no effect. Auxiliary predicate $Surrounding(x, y, d, x_1, y_1, x_2, y_2, x_3, y_3)$ can be defined by the following arithmetic formula:

$$
\begin{array}{l}
\ d = 1 \wedge (x_1, y_1) = (x+1, y) \wedge (x_2, y_2) = (x, y-1) \wedge (x_3, y_3) = (x-1, y) \\
\vee\ d = 2 \wedge (x_1, y_1) = (x, y-1) \wedge (x_2, y_2) = (x-1, y) \wedge (x_3, y_3) = (x, y+1) \\
\vee\ d = 3 \wedge (x_1, y_1) = (x-1, y) \wedge (x_2, y_2) = (x, y+1) \wedge (x_3, y_3) = (x+1, y) \\
\vee\ d = 4 \wedge (x_1, y_1) = (x, y+1) \wedge (x_2, y_2) = (x+1, y) \wedge (x_3, y_3) = (x, y-1)
\end{array}
$$

As an example, suppose given this state specification (cf. Figure 7.5, left hand side):

$$
\begin{array}{rl}
KState(S, z) \equiv & Holds(At(3, 2), z) \wedge Holds(Facing(2), z) \\
& \wedge \neg Holds(Occupied(2, 2), z) \wedge \neg Holds(Occupied(4, 2), z) \\
& \wedge [Holds(Occupied(3, 1), z) \vee Holds(Occupied(3, 3), z)] \\
& \wedge Consistent(z)
\end{array}
$$

Since $Surrounding(3, 2, 2, x_1, y_1, x_2, y_2, x_3, y_3)$ is equivalent to $(x_1, y_1) = (3, 1)$, $(x_2, y_2) = (2, 2)$, and $(x_3, y_3) = (3, 3)$, the new update axiom for *Go* entails

$$
\begin{array}{rl}
KState(S', z) \supset & Holds(At(4, 2), z) \wedge Holds(Facing(2), z) \\
& \wedge \neg Holds(Occupied(2, 2), z) \wedge \neg Holds(Occupied(4, 2), z) \\
& \wedge Consistent(z)
\end{array}
$$

where $S' = Do(Go, S)$. The exact knowledge state depends on the sensing outcome of the *Go* action, but in any case nothing follows about whether rooms $(3, 1)$ and $(3, 3)$ are still occupied, because the corresponding fluent may or may not be subtracted from the state.

---

[1] In the following, for the sake of readability we slightly stretch the formal notion of physical effect specifications by splitting the update. The axiom can be brought into the standard form by multiplying out the disjunctions and by combining, in each of the resulting disjuncts, the split update into a single equation.

```
state_update(Z1, go, Z2, [Light]) :-
   holds(at(X,Y), Z1),
   holds(facing(D), Z1),
   adjacent(X, Y, D, Xp, Yp),
   surrounding(X, Y, D, X1, Y1, X2, Y2, X3, Y3),
   update(Z1, [at(Xp,Yp)], [at(X,Y)], Za),
   ( \+ knows_not(occupied(X1,Y1), Za)
        -> cancel(occupied(X1,Y1), Za, Zb) ; Zb=Za ),
   ( \+ knows_not(occupied(X2,Y2), Zb)
        -> cancel(occupied(X2,Y2), Zb, Zc) ; Zc=Zb ),
   ( \+ knows_not(occupied(X3,Y3), Zc)
        -> cancel(occupied(X3,Y3), Zc, Z2) ; Z2=Zc ),
   light(Xp, Yp, Light, Z2).

surrounding(X, Y, D, X1, Y1, X2, Y2, X3, Y3) :-
   [X1,Y1,X2,Y2,X3,Y3] :: 0..6,
   D#=1#/\X1#=X+1#/\Y1#=Y#/\X2#=X#/\Y2#=Y-1#/\X3#=X-1#/\Y3#=Y
   #\/
   D#=2#/\X1#=X#/\Y1#=Y-1#/\X2#=X-1#/\Y2#=Y#/\X3#=X#/\Y3#=Y+1
   #\/
   D#=3#/\X1#=X-1#/\Y1#=Y#/\X2#=X#/\Y2#=Y+1#/\X3#=X+1#/\Y3#=Y
   #\/
   D#=4#/\X1#=X#/\Y1#=Y+1#/\X2#=X+1#/\Y2#=Y#/\X3#=X#/\Y3#=Y-1.
```

Figure 7.6: FLUX encoding of the nondeterministic update axioms for *Go*.

Figure 7.6 shows a FLUX encoding for the nondeterministic knowledge update axiom for *Go*. The nondeterminism is modeled by cancelling knowledge of $Occupied(x_i, y_i)$ for the three squares which surround the origin (x, y) and which are not in direction d. Cancellation is effective only in case the location in question is not known to be empty. This reflects the assumption that people may leave but not enter offices in the evening.

As an example, suppose the robot goes east from $(2, 2)$ and senses light there, as depicted in Figure 7.5 (left hand side). At least one of the two adjacent offices must therefore be occupied:

```
?- holds(at(2,2),Z1), holds(facing(2),Z1),
   not_holds(occupied(2,2),Z1), not_holds(occupied(3,2),Z1),
   not_holds(occupied(4,2),Z1), consistent(Z1),
   state_update(Z1,go,Z2,[true]).

Z2 = [at(3,2),facing(2) | Z]

Constraints:
or_holds([occupied(3,3),occupied(3,1)], Z)
...
```

However, if the cleanbot continues its way along the corridor, then this knowledge is given up, and later there may no longer be light at location $(3, 2)$:

```
?- Z2 = [at(3,2),facing(2) | Z],
   or_holds([occupied(3,3),occupied(3,1)], Z),
   state_update(Z2,go,Z3,[false]),
   state_update(Z3,turn,Z4,[]), state_update(Z4,turn,Z5,[]),
   state_update(Z5,go,Z6,[false]).

Z6 = [at(3,2),facing(4) | Z]

Constraints:
not_holds(occupied(3,3), Z)
not_holds(occupied(3,1), Z)
...
```

The control strategy for the cleanbot of Chapter 5 can be applied under the modified scenario, too. If people leave offices during the program run and the robot notices this eventually, then it has the chance to enter and clean these rooms, too. On the other hand, it may happen that these changes occur towards the end of the program run, where all choice points have already been considered and the cleanbot is on its way home. With a minor elaboration of the control program, the robot can be ordered to take the opportunity in this case, too (see Exercise 7.7). □

7.3 Bibliographical Notes

The systematic approach of [Sandewall, 1994] includes a formal definition of nondeterministic actions, as do some dialects of the so-called action description language, e.g., [Kartha and Lifschitz, 1994; Thielscher, 1994; Baral, 1995]. A classical solution to the frame problem in the presence of uncertain effects has been to exempt certain fluents from the frame assumption [Sandewall, 1994]. This approach has been adopted in event calculus [Shanahan, 1997]. In situation calculus with successor state axioms [Reiter, 1991], nondeterministic actions have been modeled indirectly by way of auxiliary predicates which may vary freely between situations [Boutilier and Friedmann, 1995; Lin, 1996]. Another approach has been to model uncertain effects by a set of alternative deterministic actions among which the agent cannot choose [Bacchus *et al.*, 1999]. Nondeterministic state update axioms in fluent calculus have been introduced in [Thielscher, 2000a].

7.4 Exercises

7.1. Let $State(S_0)$ be the state

$$At(5) \circ Carries(1,2,5) \circ Empty(2) \circ Empty(3) \circ Request(5,2)$$

Let $S_1 = Do(Deliver(1), S_0)$ and $S_2 = Do(Pickup(1), S_1)$. Use fluent calculus and the axiomatization of the nondeterministic mail delivery domain to prove that

$$Holds(Carries(1,5,2), S_1) \vee Holds(Carries(1,5,2), S_2)$$

7.2. Specify the effect of tossing a coin by a nondeterministic state update axiom. Give a sound FLUX encoding of this update axiom.

Hint: To verify the FLUX encoding, pose the query whether the agent knows heads or knows tails after executing the action.

7.3. Prove that the FLUX encoding of $Deliver(b)$ in Figure 7.1 is sound (in the sense of Definition 5.14, page 126) wrt. knowledge update axiom (7.1).

7.4. Show that the mailbot program of Figure 7.3 is sound wrt. the axiomatization of the nondeterministic mail delivery domain.

7.5. Adapt the control program for the mailbot with the improved strategy of Figure 2.8 (page 46) so as to be applicable to the nondeterministic case.

7.6. Extend the FLUX program for the mailbot of Section 7.1 by having the robot drop undeliverable packages in an additional storage room 7.

7.7. Modify the control program for the cleanbot so that it will not let the opportunity pass to clean a room which it notices to be free.

Hint: The problem with the original strategy arises when the robot backtracks to a location where it has sensed light earlier and therefore discarded the choice point.

7.8. Modify the specification of the office cleaning domain so that people may freely come and go. Encode the specification in FLUX and test it with the cleanbot program.

Hint: The formalization of the update should not imply that the robot looses its knowledge that locations in the corridor or outside of the boundaries cannot be occupied.

7.9. Modify the specification of the Wumpus World of Exercise 4.8 and 5.10 so as to model movements of the Wumpus. Assume that the Wumpus may wander to any of the adjacent squares any time the agent performs a *Go* action itself (unless, of course, the Wumpus is dead). The agent hears a scream if the Wumpus has entered a square with a pit and died from this false step. The agent hears a roar upon entering a square simultaneously with the Wumpus. Write a FLUX program for an agent in the modified world. Perceiving a roar, the agent should leave the current square immediately.

Chapter 8

Imprecision*

Imagine a robot whose task is to search a room for specific objects and to collect and carry them to specific locations. The objects are cylindrical and each has a unique color. The robot is equipped with a simple gripper by which it can lift, hold, and put down the parts. A camera and laser range finder help the robot locating itself and identifying and tracking down the parts it should collect. Figure 8.1 depicts an example scenario, in which the robot has to find four parts and deliver them to two different destinations.

What makes the task of the "collectbot" intricate is the fact that the parts can be at any of uncountably many locations on a continuous scale while the sensors, although reasonably accurate, cannot provide perfectly precise information about these positions. Generally, sensing of non-discrete values in the real world is always subject to a certain imprecision so that our robot will never know exactly where the objects are. Moreover, when the robot moves towards a specific location in the room, it will never end up at precisely the intended position due to its necessarily imprecise effectors.

Fortunately, mathematical precision is not required for robots to collect parts. It always suffices if the robot moves close enough to an object so that the gripper can securely clasp it. Moreover, the user should consider the task carried out successfully once the collectbot drops the right parts somewhere in the near vicinity of the specified destinations. From a practical point of view, imprecision, if within small boundaries, is perfectly acceptable.

Imprecise sensors require robots to handle sensed data with the appropriate caution. Reasoning robots use their cognitive faculties to logically combine uncertain information collected over time, thereby increasing the accuracy of their estimations. Imprecise effectors require robots to make cautious predictions about the effects of their physical actions. Here, the ability to reason enables a robot to ascertain, for example, that it can move close enough to an object for grasping it. This chapter is devoted to the programming of agents, in particular physical robots, with imprecise sensors and effectors. Specifically, we discuss

8.1. how to model sensors with imprecision;

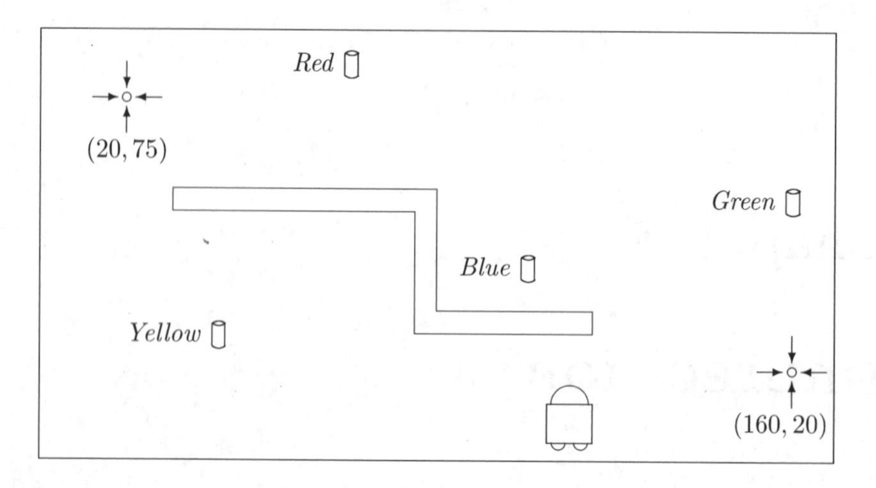

Figure 8.1: An example scenario, where the robot is ordered to bring four objects of different color to two specific locations, which are in the upper left and lower right corner, respectively. Objects may not be visible from all observation points.

8.2. how to model effectors with imprecision;

8.3. how to encode imprecise sensors and effectors in a hybrid variant of FLUX in which real-valued intervals are used to encode imprecise knowledge of the value of a parameter.

8.1 Modeling Imprecise Sensors

Real world sensors may have very different degrees of accuracy. A sonar sensor for measuring the distance to a wall, say, is relatively imprecise and can provide only a rough estimate of the actual distance. An expensive laser range finder can give much more accurate information. Still, of course, it has a limited precision, too, just like any other physical device. Sensor data always have to be interpreted in the context of a suitable **sensor model**, which specifies the accuracy and limitations of a sensor. As programmers, we need to take into account these sensor models when dealing with sensor data on a continuous scale and where small errors may matter.

In Chapter 5, we have introduced the concept of knowledge update axioms to model sensing of a value for some fluent. Thereby we have assumed that sensing such a value always results in the agent knowing the exact instance. That is to say, for each fluent F for which a value is being sensed, the assumption was that there will be an instance $F(\vec{t})$ which afterwards is true in all possible states. We therefore need to generalize the notion of a knowledge update axiom so as to also allow for imprecise sensing.

Definition 8.1 Let A be an action function in a fluent calculus signature. A **knowledge update axiom for imprecise sensing** for A is a formula

$$\begin{array}{l} Poss(A(\vec{x}), s) \supset \\ \quad (\exists \vec{y})(\forall z')\,(\,KState(Do(A(\vec{x}), s), z') \equiv \\ \qquad\qquad (\exists z)\,(KState(s, z) \wedge \Psi(z', z)) \wedge \Pi(z', Do(A(\vec{x}), s))\,) \end{array}$$

where the physical effect $\Psi(z', z)$ is as before (cf. Definition 5.7, page 112) and the cognitive effect $\Pi(z', Do(A(\vec{x}), s))$ is a formula of the form

$$\begin{array}{l} [\,\Pi_1(z') \equiv \Pi_1(Do(A(\vec{x}), s))\,] \wedge \ldots \wedge [\,\Pi_k(z') \equiv \Pi_k(Do(A(\vec{x}), s))\,] \\ \wedge \\ (\exists \vec{w})\, \Pi_{k+1}(z') \end{array} \tag{8.1}$$

where

- $k \geq 0$;
- for $i = 1, \ldots, k$, $\Pi_i(z')$ is a state formula with free variables among $\vec{x}$;
- $\Pi_{k+1}(z')$ is a state formula with variables among $\vec{w}, \vec{x}, \vec{y}$. □

As in the special case, variables $\vec{y}$ encode the acquired sensor data. In contrast to the special case, the generalized value-sensing component $\Pi_{k+1}(z')$ allows the sensed fluents to actually have values $\vec{w}$ instead of $\vec{y}$. The degree of imprecision is then determined by how much the values $\vec{w}$ may differ from the sensed values $\vec{y}$. It is easy to see that the new definition is a proper generalization of the previous one: Precise sensing is obtained as a special case where $(\exists \vec{w})\, \Pi_{k+1}(z')$ is the conjunction $Holds(F_1(\vec{t_1}), z') \wedge \ldots \wedge Holds(F_l(\vec{t_l}), z') \wedge \Pi(\vec{x}, \vec{y})$.

Example 5 For the collectbot domain we use a special sort COLOR to identify the colored parts to be collected. The location of the collectbot and the positions of the colored parts shall be modeled by two fluents with real-valued arguments:

$$\begin{array}{rll} At: & \mathbb{R} \times \mathbb{R} \mapsto \text{FLUENT} & At(x, y) \mathrel{\hat{=}} \text{robot is at } (x, y) \\ Loc: & \text{COLOR} \times \mathbb{R} \times \mathbb{R} \mapsto \text{FLUENT} & Loc(c, x, y) \mathrel{\hat{=}} \text{part of color } c \text{ is at } (x, y) \end{array}$$

The following three domain constraints stipulate that the robot must be somewhere unique and that each individual part cannot be at more than one location in the same situation:

$$\begin{array}{l} (\exists x, y)\, Holds(At(x, y), s) \\ Holds(At(x_1, y_1), s) \wedge Holds(At(x_2, y_2), s) \supset x_1 = x_2 \wedge y_1 = y_2 \\ Holds(Loc(c, x_1, y_1), s) \wedge Holds(Loc(c, x_2, y_2), s) \supset x_1 = x_2 \wedge y_1 = y_2 \end{array}$$

We assume that the collectbot can perform these two high-level sensing actions:

$$\begin{array}{rll} SelfLocate: & \text{ACTION} & SelfLocate \mathrel{\hat{=}} \text{sensing own position} \\ Locate: & \text{COLOR} \mapsto \text{ACTION} & Locate(c) \mathrel{\hat{=}} \text{locating the part of color } c \end{array}$$

Both actions are always possible:

$$\begin{aligned} Poss(SelfLocate, s) &\equiv \top \\ Poss(Locate(c), s) &\equiv \top \end{aligned}$$

Due to the imprecision of the laser range finder, the robot's actual position may deviate slightly from what is being sensed. Suppose the sensor model implies that both the x- and y-coordinate can be measured with an accuracy of ± 1.5 centimeters, then the following knowledge update axiom for imprecise sensing is an appropriate specification of the effect of self location:

$$\begin{aligned} &Poss(SelfLocate, s) \supset \\ &\quad (\exists x, y)(\forall z')(KState(Do(SelfLocate, s), z') \equiv KState(s, z') \wedge \Pi(z')) \end{aligned} \tag{8.2}$$

where $\Pi(z')$ stands for the formula

$$(\exists x', y')\,(Holds(At(x', y'), z') \wedge |x' - x| \leq 1.5 \wedge |y' - y| \leq 1.5)$$

Put in words, upon performing a *SelfLocate* action, the robot senses values x, y for its location and then considers possible only those states in which $At(x', y')$ holds for some x', y' such that $|x' - x| \leq 1.5$ and $|y' - y| \leq 1.5$. The knowledge update axiom also says that the sensing result can deviate from the robot's actual position in situation $Do(SelfLocate, s)$ by at most the same error. Suppose, for example, that a *SelfLocate* action performed in situation S_0 results in the sensed data $x = 109.77$ and $y = 11.02$, then the robot knows it is somewhere inside of a square with center $(109.77, 11.02)$ and a width of three centimeters: The instance $\{s/S_0, x/109.77, y/11.02\}$ of axiom (8.2) implies

$$\begin{aligned} &KState(Do(SelfLocate, S_0), z') \supset \\ &\quad (\exists x', y')\,(\,Holds(At(x', y'), z') \wedge \\ &\qquad\qquad 108.27 \leq x' \leq 111.27 \wedge 9.52 \leq y' \leq 12.52\,) \end{aligned} \tag{8.3}$$

Next, we consider how the collectbot tracks down objects. Provided the view of a colored part is unobstructed from the current position, the sensors return an estimation for the pose of the object, that is, values for direction and distance relative to the robot. Figure 8.2 shows how to use these data in combination with the knowledge of the robot's own position to calculate the actual location of the part. Again, however, sensing is imprecise. Suppose the sensor model implies that the orientation is measured with an accuracy of ± 0.01 (where angles are expressed as values between 0 and $2 \cdot \pi$). The distance d shall be measurable with an accuracy of $\pm 0.02 \cdot d$, that is, the farther away the object the less precise the measurement. Moreover, the sensing apparatus of the robot may not be able to locate a part at all, e.g., if the view is obstructed. If so, the sensors return the value 0.0 for the distance. Put together, the effect of *SenseLoc* on the knowledge of the collectbot is specified by the following knowledge update axiom for imprecise sensing:

$$\begin{aligned} &Poss(Locate(c), s) \supset \\ &\quad (\exists d, o)(\forall z')(KState(Do(Locate(c), s), z') \equiv KState(s, z') \wedge \Pi(z')) \end{aligned} \tag{8.4}$$

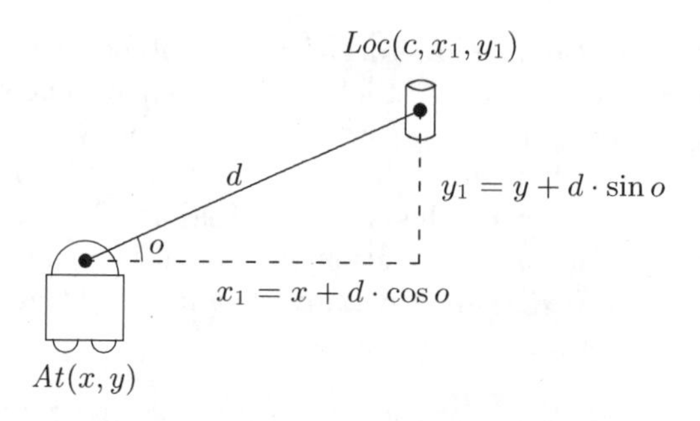

Figure 8.2: How to calculate the location (x_1, y_1) of an object c given its pose, that is, the orientation o and the distance d relative to the position (x, y) of the robot.

---

where $\Pi(z')$ stands for the formula

$$
\begin{array}{l}
(\exists d', o', x, y, x_1, y_1)\,(\,\mathit{Holds}(\mathit{At}(x,y), z') \wedge \mathit{Holds}(\mathit{Loc}(c, x_1, y_1), z') \wedge \\
\qquad |o' - o| \leq 0.01 \wedge |d' - d| \leq 0.02 \cdot d \wedge \\
\qquad x_1 = x + d' \cdot \cos o' \wedge y_1 = y + d' \cdot \sin o' \\
\qquad \vee\ d = 0.0\,)
\end{array}
$$

Suppose, for example, after having located itself the robot sees the green part at a distance $d = 68.2$ in direction $o = 0.7922$. Since $d \neq 0.0$, the instance $\{s/S_1, c/\mathit{Green}, d/68.2, o/0.7922\}$ of axiom (8.4) implies, along with the knowledge of the robot's own location given by axiom (8.3),

$$
\begin{array}{l}
\mathit{KState}(\mathit{Do}(\mathit{Locate}(\mathit{Green}), S_1), z') \supset (\exists d', o', x, y, x_1, y_1) \\
\quad (\,\mathit{Holds}(\mathit{At}(x,y), z') \wedge \mathit{Holds}(\mathit{Loc}(\mathit{Green}, x_1, y_1), z') \wedge \\
\quad 108.27 \leq x \leq 111.27 \ \wedge\ 9.52 \leq y \leq 12.52 \wedge \\
\quad 66.836 \leq d' \leq 69.564 \ \wedge\ 0.7822 \leq o' \leq 0.8022 \wedge \\
\quad x_1 = x + d' \cdot \cos o' \ \wedge\ y_1 = y + d' \cdot \sin o'\,)
\end{array}
\tag{8.5}
$$

where $S_1 = \mathit{Do}(\mathit{SelfLocate}, S_0)$. The given inequations imply boundary conditions for the position of the green part: From $o' \leq 0.8022$, $d' \geq 66.836$, $x \geq 108.27$, and the equation for x_1 it follows that $x_1 \geq 154.73$. On the other hand, from $o' \geq 0.7822$, $d' \leq 69.564$, and $x \leq 111.27$ it follows that $x_1 \leq 160.62$. In a similar fashion, the equation for y_1 and the boundaries for the parameters imply that $56.63 \leq y_1 \leq 62.53$. Hence, the robot has a pretty good idea of where to find the green part in situation $S_2 = \mathit{Do}(\mathit{Locate}(\mathit{Green}), S_1)$:

$$
\begin{array}{l}
\mathit{KState}(S_2, z') \supset \\
\quad (\exists x_1, y_1)\,(\,\mathit{Holds}(\mathit{Loc}(\mathit{Green}, x_1, y_1), z') \wedge \\
\qquad\qquad 154.73 \leq x_1 \leq 160.62 \ \wedge\ 56.63 \leq y_1 \leq 62.53\,)
\end{array}
\tag{8.6}
$$

□

Knowledge update axioms allow to logically combine sensor data acquired over time. Agents can thus enhance the accuracy of their knowledge either by using different sensors to measure the same parameter, or by repeatedly applying the same sensing method but under different conditions, like tracking down an object from different positions. This so-called **fusion** of sensor data can only increase the accuracy of a value estimation. Sensor fusion is made possible by a general property of knowledge update axioms, namely, that agents never loose knowledge by applying a pure sensing action:

Proposition 8.2 *Consider a knowledge update axiom for imprecise sensing with formula* $\Psi(z', z)$ *in Definition 8.1 being* $z' = z$. *For any knowledge expression* ϕ,

$$Poss(A(\vec{x}), s) \models Knows(\phi, s) \supset Knows(\phi, Do(A(\vec{x}), s))$$

Proof: By definition, $Knows(\phi, s)$ implies $KState(s, z) \supset HOLDS(\phi, z)$. From the knowledge update axiom and the assumption that $\Psi(z', z)$ is $z' = z$ it follows that $KState(Do(A(\vec{x}), s), z') \supset (\exists z)\,(KState(s, z) \wedge z' = z)$, which implies $KState(Do(A(\vec{x}), s), z') \supset HOLDS(\phi, z')$; hence, $Knows(\phi, Do(A(\vec{x}), s))$ by definition. ■

Another property of knowledge update axioms, which is relevant for sensor fusion, is that no matter in which order different means of sensing are applied, the resulting knowledge states will be the same. This order-independence is just as easy to verify (see Exercise 8.2).

Example 5 (continued) Suppose, for the sake of argument, that the collectbot had additional means to measure the distance of an object to the left wall.[1] The following update axiom for $SenseWallDist$: COLOR $\mapsto$ ACTION specifies that the returned value may be imprecise up to ± 1.8 centimeters, provided the view of the object is unobstructed:

$$\begin{array}{l} Poss(SenseWallDist(c), s) \supset \\ \quad (\exists x)(\forall z')(KState(Do(SenseWallDist(c), s), z') \equiv \\ \qquad\qquad KState(s, z') \wedge \Pi(z')) \end{array} \tag{8.7}$$

where $\Pi(z')$ stands for the formula

$$(\exists x_1, y_1)\,(\,Holds(Loc(c, x_1, y_1), z') \wedge |x_1 - x| \leq 1.8 \;\vee\; x = 0.0\,)$$

The new action in combination with $SenseLoc(c)$ may help a robot to get a better estimate of the location of an object than would be possible with each action alone. Recall, for instance, the first measurement of the position of the green part, (8.5). Let $S_2 = Do(Locate(Green), S_1)$, and suppose that $SenseWallDist(Green)$ is possible in S_2 and results in the sensed value $x = 161.4$. The instance $\{s/S_2, c/Green, x/161.4\}$ of axiom (8.7) implies that

[1] This may be possible, say, because the floor exhibits a regular vertical pattern.

$Holds(Loc(Green, x_1, y_1), z')$ such that $159.6 \leq x_1 \leq 163.2$ in all possible states z' of situation $Do(SenseWallDist(Green), S_2)$.

But there is more that can be inferred. The additional measurement of the x-coordinate for the green part even allows to derive a better estimation for the y-coordinate! This is so because from $x_1 \geq 159.6$ and the given equations and inequations in (8.5) it follows that $d' \geq 68.13$, where d' is the distance of the part wrt. the location of the robot. In turn, this implies that $y_1 \geq 57.54$, which is a new and stronger lower bound for the y-coordinate of the green object. This shows the power of sensor fusion: Reasoning robots are able to infer additional information by logically combining sensor data that have been collected over time. □

Agents normally do not know the exact value of a fluent with noisy parameters. The standard predicate $KnowsVal(\vec{x}, f(\vec{x}), s)$ is therefore rarely true, even if the agent knows the value quite accurately. The usual way of verifying sufficient knowledge of an uncertain fluent is to use $Knows((\exists \vec{x})\,(f(\vec{x}) \wedge \phi(\vec{x})), s)$, where ϕ defines the desired property of the parameters. For example, without knowing its exact position after locating itself, our collectbot does know for sure that it is less than 20 centimeters away from the southern wall of the room: Axiom (8.3) entails

$$Knows((\exists x, y)\,(Holds(At(x, y) \wedge y < 20.0)), Do(SelfLocate, S_0))$$

Such knowledge statements, where a noisy parameter is verified to satisfy a given condition, are typical for agent programs which process imprecise data.

8.2 Modeling Imprecise Effectors

Acting with perfect accuracy in the real world is as impossible as sensing with absolute precision. When real robots move towards a specific point in space, then depending on the quality of the physical apparatus they will end up more or less close to the goal location. This natural uncertainty need not hinder a robot from successfully completing its task, e.g., moving near an object and picking it up.

The notion of nondeterminism introduced in Chapter 7 can be readily used to model imprecise effectors. In so doing, imprecision is reflected by the uncertainty of the update. Let, for example, action $A(\vec{x})$ be the attempt to achieve value $\vec{x}$ for a fluent F. Suppose the action has an **imprecise effect**, then this can be specified by the following nondeterministic update from state z to z':

$$(\exists \vec{v}, \vec{y})\,(Holds(F(\vec{y}), z) \wedge \Delta(\vec{v}, \vec{x}) \wedge z' = z - F(\vec{y}) + F(\vec{v})\,)$$

Here, formula $\Delta(\vec{v}, \vec{x})$ specifies the degree of imprecision by which the actual resulting instance $F(\vec{v})$ differs from the intended instance $F(\vec{x})$. Unless Δ fully determines $\vec{v}$ from $\vec{x}$, this update axiom is nondeterministic because the successor state cannot be predicted even for ground states z.

Example 5 (continued) Two further fluents in the collectbot domain indicate which part the robot is currently carrying, if any, and which parts are being requested to specific locations:

$$\begin{array}{rl} \mathit{Carries}: & \text{COLOR} \mapsto \text{FLUENT} \\ & \mathit{Carries}(c) \mathrel{\hat{=}} \text{robot carries part of color } c \\ \mathit{Request}: & \text{COLOR} \times \mathbb{R} \times \mathbb{R} \mapsto \text{FLUENT} \\ & \mathit{Request}(c,x,y) \mathrel{\hat{=}} \text{part } c \text{ requested at } (x,y) \end{array}$$

Besides the two sensing actions introduced in the previous section, the robot can perform three actions with actual physical effects, namely, moving to a specific location and picking up and dropping the colored parts:

$$\begin{array}{rll} \mathit{Go}: & \mathbb{R} \times \mathbb{R} \mapsto \text{ACTION} & \mathit{Go}(x,y) \mathrel{\hat{=}} \text{move to position } (x,y) \\ \mathit{Pickup}: & \text{COLOR} \mapsto \text{ACTION} & \mathit{Pickup}(c) \mathrel{\hat{=}} \text{pick up part of color } c \\ \mathit{Drop}: & \text{ACTION} & \mathit{Drop} \mathrel{\hat{=}} \text{drop part held in the gripper} \end{array}$$

The intended result of action $Go(x,y)$ is to have the center of the gripper of the robot end up above point (x,y). For the sake of simplicity, we assume that moving to a specific location is always possible:[2]

$$\mathit{Poss}(\mathit{Go}(x,y),s) \equiv \top$$

The action is, however, imprecise, as the actually reached position may deviate from the intended location. With the following nondeterministic knowledge update axiom we assume that the low-level control is good enough for the collectbot to reach any position with an accuracy of ± 2 centimeters in both the x- and y-coordinate.

$$\begin{array}{l} \mathit{Poss}(\mathit{Go}(x,y),s) \supset \\ \quad [\mathit{KState}(\mathit{Do}(\mathit{Go}(x,y),s),z') \equiv (\exists z)\,(\mathit{KState}(s,z) \land \\ \qquad (\exists x_0,y_0,x',y')\,(\,\mathit{Holds}(\mathit{At}(x_0,y_0),z) \land \\ \qquad\qquad |x'-x| \leq 2.0 \land |y'-y| \leq 2.0 \land \\ \qquad\qquad z' = z - \mathit{At}(x_0,y_0) + \mathit{At}(x',y')\,)\,)] \end{array} \tag{8.8}$$

Put in words, rather than always being able to reach the exact goal position (x,y), the robot ends up at some location (x',y') in the proximity of (x,y). As an example, recall the knowledge of our collectbot after self location and locating the green part (cf. axiom (8.5)). Based on the derived boundary for the location of the green object (cf. axiom (8.6)), the robot decides to move to position $(157.68, 59.58)$, which is the median of the area inside of which the green object is known to stand. Knowledge update axiom (8.8) then implies

$$\begin{array}{l} \mathit{KState}(S_3,z') \supset \\ \quad (\exists x',y',x_1,y_1)\,(\,\mathit{Holds}(\mathit{At}(x',y'),z') \land \\ \qquad \mathit{Holds}(\mathit{Loc}(\mathit{Green},x_1,y_1),z') \land \\ \qquad 155.68 \leq x' \leq 159.68 \land 57.58 \leq y' \leq 61.58 \land \\ \qquad 154.73 \leq x_1 \leq 160.62 \land 56.63 \leq y_1 \leq 62.53\,) \end{array} \tag{8.9}$$

[2] Of course, the robot is supposed to not bump into the colored objects or any other obstacle. We assume that the low-level control takes care of this and that any position can be safely reached within the error range specified by the update axiom for *Go*.

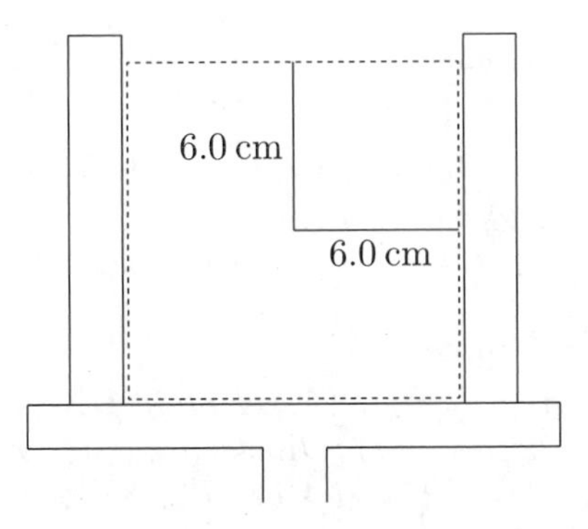

Figure 8.3: Any object within the marked square can be grasped by lowering and closing the gripper.

where $S_3 = Do(Go(157.68, 59.58), S_2)$.

In order for the collectbot to grasp a colored part, the gripper needs to be empty and the object in question needs to be within reach. Suppose the latter holds whenever the part is inside of a 6×6cm square around the center of the gripper, as shown in Figure 8.3. The precondition for picking up an object can then be formalized as follows:

$$\begin{array}{l} Poss(Pickup(c), s) \equiv (\forall c')\,\neg Holds(Carries(c'), s) \wedge \\ \quad (\exists x, y, x_1, y_1)\,(\,Holds(At(x, y), s) \wedge Holds(Loc(c, x_1, y_1), s) \wedge \\ \qquad |x_1 - x| < 6.0 \wedge |y_1 - y| < 6.0\,) \end{array} \tag{8.10}$$

The knowledge of our robot in the situation after it moved to the green part suffices to ensure that this object can be picked up: The boundaries in axiom (8.9) for the location of the robot on the one hand and of the object on the other hand imply that the object is within a distance of $\Delta x \leq 4.95$ and $\Delta y \leq 4.95$ from the center of the gripper. Hence it follows that

$$\begin{array}{l} Knows((\exists x, y, x_1, y_1)\,(\,At(x, y) \wedge Loc(Green, x_1, y_1) \wedge \\ \qquad |x_1 - x| < 6.0 \wedge |y_1 - y| < 6.0\,),\ S_3\,) \end{array}$$

In a similar fashion, we allow the robot to drop a part only if the center of the gripper is less than 10 centimeters away from the requested destination:

$$\begin{array}{l} Poss(Drop, s) \equiv (\exists c, x, y, x_1, y_1) \\ \quad (\,Holds(Carries(c), s) \wedge Holds(At(x, y), s) \wedge \\ \quad Holds(Request(c, x_1, y_1), s) \wedge \sqrt{(x_1 - x)^2 + (y_1 - y)^2} < 10.0\,) \end{array} \tag{8.11}$$

The update axioms for picking up and dropping parts complete the axiomatization of the collectbot domain:

$$
\begin{array}{l}
Poss(Pickup(c), s) \supset \\
\quad [KState(Do(Pickup(c), s), z') \equiv (\exists z)\,(KState(s, z) \wedge \\
\qquad (\exists x, y)\,(\,Holds(Loc(c, x, y), z) \wedge \\
\qquad\qquad z' = z - Loc(c, x, y) + Carries(c)\,)\,)\,] \\
\\
Poss(Drop, s) \supset \\
\quad [KState(Do(Drop, s), z') \equiv (\exists z)\,(KState(s, z) \wedge \\
\qquad (\exists c, x, y)\,(\,Carries(c), z) \wedge Holds(Request(c, x, y), z) \wedge \\
\qquad\qquad z' = z - Carries(c) \circ Request(c, x, y)\,)\,)\,]
\end{array}
\tag{8.12}
$$

□

8.3 Hybrid FLUX

Programming agents with imprecise sensors and effectors requires to use a hybrid extension of FLUX in which fluents can have both discrete and real-valued arguments. With the help of the standard Eclipse constraint system *IC* (for: *interval constraints*), a noisy parameter is encoded by a real-valued **interval**. This interval represents the incomplete knowledge the agent has regarding the actual value. An extensive interval indicates a high degree of uncertainty while a narrow interval means that the agent has fairly accurate knowledge of the parameter.

The IC-constraint language allows to formulate constraints over the reals using standard mathematical functions, including the trigonometric ones, along with equality, inequality, and ordering predicates. By combining constraints, the solver is able to infer new boundaries for the variables. Here is a simple example:

```
?- reals([X]), X*X >= 2.5+X, X*X =< 7-X.

X = X{2.1583123807918478 .. 2.2003835163916667}
```

The standard IC-predicate $Reals(\vec{x})$ defines the variables in list $\vec{x}$ to be of sort $\mathbb{R}$. The example shows that the system can solve non-linear constraints. We need such a powerful solver for handling constraints which define trigonometric relations.

Using the FD- and IC-solver simultaneously, hybrid FLUX allows to combine discrete and continuous arguments in fluents. Real-valued parameters are identified with the help of the library predicate $IsIcVar(x)$ and can thus be distinguished from finite-domain arguments (which are identified by $IsDomain(x)$ as before). The Eclipse IC-system also provides the Boolean operators *And* and *Or*, which allow for logical combinations of both FD- and IC-constraints. These additions are needed to extend the constraint kernel of FLUX to hybrid domains as shown in Figure 8.4. The semantics of the defined predicates is as in discrete FLUX (cf. Figure 4.2 and 4.3 on pages 80 and 83, respectively). For

```
:- lib(fd).
:- lib(ic).

:- import (::)/2,(#=)/2,(#\=)/2,(#>)/2,(#<)/2,(#>=)/2,(#=<)/2,
          indomain/1 from fd.
:- import (=:=)/2,(=\=)/2,(>=)/2,(=<)/2 from ic.

or_neq(_,[],[],(0=\=0)).
or_neq(Q,[X|X1],[Y|Y1],D) :-
  or_neq(Q,X1,Y1,D1),
  ( Q=forall, var(X), \+ is_domain(X), \+ is_ic_var(X) ->
       ( binding(X,X1,Y1,YE) ->
            ( is_ic_var(Y) -> D = (abs(Y-YE)>=1e-8 or D1) ;
              real(Y)      -> D = (abs(Y-YE)>=1e-8 or D1) ;
                              D = (fd:(Y#\=YE) or D1) )
        ; D=D1 ) ;
    is_ic_var(Y) -> D = (abs(X-Y)>=1e-8 or D1) ;
    real(Y)      -> D = (abs(X-Y)>=1e-8 or D1) ;
                    D = (fd:(X#\=Y) or D1) ).

and_eq([],[],(0=:=0)).
and_eq([X|X1],[Y|Y1],D) :-
   and_eq(X1,Y1,D1), ( is_ic_var(Y) -> D = ((X=:=Y) and D1) ;
                       real(Y)      -> D = ((X=:=Y) and D1) ;
                                       D = (fd:(X#=Y) and D1) ).

or_and_eq([],(0=\=0)).
or_and_eq([eq(X,Y)|Eq],(D1 or D2)) :-
   and_eq(X, Y, D1), or_and_eq(Eq, D2).
```

Figure 8.4: Extended constraint kernel of hybrid FLUX.

pragmatic reasons, two real-valued arguments are considered unequal only if they differ by at least 10^{-8}.

As an example for the added expressiveness of the constraint solver, let the colors *Yellow*, *Red*, *Green*, *Blue* be encoded, respectively, by the natural numbers 1–4. The following query is proved unsatisfiabile:

```
?- Z = [loc(2,40.0,80.0)],
   reals([X,Y]), or_holds([loc(C,_,_),loc(_,X,Y)],Z),
   C :: [1,4],
   sqrt(X*X+Y*Y) =< 50.0.

no (more) solution.
```

Here, fluent $Loc(c, \_, \_)$ is derived to not hold in z by the FD-solver using the fact that $c \neq Red$ (because c is either *Yellow* (1) or *Blue* (4)) while fluent $Loc(\_, x, y)$ is derived to not hold in z by the IC-solver using the fact that $\sqrt{x^2 + y^2} \leq 50 < \sqrt{40^2 + 80^2}$.

When a real-valued interval is used to encode imprecise information, the agent will hardly ever know the exact value of that parameter. This implies that the standard FLUX predicate *KnowsVal* usually fails when applied to such an argument:

```
?- reals([X,Y]), 108.27=<X, X=<111.27, 9.52=<Y, Y=<12.52,
   Z=[at(X,Y)|_], knows_val([Y],at(_,Y),Z)

no (more) solution.
```

Fortunately, agents never really need to know exact values of noisy parameters. It usually suffices to check whether the parameter is known to satisfy certain conditions. As always in FLUX, conditions are proved to be known in the indirect way, that is, by ensuring that the negation of a condition is unsatisfiabile. In the example just mentioned, for instance, the robot may not know its exact position but it does know, say, that it is less than 20 centimeters away from the southern wall of the room. This is easily verified thus:

```
?- reals([X,Y]), 108.27=<X, X=<111.27, 9.52=<Y, Y=<12.52,
   Z=[at(X,Y)|_],
   reals([V]), holds(at(_,V),Z), \+ V >= 20.

yes.
```

This coding principle for conditions on noisy arguments is used in the precondition axioms for the actions of the collectbot as depicted in Figure 8.5.

Knowledge update axioms for imprecise sensing are encoded using real-valued constraints to restrict the sensed parameters so as to comply with the sensing result. The first two clauses in Figure 8.6 encode knowledge update axioms (8.2) and (8.4) for self location and locating an object, respectively.

```
poss(pickup(C),Z) :-
   knows_not(carries(_),Z),
   reals([X,X1,Y,Y1]), holds(at(X,Y),Z), holds(loc(C,X1,Y1),Z),
   \+ (abs(X1-X) >= 6.0 ; abs(Y1-Y) >= 6.0).

poss(drop,Z) :-
   knows_val([C],carries(C),Z),
   reals([X,X1,Y,Y1]),
   holds(at(X,Y),Z), holds(request(C,X1,Y1),Z),
   \+ sqrt((X1-X)*(X1-X)+(Y1-Y)*(Y1-Y)) >= 10.0.
```

Figure 8.5: FLUX encoding of precondition axioms (8.10) and (8.11) in the collectbot domain.

As in the fluent calculus axiomatization, trigonometric functions constrain the coordinates of a sensed object according to the measured pose.

Nondeterministic update axioms specifying imprecise effectors are encoded in a similar fashion. Real-valued constraints determine new boundaries for the updated parameters of affected fluents. The third axiom in Figure 8.6 describes the effect of the collectbot going to a specific location in the room. The last two update axioms in the program for the collectbot, which are for picking up and dropping parts, are straightforward encodings of the respective fluent calculus axioms (8.12).

Based on hybrid FLUX and in particular the IC-constraint solver, the encoded domain axioms enable the collectbot to interpret imprecise sensing results in the context of its current world model. Suppose, for example, the robot initially only knows that its gripper is empty and that both its own position and the position of the green part must be unique. Suppose further that the low-level sensing module reads values $x = 109.77$ and $y = 11.02$ as an approximation of the current position of the collectbot along with an estimated distance $d = 68.2$ and orientation $o = 0.7922$ regarding the pose of the green part. Similar to our axiomatic derivation in Section 8.1, the respective knowledge update axioms imply boundaries for the actual values:

```
init(Z0) :-
   reals([X,Y,X1,Y1]), Z0=[at(X,Y),loc(3,X1,Y1)|Z],
   not_holds_all(carries(_),Z), not_holds_all(at(_,_),Z),
   not_holds_all(loc(3,_,_),Z).

?- init(Z0),
   state_update(Z0,self_locate,Z0,[109.77,11.02]),
   state_update(Z0,locate(3),Z0,[68.2,0.7922]).

X = X{108.27..111.27}
Y = Y{9.52..12.52}
```

```
state_update(Z,self_locate,Z,[X,Y]) :-
   reals([X1,Y1]), holds(at(X1,Y1),Z),
   abs(X1-X) =< 1.5, abs(Y1-Y) =< 1.5.

state_update(Z,locate(C),Z,[D,O]) :-
   D =\= 0.0 ->
      reals([X,X1,Y,Y1]), holds(at(X,Y),Z), holds(loc(C,X1,Y1),Z),
      abs(O1-O) =< 0.01, abs(D1-D) =< 0.02*D,
      X1=:=X+D1*cos(O1), Y1=:=Y+D1*sin(O1)
   ; true.

state_update(Z1,go(X1,Y1),Z2,[]) :-
   reals([X,X1,Y,Y1]), holds(at(X,Y),Z1),
   abs(X2-X1) =< 2.0, abs(Y2-Y1) =< 2.0,
   update(Z1,[at(X2,Y2)],[at(X,Y)],Z2).

state_update(Z1,pickup(C),Z2,[]) :-
   reals([X,Y]), holds(loc(C,X,Y),Z1),
   update(Z1,[carries(C)],[loc(C,X,Y)],Z2).

state_update(Z1,drop,Z2,[]) :-
   reals([X,Y]), holds(carries(C),Z1), holds(request(C,X,Y),Z1),
   update(Z1,[],[carries(C),request(C,X,Y)],Z2).
```

Figure 8.6: FLUX encoding of update axioms for the collectbot.

```
X1 = X1{154.7294975578213..160.61623932156417}
Y1 = Y1{56.62880158469283..62.528662810311559}
```

Thanks to the constraint solving abilities, the fusion of sensor data from different sources comes for free in hybrid FLUX. As in Section 8.1, suppose, for the sake of argument, that our robot had additional means to estimate the distance of the green part from the left wall. The extra information can then be logically combined with the constraints from above, thus yielding a better approximation of the location of the object. The following computation shows what can be inferred if the additional sensing action reveals that the x-coordinate of the green part lies between 159.6 and 163.2:

```
?- init(Z0),
   state_update(Z0,self_locate,Z0,[109.77,11.02]),
   state_update(Z0,locate(3),Z0,[68.2,0.7922]),
   159.6 =< X1, X1 =< 163.2.

X = X{110.25376067843588..111.27}
Y = Y{9.52..12.52}
X1 = X1{159.6..160.61623932156417}
Y1 = Y1{57.541849989474649..62.528662810311559}
```

Again sensor fusion allows to improve not only the estimation of parameter x_1 but also, indirectly, of y_1. As explained in Section 8.1, this follows from the fact that $159.6 \leq x_1$ constrains further the distance to the object, which in turn implies a stronger lower bound for its y-coordinate.

As an example for calculating the update caused by imprecise effectors, consider the collectbot moving to position $(157.68, 59.58)$ after having performed self location followed by locating the green part. With the help of the specification of the *Go* action, the robot can infer that it ends up in the vicinity of the position it is aiming at:

```
?- init(Z0),
   state_update(Z0,self_locate,Z0,[109.77,11.02]),
   state_update(Z0,locate(3),Z0,[68.2,0.7922]),
   state_update(Z0,go(157.68,59.58),Z1,[]).

Z1 = [at(X2,Y2),loc(3,X1,Y1)|Z]
X2 = X2{155.68..159.68}
Y2 = Y2{57.58..61.58}
X1 = X1{154.7294975578213..160.61623932156417}
Y1 = Y1{56.62880158469283..62.528662810311559}
```

With regard to the preconditions of picking up an object (Figure 8.5), the resulting constraints suffice to conclude that by lowering and closing its gripper the robot can now grab the green part:

```
?- init(Z0),
```

```
init(Z0) :- Z0 = [request(1,20.0,75.0),
                  request(2,160.0,20.0),
                  request(3,160.0,20.0),
                  request(4,20.0,75.0) | Z],
            not_holds_all(request(_,_,_),Z),
            not_holds_all(carries(_),Z),
            consistent(Z0).

consistent(Z) :-
   reals([X,Y]),
   ( holds(at(X,Y),Z,Z1) -> not_holds_all(at(_,_),Z1) ),
   location_unique(Z),
   duplicate_free(Z).

location_unique([F|Z]) <=>
   ( F = loc(C,X,Y) -> not_holds_all(loc(C,_,_),Z) ; true ),
   location_unique(Z).
```

Figure 8.7: Sample initial task specification for the collectbot.

```
    state_update(Z0,self_locate,Z0,[109.77,11.02]),
    state_update(Z0,locate(3),Z0,[68.2,0.7922]),
    state_update(Z0,go(157.68,59.58),Z1,[]),
    poss(pickup(3),Z1).

 yes.
```

Having specified the actions of the collectbot, we are now in a position to write control programs for the robot by which it searches for objects and delivers them to any specified location. An example task specification is given by the first clause of the program shown in Figure 8.7. The robot is requested to bring four parts of different color to two goal locations. The collectbot knows that it starts with the gripper empty. It has no information about its own position nor about that of any of the requested parts. The domain constraints state that both the robot and all objects can only be at one location at a time. Specifically, by the auxiliary constraint *LocationUnique*(z), state z is required to contain only one instance of fluent *Loc*(c, x, y) for any c. This encodes the fact that all objects are of different color.

The main program for the collectbot depicted in Figure 8.8 implements a simple search and delivery strategy. Unless all requests have been carried out, by predicate *LocateAll* the robot first locates itself, followed by trying to spot all objects that still need to be delivered. Next, if the robot happens to be close enough to some part, it picks it up and brings it to the requested destination. Otherwise, the collectbot selects one of the objects and moves to its estimated location. The selection is defined via the clause for predicate *SelectLocation*. Using the standard IC-predicate *Delta*(x, δ), which defines δ to be the length of

```
main :- init(Z), main_loop(Z).

main_loop(Z) :-
   not_holds_all(request(_,_,_),Z) -> true ;
   locate_all(Z),
   ( knows_val([C,X,Y],request(C,X,Y),Z), poss(pickup(C),Z)
        -> execute(pickup(C),Z,Z1), execute(go(X,Y),Z1,Z2),
           execute(drop,Z2,Z3)
     ;
     select_location(X,Y,Z) -> execute(go(X,Y),Z,Z3) ),
   main_loop(Z3).

locate_all(Z) :-
   execute(self_locate,Z,Z),
   findall(C,knows_val([C],request(C,_,_),Z),Cs),
   locate_all(Cs,Z).

locate_all([],_).
locate_all([C|Cs],Z) :-
   execute(locate(C),Z,Z), locate_all(Cs,Z).

select_location(MX,MY,Z) :-
   reals([X,X1,Y,Y1]),
   knows_val([C],request(C,_,_),Z), holds(loc(C,X,Y),Z),
   delta(X,DX), delta(Y,DY),
   \+ ( knows_val([C1],request(C1,_,_),Z), holds(loc(C1,X1,Y1),Z),
        delta(X1,DX1), delta(Y1,DY1), DX1+DY1 =< DX+DY-0.1 ),
   get_ic_bounds(X,X1,X2), MX is X1+(X2-X1)/2,
   get_ic_bounds(Y,Y1,Y2), MY is Y1+(Y2-Y1)/2.
```

Figure 8.8: A FLUX strategy for the collectbot.

the current interval for variable x, the robot selects among the requested objects the one c for whose position (x, y) it has the best approximation. With the help of the IC-predicate $GetIcBounds(x, l, r)$, which defines l to be the left and r the right boundary of the current interval for x, the collectbot infers the median (m_x, m_y) of the estimated location with the intention to move there.

After each *Go* action, the collectbot once more senses the positions of all objects that remain to be delivered. So doing allows it to find parts it has not yet located because they were hidden from the view. Moreover, by sensing the same object from different points of observation, the robot increases the accuracy of its knowledge by sensor fusion.

In the following, we give the trace of an actual run of the program with the initial state as shown in Figure 8.7. The values in brackets are the observed values when executing a sensing action, that is, coordinates in case of self location and distance plus orientation in case of locating a colored part. A distance of 0.0 means that the robot failed to locate the respective object:

SelfLocate	[109.77, 11.02]
Locate(1)	[77.84, 2.879]
Locate(2)	[0.0, 0.0]
Locate(3)	[68.2, 0.792]
Locate(4)	[0.0, 0.0]
Go(157.68, 59.58)	
SelfLocate	[160.92, 59.97]
Locate(1)	[0.0, 0.0]
Locate(2)	[122.30, 2.983]
Locate(3)	[1.33, 3.799]
Locate(4)	[58.61, 3.414]
Pickup(3)	
Go(160.0, 20.0)	
Drop	
SelfLocate	[160.8, 19.48]
⋮	
SelfLocate	[159.18, 18.34]
Locate(1)	[0.0, 0.0]
Go(35.77, 29.90)	
SelfLocate	[37.91, 28.21]
Locate(1)	[2.51, 2.609]
Pickup(1)	
Go(20.0, 75.0)	
Drop	

Notice that the robot has remembered from the very first round of sensing the whereabouts of the yellow cylinder (color 1), which later on was hidden from the view until the robot moved next to it.

8.4 Bibliographical Notes

High-level reasoning and planning with noisy actions has received little attention in literature. A special planning algorithm for domains with noisy sensors has been developed in [Kushmerick *et al.*, 1995]. [Bacchus *et al.*, 1999] have presented an extension of situation calculus where noisy effects have been modeled as the nondeterministic selection among actions with deterministic effects. The use of nondeterministic update axioms to model noisy data in fluent calculus and FLUX has been introduced in [Thielscher, 2001b].

8.5 Exercises

8.1. The farther away an object, the less precise is the estimation of its pose when our collectbot locates it, so that it is sometimes necessary to iterate the process of spotting an object and moving towards its estimated position. For the following, assume that the robot knows it starts at location $(0, 0)$.

   (a) Suppose that a colored part c is at a distance of 40 centimeters (and visible). Show that after locating c and going to the median of the position estimation, the robot knows it can pick up the part. (Hint: Prove that after locating c, both its x- and y-coordinate are known with an accuracy better than ± 4.0 centimeters, and show that this implies the claim.)

   (b) Suppose that a colored part c is 100 centimeters away (and visible). Show that after locating c and going to the median of the position estimation, the robot does not know that it can pick up the part. Show also that the robot knows it can pick up the part after locating c again (assuming it is still visible) and going to the median of the new position estimation.

8.2. Consider a collection of pure sensing actions $\alpha_1, \ldots, \alpha_n$. Formulate and prove a theorem that says that, as far as the resulting knowledge state is concerned, the order is irrelevant in which these actions are applied.

8.3. Extend the fluent calculus axiomatization of the collectbot domain by a fluent $Time(t)$ denoting the actual time that has elapsed since the initial situation.

   (a) The time it takes for the collectbot to execute an action is uncertain. Suppose that a *Pickup* action takes between 8 and 10 seconds, a *Drop* action requires 2 to 3 seconds, and each *Go* action is performed with an average speed of 20–21 centimeters per second. Extend the update axioms accordingly.

   (b) Consider the situation

$$
\begin{aligned}
&Do(Drop, Do(Go(160, 20),\\
&\quad Do(Pickup(Green), Do(Go(157.68, 59.58),\\
&\qquad Do(Locate(Green), Do(SelfLocate, S_0))))))
\end{aligned}
$$

   Use the update axioms to infer both a lower and an upper bound for the time it takes for the robot to execute these actions. Suppose, to this end, that the sensing actions result in the same values as in Section 8.1.

8.4. Add a strategy to the FLUX program of Figure 8.8 which has the collectbot systematically search the room for the parts it fails to spot in the main loop of the program of Figure 8.8.

8.5. Extend the FLUX specification of the collectbot domain by time following Exercise 8.3. Modify the control program to have the robot deliver the parts in such a way as to minimize the execution time.

8.6. This exercise is concerned with lifting the restriction that the parts in the collectbot domain are uniquely colored. The main problem here is to decide after seeing twice a green cylinder, say, whether there are really two such objects or whether the same object has been viewed from different angles.

(a) Under the assumption that two parts need to be at least 5 centimeters apart, modify the axiomatization of the collectbot domain so as to allow for several parts having the same color.

(b) Modify and extend the FLUX program for the collectbot to account for the possibility that objects have the same color and that the robot is requested to collect a specific number of parts of the same color.

Chapter 9

Indirect Effects: Ramification Problem*

Imagine a robotic agent whose task is to control the water pumps and output of a steam boiler. The pumps can be on or off and switched to either low or high capacity. Some of the individual water pump controllers are pairwise connected so that if one pump is turned off, the other one changes automatically to high capacity. The boiler is also equipped with a valve that can be opened to let off a small quantity of steam in order to relieve the pressure. The valve opens automatically whenever a running pump is changed to high capacity. Figure 9.1 gives a schematic illustration of such a boiler with four water pumps. The task is to program an agent to control the settings of the boiler according to requested quantities of steam output. Whenever the demand changes, the agent needs to adjust the current setting of the water pumps and the valve so as to increase or reduce the output of the boiler.

As environment of an agent, the steam boiler exhibits the interesting property of being semi-automatic in a certain sense. In environments of this kind, actions sometimes invoke processes which may then automatically cause a variety of **indirect** or delayed changes. Turning off a pump of the steam boiler, for instance, may trigger a whole chain of indirect effects according to the specification: The adjacent pump, if running low, automatically changes to high capacity. In turn, this may cause the valve to open. Moreover, every one of these changes has a specific effect on the water level in the boiler, which in turn affects the quantity of steam that is produced. An agent controlling the steam boiler must take into account all of these side effects when inferring the result of its actions.

The fundamental challenge raised by indirect effects is that they question the justification for solving the fundamental frame problem (cf. Chapter 1). This problem requires theories of actions to account for the ubiquitous inertia in our world. Physical objects and their relations have the tendency to be stable unless affected by some external force. Solutions to the frame problem therefore

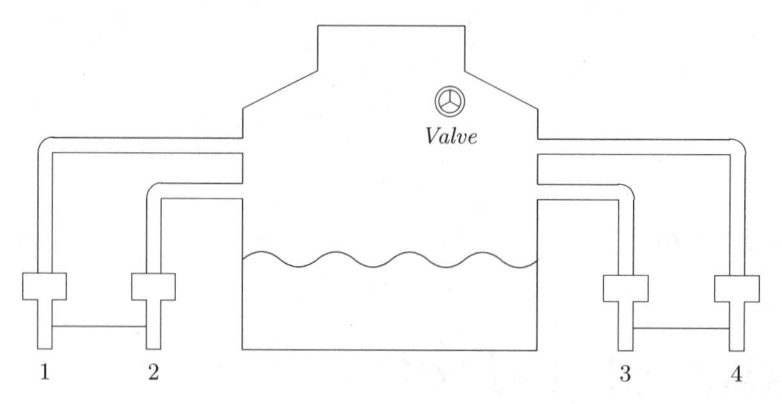

Figure 9.1: A steam boiler with four water pumps. Running low, pumps 1 and 2 have a capacity of 40 liters per second with an uncertainty of $\pm 2\,l/\,\mathrm{sec}$. The lower capacity of pumps 3 and 4 is $100 \pm 3\,l/\,\mathrm{sec}$. When running with high capacity, each pump provides an additional $25\,l/\,\mathrm{sec}$. The controllers of pumps 1 and 2 are paired together, as are those of pumps 3 and 4, to automatically change to high capacity one water pump when the other pump is turned off. If the steam boiler is provided with L liters of water per second, then it produces a quantity of $L - 20\,l/\,\mathrm{sec}$ steam. Of this, the valve, if open, lets off 10 liters per second.

---

define successor states as identical to the preceding state modulo the explicit changes effected by an action. The possibility of additional, Implicit effects threatens the applicability of this definition to semi-automatic systems. This is known as the **ramification problem**, so named because it is concerned with the possible ramifications of actions.

In fluent calculus, the ramification problem manifests in the state update equations which are being used to solve the frame problem. An equation of the form $State(Do(a, s)) = State(s) - \vartheta^- + \vartheta^+$ does not allow for any changes besides the direct effects ϑ^- and ϑ^+ of an action a in situation s. Hence, this simple definition of state update is not suitable as a model for environments where the direct effects of an action may trigger further, indirect effects.

Superficially, it seems that the ramification problem can be circumvented simply by adding to the effect terms ϑ^- and ϑ^+ all indirect effects of an action. However, reasonably complex semi-automatic systems consist of many connections among their components. The way these components take influence on each other can often be encoded by a specification of about the same size as the system itself since the connections tend to be local. To be more precise, if each of n components depends directly on at most k other components such that k is much smaller than n, then the size of a specification of these dependencies is approximately linear in n. The attempt to specify all indirect effects as direct ones, however, requires to turn the local dependencies into global ones, which usually leads to specifications that are much larger. Specifically, in

Figure 9.2: An example for cascades of indirect effects.

order to describe all global dependencies among n components, n functions need to be specified each of which has about n arguments. This leads to specifications which are at least quadratic in size.

There is also a more principled objection against the attempt to simply specify away the ramification problem. Indirect effects may have a recursive structure, in which case an action can have an unbounded number of effects. As an illustrative example, picture a masterfully composed pattern of erected dominoes (see Figure 9.2). Pushing one of them will cause all dominoes next to it to fall over as well, which in turn causes further dominoes to fall, and so on. In a conventional specification, in which all effects are given directly, one would have to set a limit to the number of dominoes that are eventually affected by the action of pushing one. No such boundary will be necessary in a recursive specification using the notion of indirect effects.

Delayed effects are yet another phenomenon of semi-automatic environments which is difficult to capture with standard update axioms. These effects occur in systems in which internal processes may be initiated, which then run automatically and eventually trigger additional effects. This often takes time, so that an agent can possibly intervene with a further action. Under such circumstances, the occurrence of an effect depends on the absence of a later action.

In this chapter, we will extend our formal framwork and programming system to be able to model

9.1. the causal relationships of a semi-dynamic system;

9.2. the ramifications determined by the causal relationships;

9.3. causality and ramification in FLUX.

9.1 Causal Relationships

The ramifications of an action in a semi-automatic environment are due to causal connections among the components. For example, if pump 1 of our steam boiler is running low, then turning off pump 2 **causes** pump 1 to change to high capacity as indirect effect. To provide means for specifying causal relations among the components of a dynamic system, the signature of fluent calculus is extended by a new standard predicate.

Definition 9.1 A tuple $\mathcal{S} \cup \langle Causes \rangle$ is a **fluent calculus signature for ramifications** if $\mathcal{S}$ is a fluent calculus signature for knowledge and

$$Causes : \text{STATE}^6 \times \text{SIT}$$

□

In general, the occurrence of an indirect effect depends on the current state, the effects that have already occurred, and the current situation. This explains the structure of the causality predicate: An instance $Causes(z, p, n, z', p', n', s)$ means that in situation s, the occurred positive and negative effects p and n, respectively, possibly cause an automatic update from state z to state z' with positive and negative effects p' and n', respectively. A causal relationship thus relates a state-effects triple $\langle z, p, n \rangle$ to another state-effects triple $\langle z', p', n' \rangle$ with respect to a situation s. The arguments p, n represent a kind of **momentum** which is responsible for the occurrence of indirect effects.

Example 6 Before we start formalizing the steam boiler, let us consider a very simple example of an indirect effect. Suppose that a light bulb is controlled by two switches as depicted in Figure 9.3. Whenever one of the switches is closed while the other one happens to be closed already, the light bulb turns on. Conversely, if one of the switches is opened, the light goes off. The states in this scenario shall be described by the two fluents

$$\begin{array}{rll} Closed\colon & \{1,2\} \mapsto \text{FLUENT} & Closed(x) \mathrel{\hat{=}} \text{switch } x \text{ is closed} \\ LightOn\colon & \text{FLUENT} & LightOn \mathrel{\hat{=}} \text{light is on} \end{array}$$

The only action to consider is

$$Alter\colon \ \{1,2\} \mapsto \text{ACTION} \quad Alter(x) \mathrel{\hat{=}} \text{toggle switch } x$$

Our intention is to model any change of the status of the light bulb as indirect effect of the action. Whether light is on or off obviously depends on the state of the switches as expressed by the following domain constraints:

$$\begin{array}{l} Holds(Closed(1), s) \wedge Holds(Closed(2), s) \supset Holds(LightOn, s) \\ \neg Holds(Closed(1), s) \supset \neg Holds(LightOn, s) \\ \neg Holds(Closed(2), s) \supset \neg Holds(LightOn, s) \end{array} \tag{9.1}$$

At first glance, these constraints seem to convey exactly the information needed to derive the effect it has on the light if the position of a switch is changed. In fact, however, as static laws these axioms do not contain sufficient causal

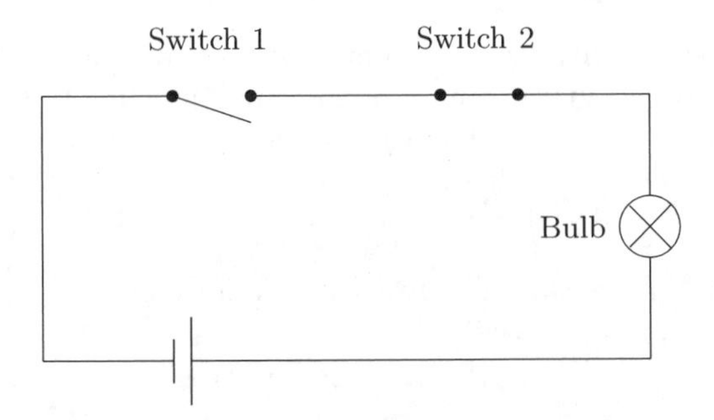

Figure 9.3: An electric circuit consisting of two switches and a light bulb, which is on if and only if both switches are closed.

---

information. To see why, consider the following formulas, which are logically equivalent to (9.1):

$$\begin{array}{l} Holds(Closed(1), s) \wedge \neg Holds(LightOn, s) \supset \neg Holds(Closed(2), s) \\ Holds(LightOn, s) \supset Holds(Closed(1), s) \\ Holds(LightOn, s) \supset Holds(Closed(2), s) \end{array}$$

Mistaking, say, the first logical implication as a causal rule would mean to consider $\neg Closed(2)$ an indirect effect of the direct effect $Closed(1)$ whenever light is off—which obviously does not correspond to the actual behavior of the circuit.

An adequate representation of what happens when switch 1 gets closed with switch 2 being closed is given by this instance of the new *Causes* predicate:

$$\begin{array}{lll} Causes(Closed(1) \circ Closed(2) & , Closed(1) & , \emptyset, \\ \quad Closed(1) \circ Closed(2) \circ LightOn & , Closed(1) \circ LightOn & , \emptyset, s) \end{array} \tag{9.2}$$

Put in words, in the intermediate state $Closed(1) \circ Closed(2)$ reached by positive effect $Closed(1)$ an automatic transition occurs, which results in the state $Closed(1) \circ Closed(2) \circ LightOn$, where the additional positive effect *LightOn* has been effected. Likewise,

$$\begin{array}{lll} Causes(Closed(1) \circ Closed(2) & , Closed(2) & , \emptyset, \\ \quad Closed(1) \circ Closed(2) \circ LightOn & , Closed(2) \circ LightOn & , \emptyset, s) \end{array} \tag{9.3}$$

Conversely, if either switch gets opened while the other one is still closed, then light goes off as indirect effect:

$$\begin{array}{lll} Causes(Closed(2) \circ LightOn & , \emptyset, & Closed(1), \\ \quad Closed(2) & , \emptyset, & Closed(1) \circ LightOn, s) \\ Causes(Closed(1) \circ LightOn & , \emptyset, & Closed(2), \\ \quad Closed(1) & , \emptyset, & Closed(2) \circ LightOn, s) \end{array} \tag{9.4}$$

□

Causal transitions like (9.2)–(9.4) are instances of general causal dependencies that characterize the dynamics of the environment. With the help of the causality predicate, these relations are formalized as a new kind of domain axiom. Each such axiom describes a particular causal law that holds among some components of the underlying system.

Definition 9.2 A **causal relationship** is a formula

$$(\forall)(\Gamma \supset Causes(z, p, n, z', p', n', s))$$

where Γ is a first-order formula without predicate *Causes* and in which no situation occurs besides s. □

Example 6 (continued) The following causal relationships characterize the causal behavior of our electric ciurcuit:[1]

$$\begin{array}{l} Holds(Closed(x), p) \wedge Holds(Closed(y), z) \wedge x \neq y \wedge \neg Holds(LightOn, z) \\ \quad \supset Causes(z, p, n, z \circ LightOn, p \circ LightOn, n, s) \\ \\ Holds(Closed(x), n) \wedge Holds(LightOn, z) \\ \quad \supset Causes(z, p, n, z - LightOn, p, n \circ LightOn, s) \end{array} \tag{9.5}$$

It is easy to verify that transitions (9.2)–(9.4) are logical consequences of these axioms under the foundational axioms of fluent calculus. □

Strictly speaking, our simple circuit would not require to represent a point in the transition phase by triples $\langle z, p, n \rangle$, because the causal behavior can actually be determined from the intermediate state z alone. But consider the example of a dynamic system depicted in Figure 9.4. The intermediate states in (b) and (e), taken as snapshots, are identical and yet the states that follow, viz. (c) and (f), are quite distinct. This difference can be captured if the momentum is taken into account: The state in (b) has been reached by the effect of the leftmost ball joining the line whereas (e) has been reached by the effect that the rightmost ball came down. Formally, these two situations can be distinguished by defining causal relationships $Causes(z, p, n, z', p', n', s)$ with identical state argument z but with different updated states z'.

Example 7 Recall the steam boiler of Figure 9.1. A state of this system shall be represented using these six fluents:

PumpOn :	$\mathbb{N} \mapsto$ FLUENT	this pump is running
PumpHigh :	$\mathbb{N} \mapsto$ FLUENT	this pump is set to high capacity
ValveOpen :	FLUENT	the valve is open
WaterLevel :	$\mathbb{N} \mapsto$ FLUENT	current water level in liters
SteamOutput :	$\mathbb{N} \mapsto$ FLUENT	current steam output in l/sec
RequestedOutput :	$\mathbb{N} \mapsto$ FLUENT	requested steam output in l/sec

[1] For the sake of simplicity, we use a compact notation where an equation that holds between two arguments of *Causes* is compiled into one of the arguments.

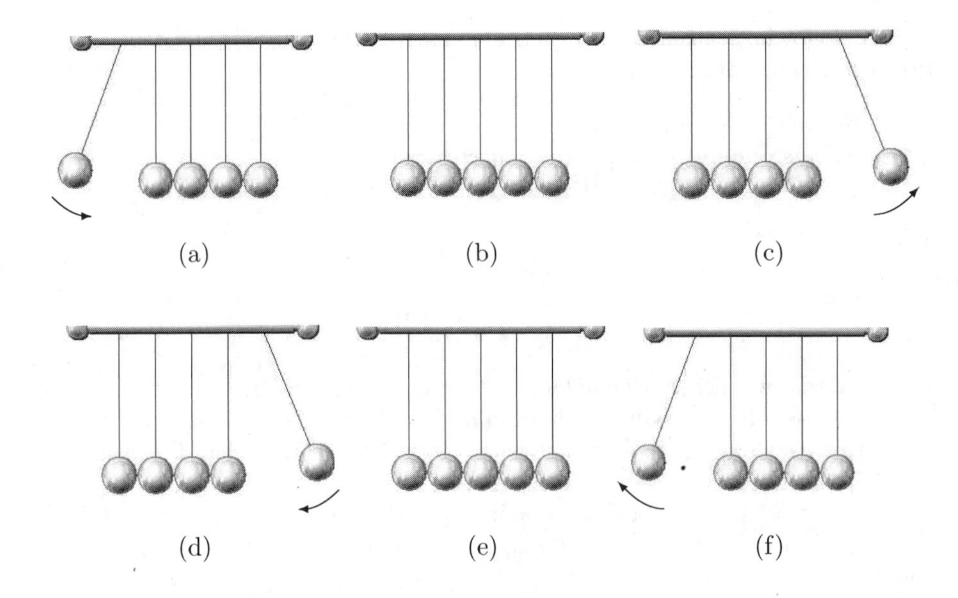

Figure 9.4: Swinging balls: An example of a dynamic system where the momentum matters.

Consider, for instance, a snapshot of the steam boiler where pump 1 is active, all pumps are switched to low capacity, the valve is closed, the water level is $107\,l$, the current steam output is $87\,l/\text{sec}$, and no other fluent holds:

$$Z = \mathit{PumpOn}(1) \circ \mathit{WaterLevel}(107) \circ \mathit{SteamOutput}(87) \tag{9.6}$$

Suppose that this state has just been reached by switching off pump 2. This moment is modeled by the state-effects triple

$$\langle Z, \emptyset, \mathit{PumpOn}(2) \rangle \tag{9.7}$$

Here, the second argument is empty because no positive effect has (yet) occurred while the third argument is the (only) negative effect that has brought about state Z. According to the specification of the steam boiler, this momentum causes pump 1 to automatically switch to higher capacity. This indirect effect brings the system from the momentary status (9.7) into the following state-effects triple:

$$\langle Z \circ \mathit{PumpHigh}(1), \mathit{PumpHigh}(1), \mathit{PumpOn}(2) \rangle$$

Thus $\mathit{PumpHigh}(1)$ has been caused as a side-effect of the prior negative effect that $\mathit{PumpOn}(2)$ became false. Suppose that this transition occurs in a particular situation, say S_{17}, then the axiomatic representation of this instance of a cause-effect relation is given by

$$\begin{array}{lll} \mathit{Causes}(Z & , \emptyset & , \mathit{PumpOn}(2), \\ \quad Z \circ \mathit{PumpHigh}(1) & , \mathit{PumpHigh}(1) & , \mathit{PumpOn}(2), S_{17}\,) \end{array} \tag{9.8}$$

The following causal relationship axiomatizes the causal dependencies between connected pumps:

$$\begin{aligned}&Holds(PumpOn(y), n) \wedge \\ &\neg Holds(PumpHigh(x), z) \wedge Connected(x, y) \\ &\quad \supset Causes(z, p, n, z + PumpHigh(x), p + PumpHigh(x), n, s)\end{aligned} \tag{9.9}$$

where

$$\begin{aligned}&Connected(x, y) \stackrel{\text{def}}{=} \\ &\quad x = 1 \wedge y = 2 \vee x = 2 \wedge y = 1 \vee x = 3 \wedge y = 4 \vee x = 4 \wedge y = 3\end{aligned}$$

Put in words, whenever $PumpOn(y)$ is a negative effect, that is, pump y has been switched off, and pump x is running low, then this causes the (positive) effect that x changes to high capacity. It is easy to see that the example transition in (9.8) follows from this general causal relationship.

The next causal relationship formalizes the fact that if an active pump x is switched to high with the valve being closed, then as a consequence the valve opens:

$$\begin{aligned}&Holds(PumpHigh(x), p) \wedge \\ &Holds(PumpOn(x), z) \wedge \neg Holds(ValveOpen, z) \\ &\quad \supset Causes(z, p, n, z + ValveOpen, p + ValveOpen, n, s)\end{aligned} \tag{9.10}$$

Recall, for example, state-effects triple $\langle Z', PumpHigh(1), PumpOn(2)\rangle$ with $Z' = Z \circ PumpHigh(1)$ which occurs as the updated snapshot in atom (9.8). According to the following instance of causal relationship (9.10), this triple causes a further transition:

$$\begin{aligned}Causes(&Z' \quad\quad\quad\quad\quad\;\;, PumpHigh(1) \quad\quad\quad\quad\quad\;\;, PumpOn(2), \\ &Z' \circ ValveOpen, PumpHigh(1) \circ ValveOpen, PumpOn(2), S_{17})\end{aligned}$$

The specification of the indirect change of the water level is trickier because the level may be simultaneously affected by multiple changes. We therefore introduce the additional effect fluent

$$LevelChange : \mathbb{N} \times \mathbb{Z} \mapsto \text{FLUENT}$$

An instance $LevelChange(x, d)$ arises in consequence of some change in the settings of pump x, which causes the water level to change by d liters per second. The new fluent does not describe a fact about the state of the steam boiler. It is merely a property of the momentum. As such, the fluent *LevelChange* arises only in the effect components of state-effects triples. The effect it has on the water level to change a pump can thus be specified by these two causal relationships:

$$\begin{aligned}&Holds(PumpOn(x), p) \wedge (\forall d')\, \neg Holds(LevelChange(x, d'), p) \wedge \\ &Capacity(x, d, z) \\ &\quad \supset Causes(z, p, n, z, p + LevelChange(x, d), n, s)\end{aligned} \tag{9.11}$$

$$\begin{array}{l} Holds(PumpOn(x), n) \wedge (\forall d') \neg Holds(LevelChange(x, d'), p) \wedge \\ Capacity(x, d, z) \\ \quad \supset Causes(z, p, n, z, p + LevelChange(x, -d), n, s) \end{array} \quad (9.12)$$

where

$$\begin{array}{l} Capacity(x, d, z) \stackrel{\text{def}}{=} \\ \quad \neg Holds(PumpHigh(x), z) \wedge [x = 1 \vee x = 2] \wedge 38 \leq d \leq 42 \vee \\ \quad \neg Holds(PumpHigh(x), z) \wedge [x = 3 \vee x = 4] \wedge 97 \leq d \leq 103 \vee \\ \quad Holds(PumpHigh(x), z) \wedge [x = 1 \vee x = 2] \wedge 63 \leq d \leq 67 \vee \\ \quad Holds(PumpHigh(x), z) \wedge [x = 3 \vee x = 4] \wedge 122 \leq d \leq 128 \end{array}$$

Put in words, if a pump is switched on (or off, respectively), then this causes a specific increase (or decrease, respectively) of the water level, depending on whether the pump is running high or low. The inequations in the definition of *Capacity* reflect the uncertainty of the output of the pumps. As a consequence, there are several instances of a causal relationship (9.11) or (9.12) for the same state-effects triple $\langle z, p, n \rangle$. Hence, several transitions are possible. This is similar to the existence of multiple instances of an update equation in a nondeterministic state update axiom. Recall, for example, the status of the steam boiler with state Z as in (9.6) and the only effect that $PumpOn(2)$ became false. Causal relationship (9.12) implies that the following atom holds for all d such that $38 \leq d \leq 42$:

$$\begin{array}{llll} Causes(Z\,, & \emptyset & , PumpOn(2), & \\ & Z\,, & WaterLevel(2, d)\,, PumpOn(2), & S_{17}\,) \end{array}$$

The water level is also affected by a capacity change of an active pump. If a pump switches from low to high (or from high to low, respectively), then this causes an increase (a decrease, respectively) of the water level by $25\,l$:

$$\begin{array}{l} Holds(PumpHigh(x), p) \wedge Holds(PumpOn(x), z) \wedge \\ \neg Holds(LevelChange(x, 25), p) \\ \quad \supset Causes(z, p, n, z, p + LevelChange(x, 25), n, s) \end{array} \quad (9.13)$$

$$\begin{array}{l} Holds(PumpHigh(x), n) \wedge Holds(PumpOn(x), z) \wedge \\ \neg Holds(LevelChange(x, -25), p) \\ \quad \supset Causes(z, p, n, z, p - LevelChange(x, -25), n, s) \end{array} \quad (9.14)$$

The effect of an instance of $LevelChange(x, d)$ on the water level needs to be axiomatized in such a way that the level actually changes by d due to pump x only if it has not changed already due to this cause. Otherwise, the change would apply repeatedly, causing the water level to rise or fall forever. Therefore, the momentum $LevelChange(x, d)$ is specified as a negative effect of the respective change of the water level:

$$\begin{array}{l} Holds(LevelChange(x, d), p) \wedge \neg Holds(LevelChange(x, d), n) \wedge \\ Holds(WaterLevel(l_1), z) \wedge l_2 = l_1 + d \\ \quad \supset Causes(z, p, n,\; z - WaterLevel(l_1) + WaterLevel(l_2), \\ \qquad\qquad\qquad\qquad\; p - WaterLevel(l_1) + WaterLevel(l_2), \\ \qquad\qquad\qquad\qquad\; n + LevelChange(x, d), s) \end{array} \quad (9.15)$$

Put in words, if $LevelChange(x, d)$ occurs as a positive effect but not yet as a negative effect, then the water level is adjusted accordingly.

Each change in the water level in turn causes a change of the steam output, as axiomatized by the following causal relationship:

$$\begin{array}{l} Holds(WaterLevel(l), p) \wedge \\ [\, Holds(ValveOpen, z) \supset o_2 = l - 30\,] \wedge \\ [\, \neg Holds(ValveOpen, z) \supset o_2 = l - 20\,] \wedge \\ Holds(SteamOutput(o_1), z) \wedge o_1 \neq o_2 \\ \quad \supset Causes(z, p, n,\ z - SteamOutput(o_1) + SteamOutput(o_2), \\ \qquad\qquad\qquad p - SteamOutput(o_1) + SteamOutput(o_2), n, s) \end{array} \tag{9.16}$$

A new steam level may also be caused by changing the status of the valve. This indirect effect is accounted for by the final two causal relationships of the steam boiler domain:

$$\begin{array}{l} Holds(ValveOpen, p) \wedge \\ Holds(WaterLevel(l), z) \wedge o_2 = l - 30 \wedge \\ Holds(SteamOutput(o_1), z) \wedge o_1 \neq o_2 \\ \quad \supset Causes(z, p, n,\ z - SteamOutput(o_1) + SteamOutput(o_2), \\ \qquad\qquad\qquad p - SteamOutput(o_1) + SteamOutput(o_2), n, s) \end{array} \tag{9.17}$$

$$\begin{array}{l} Holds(ValveOpen, n) \wedge \\ Holds(WaterLevel(l), z) \wedge o_2 = l - 20 \wedge \\ Holds(SteamOutput(o_1), z) \wedge o_1 \neq o_2 \\ \quad \supset Causes(z, p, n,\ z - SteamOutput(o_1) + SteamOutput(o_2), \\ \qquad\qquad\qquad p - SteamOutput(o_1) + SteamOutput(o_2), n, s) \end{array} \tag{9.18}$$

To illustrate the causal relationships of the steam boiler at work, Figure 9.5 depicts one possible chain of indirectly affected changes starting with the single direct effect of switching off pump 2. For the same situation, Figure 9.6 shows another way of concatenating state-effects triples using the underlying causal relationships. In this chain, the steam output is adjusted several times, which, however, has no influence on the state that is finally obtained. A further difference is that the decrease of the water level due to the change in pump 2 is higher than in Figure 9.5, which leads to a slightly different water level and steam output in the end. This is a consequence of causal relationship (9.12) not being fully deterministic. □

9.2 Inferring Ramifications of Actions

The causal relationships of a domain provide the basis for inferring the ramifications of an action. The valid instances of *Causes* can be viewed as inducing a situation-dependent relation on state-effects triples. As such, the relation can be depicted as a graph in which each edge represents the occurrence of an indirect effect which transforms one system snapshot into another one; see Figure 9.7.

$z = PumpOn(1) \circ WaterLevel(107) \circ SteamOutput(87),\ p = \emptyset,\ n = PumpOn(2)$

	z	p	n
(9.9)	$+PumpHigh(1)$	$+PumpHigh(1)$	
(9.10)	$+ValveOpen$	$+ValveOpen$	
(9.13)		$+LevelChange(1,25)$	
(9.12)		$+LevelChange(2,-38)$	
(9.15)	$-WaterLevel(107)$ $+WaterLevel(132)$	 $+WaterLevel(132)$	$+LevelChange(1,25)$
(9.15)	$-WaterLevel(132)$ $+WaterLevel(94)$	$-WaterLevel(132)$ $+WaterLevel(94)$	$+LevelChange(2,-38)$
(9.16)	$-SteamOutput(87)$ $+SteamOutput(64)$	 $+SteamOutput(64)$	

Figure 9.5: A chain of indirect effects, starting in state z with the only direct effect that $PumpOn(2)$ became false. Each row shows the changes of the three components of the state-effects triple which are obtained by applying an instance of the causal relationship in the first column.

$z = PumpOn(1) \circ WaterLevel(107) \circ SteamOutput(87),\ p = \emptyset,\ n = PumpOn(2)$

	z	p	n
(9.9)	$+PumpHigh(1)$	$+PumpHigh(1)$	
(9.10)	$+ValveOpen$	$+ValveOpen$	
(9.17)	$-SteamOutput(87)$ $+SteamOutput(77)$	 $+SteamOutput(77)$	
(9.12)		$+LevelChange(2,-41)$	
(9.15)	$-WaterLevel(107)$ $+WaterLevel(66)$	 $+WaterLevel(66)$	$+LevelChange(2,-41)$
(9.16)	$-SteamOutput(77)$ $+SteamOutput(36)$	$-SteamOutput(77)$ $+SteamOutput(36)$	
(9.13)		$+LevelChange(1,25)$	
(9.15)	$-WaterLevel(66)$ $+WaterLevel(91)$	 $+WaterLevel(91)$	$+LevelChange(1,25)$
(9.16)	$-SteamOutput(36)$ $+SteamOutput(61)$	$-SteamOutput(36)$ $+SteamOutput(61)$	

Figure 9.6: A different chain of indirect effects, originating from the same state-effects triple as in Figure 9.5.

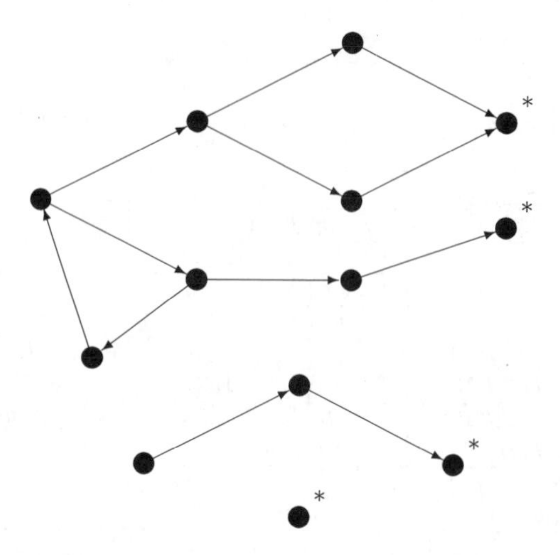

Figure 9.7: The (situation-dependent) relation induced by a set of causal relationships depicted as graph. Each node stands for a particular state-effects triple $\langle z, p, n \rangle$. A directed edge from one such node to another one $\langle z', p', n' \rangle$ indicates a valid instance $Causes(z, p, n, z', p', n', s)$. The nodes marked with $*$ represent states in which no more causal relationships apply.

So-called sinks are nodes in a graph which have no outgoing edges. In the causality graph, this is the case for state-effects triples that are not related to any other triple. Such a triple admits no (further) indirect effect, so that it contains a state in which all ramifications of an action have occurred. Inferring indirect effects thus amounts to finding sinks starting in the node which represents the status of the environment after the direct effect of the respective action. Then each reachable sink corresponds to a possible successor state in which all ramifications are realized. With regard to our example of the steam boiler domain, the ramifications described in Figures 9.5 and 9.6, for instance, determine two out of a total of five possible successor states wrt. the given initial state-effects triple. These states are identical but for a slightly varying water level and steam output due to the implicit uncertainty in causal relationship (9.12).

A sink in the causality graph is a state-effects triples $\langle z, p, n \rangle$ for which $\neg(\exists z', p', n')\, Causes(z, p, n, z', p', n', s)$ holds in the current situation s. It is important to realize, however, that a collection of causal relationships alone does not suffice to entail the property of being a sink. The reason is that causal relationships define positive instances of *Causes* only. It is therefore necessary to consider the following closure axiom of a set of causal relationships.

Definition 9.3 Let $\{\Gamma_i \supset Causes(z, p, n, z', p', n', s)\}_{i \leq n}$ be a finite set of causal relationships ($n \geq 0$). A **causal relation** Σ_{cr} is an axiom

$$Causes(z, p, n, z', p', n', s) \equiv \bigvee_{i=1}^{n} \Gamma_i$$

□

As an example, consider a state-effects triple of the form $\langle z_1, \mathit{ValveOpen}, \emptyset \rangle$, that is, where *ValveOpen* is the only effect which occurred. Of all causal relationships for the steam boiler, only (9.17) applies in this case. Therefore, the causal relation which is determined by the relationships (9.9)–(9.18) of the steam boiler domain, along with the usual unique-name axiom and the foundational axioms, entails

$$\begin{array}{l} p_1 = \mathit{ValveOpen} \wedge n_1 = \emptyset \supset \\ \quad Causes(z_1, p_1, n_1, z_2, p_2, n_2, s) \supset \\ \quad\quad (\exists o)\,(\, z_2 = z_1 - \mathit{SteamOutput}(o) + \mathit{SteamOutput}(o - 10) \wedge \\ \quad\quad\quad p_2 = p_1 + \mathit{SteamOutput}(o - 10) \wedge n_2 = n_1 \,) \end{array}$$

Hence, the only possible indirect effect is that the steam output gets reduced by 10 liters per second. Once this effect is incorporated, a sink has been reached: Since none of the given causal relationships applies, our causal relation for the steam boiler entails

$$\begin{array}{l} (\exists o')\, p_2 = \mathit{ValveOpen} \circ \mathit{SteamOutput}(o') \wedge n_2 = \emptyset \supset \\ \quad \neg(\exists z_3, p_3, n_3)\, Causes(z_2, p_2, n_2, z_3, p_3, n_3, s) \end{array}$$

The formal definition of successor states as sinks requires to axiomatize the transitive closure of a given causal relation. As this cannot be axiomatized in first-order logic (see Exercise 9.1), we use a standard second-order axiom stating that $(\vec{x}, \vec{y}, s)$ belongs to the transitive closure (meaning that $\vec{y}$ is reachable from $\vec{x}$) of a situation-dependent relation R:

$$(\forall P) \left\{ \begin{array}{l} (\forall \vec{u})\, P(\vec{u}, \vec{u}, s) \wedge (\forall \vec{u}, \vec{v}, \vec{w})\,(\, P(\vec{u}, \vec{v}, s) \wedge R(\vec{v}, \vec{w}, s) \supset P(\vec{u}, \vec{w}, s)\,) \\ \qquad\qquad\qquad\qquad \supset P(\vec{x}, \vec{y}, s) \end{array} \right\}$$

This axiom says that *all* relations P which are reflexive and transitive through R must satisfy $P(\vec{x}, \vec{y}, s)$. If this holds, then indeed $(\vec{x}, \vec{y}, s)$ must be in the transitive closure of R. On the other hand, $\vec{y}$ cannot be reached from $\vec{x}$ if there is a relation P which is reflexive and transitive through R but which does not satisfy $P(\vec{x}, \vec{y}, s)$.

In fluent calculus, the second-order characterization of transitive closure is used to define a macro $Ramify(z, p, n, z', s)$. The intended meaning is that z is a state which gets updated by positive and negative effects p and n, respectively, and then leads to a sink z' as a possible successor state for situation s. In other words, z' can be reached via the underlying causal relation from the state-effects triple $\langle z - n + p, p, n \rangle$. Formally,

$$\begin{array}{l} Ramify(z, p, n, z', s) \stackrel{\text{def}}{=} \\ \quad (\exists p', n')\,(\, Causes^*(z - n + p, p, n, z', p', n', s) \wedge \\ \quad\quad \neg(\exists z_2, p_2, n_2)\, Causes(z', p', n', z_2, p_2, n_2, s)\,) \end{array} \qquad (9.19)$$

Here, $Causes^*(z_1, p_1, n_1, z_2, p_2, n_2, s)$ stands for the second-order schema just mentioned with R replaced by *Causes* and $\vec{x}$ and $\vec{y}$ replaced by z_1, p_1, n_1 and z_2, p_2, n_2, respectively.

We do not attempt to give step-wise proofs using the second-order axiom here, but see Exercise 9.1 for a small example. We rather base formal arguments on the fact that this axiom defines the transitive closure of a given causal relation. Recall, for instance, the example shown in Figures 9.5 and 9.6. The relation determined by the causal relationships of the steam boiler domain entail

$$
\begin{array}{l}
z = PumpOn(1) \circ PumpOn(2) \circ WaterLevel(107) \circ SteamOutput(87) \supset \\
\quad [\, Ramify(z, \emptyset, PumpOn(2), z', s) \equiv \\
\qquad (\exists l)\,(\, z' = PumpOn(1) \circ PumpHigh(1) \circ ValveOpen \\
\qquad\qquad\qquad \circ\, WaterLevel(l) \circ SteamOutput(l-30) \\
\qquad\quad \wedge\, 90 \leq l \leq 94\,)\,]
\end{array}
$$

That is to say, switching off pump 2 in the given state z may lead to state z' if and only if z' satisfies that pump 1 is now running high, the valve is open, the water level l is between 90 and 94 liters, and the steam output is $l - 30$ liters per second.

The definition of ramification as the reflexive and transitive closure of a causal relation paves the way for a generalized definition of update axioms which account for indirect effects of actions. Rather than simply being the result of some direct effects, possible successor states must satisfy the ramification relation wrt. the original state modulo the direct effects. Specifically, the concept of knowledge update axioms is generalized to indirect effects as follows.

Definition 9.4 Let A be an action function in a fluent calculus signature. A **knowledge update axiom with ramification** (and a **knowledge update axiom with ramification and imprecise sensing**, respectively) for A is a formula

$$
\begin{array}{l}
Poss(A(\vec{x}), s) \supset \\
\quad (\exists \vec{y})(\forall z')\,(\, KState(Do(A(\vec{x}), s), z') \equiv \\
\qquad\qquad (\exists z)\,(KState(s, z) \wedge \Psi(z', z)) \wedge \Pi(z', Do(A(\vec{x}), s))\,)
\end{array}
\tag{9.20}
$$

where the cognitive effect $\Pi(z', Do(A(\vec{x}), s))$ is as in Definition 5.7 on page 112 (respectively, Definition 8.1 on page 193, for imprecise sensing), and the physical effect $\Psi(z', z)$ is a formula of the form

$$
\begin{array}{l}
(\exists \vec{y}_1)\,(\Delta_1(z) \wedge Ramify(z, \vartheta_1^+, \vartheta_1^-, z', Do(A(\vec{x}), s))\,) \\
\vee \ldots \vee \\
(\exists \vec{y}_n)\,(\Delta_n(z) \wedge Ramify(z, \vartheta_n^+, \vartheta_n^-, z', Do(A(\vec{x}), s))\,)
\end{array}
\tag{9.21}
$$

where

- $n \geq 1$;
- for $i = 1, \ldots, n$,

– $\Delta_i(z)$ is a state formula with free variables among $\vec{x}, \vec{y}_i, z$;

– $\vartheta_i^+, \vartheta_i^-$ are finite states with variables among $\vec{x}, \vec{y}_i$. □

Example 6 (continued) The action of toggling a switch in the electric circuit of Figure 9.3 can be described by the following knowledge update axiom with ramification, tacitly assuming that the action is always possible:

$$\begin{array}{l} KState(Do(Alter(x), s), z') \equiv (\exists z)\,(KState(s, z) \wedge \\ \quad Holds(Closed(x), z) \wedge Ramify(z, \emptyset, Closed(x), z', Do(Alter(x), s)) \\ \quad \vee \\ \quad \neg Holds(Closed(x), z) \wedge Ramify(z, Closed(x), \emptyset, z', Do(Alter(x), s))\,) \end{array} \qquad (9.22)$$

Take, e.g., the state depicted in Figure 9.3, where switch 1 is open while switch 2 is closed (and, of course, light is off):

$$KState(S_0, z) \;\equiv\; z = Closed(2)$$

Consider the action of altering switch 1. According to the knowledge update axiom for $Alter(1)$, the only direct effect in the initial state is $+Closed(1)$. The causal relation determined by the causal relationships for the circuit, axioms (9.5), entail

$$\begin{array}{l} Causes(Closed(2) \circ Closed(1), Closed(1), \emptyset, z', p', n', s) \\ \equiv \\ z' = Closed(1) \circ Closed(2) \circ LightOn \wedge p' = Closed(1) \circ LightOn \wedge n' = \emptyset \end{array}$$

Moreover,

$$\begin{array}{l} Causes(Closed(1) \circ Closed(2) \circ LightOn, Closed(1) \circ LightOn, \emptyset, z', p', n', s) \\ \equiv \bot \end{array}$$

Hence, by the definition of $Ramify$,

$$KState(Do(Alter(1), S_0), z) \equiv z = Closed(1) \circ Closed(2) \circ LightOn$$

Now suppose that switch 2 is opened. According to the knowledge update axiom for $Alter(2)$, the only direct effect in the state in $Do(Alter(1), S_0)$ is $-Closed(2)$. The causal relation determined by the causal relationships for the circuit entail

$$\begin{array}{l} Causes(Closed(1) \circ LightOn, \emptyset, Closed(2), z', p', n', s) \\ \equiv \\ z' = Closed(1) \wedge p' = \emptyset \wedge n' = Closed(2) \circ LightOn \end{array}$$

Moreover,

$$Causes(Closed(1), \emptyset, Closed(2) \circ LightOn, z', p', n', s) \equiv \bot$$

Hence, by the definition of $Ramify$,

$$KState(Do(Alter(2), Do(Alter(1), S_0)), z) \equiv z = Closed(1)$$

□

The following proposition shows that these non-standard update axioms reduce to the standard case under the empty set of causal relationships.

Proposition 9.5 *Let Σ_{cr} be the causal relation determined by the set $\{\}$ of causal relationships according to Definition 9.3, then Σ_{cr} entails that (9.21) is equivalent to*

$$
\begin{array}{l}
(\exists \vec{y}_1)\,(\Delta_1(z) \wedge z' = z - \vartheta_1^- + \vartheta_1^+) \\
\vee \ldots \vee \\
(\exists \vec{y}_n)\,(\Delta_n(z) \wedge z' = z - \vartheta_n^- + \vartheta_n^+)
\end{array}
$$

Proof: From Definition 9.3, $\Sigma_{cr} = \{Causes(z_1, p_1, n_1, z_2, p_2, n_2, s) \equiv \bot\}$; hence, $Causes^*(z_1, p_1, n_1, z_2, p_2, n_2, s) \equiv z_1 = z_2 \wedge p_1 = p_2 \wedge n_1 = n_2$. Then by (9.19),

$$
\begin{array}{l}
Ramify(z, \vartheta_i^+, \vartheta_i^-, z', Do(A(\vec{x}), s)) \\
\equiv (\exists p', n')\,(z' = z - \vartheta_i^- + \vartheta_i^+ \wedge p' = \vartheta_i^+ \wedge n' = \vartheta_i^-)
\end{array}
$$

for each $1 \leq i \leq n$, which proves the claim. ∎

Example 7 (continued) Suppose the control agent for the steam boiler can perform these eight actions:

$TurnOn: \mathbb{N} \mapsto$ ACTION		turn on this pump
$TurnOff: \mathbb{N} \mapsto$ ACTION		turn off this pump
$SetHi: \mathbb{N} \mapsto$ ACTION		switch this pump to high capacity
$SetLo: \mathbb{N} \mapsto$ ACTION		switch this pump to low capacity
$OpenValve:$ ACTION		open the valve
$CloseValve:$ ACTION		close the valve
$CheckSteamOutput:$ ACTION		sense current steam output
$AskRequestedOutput:$ ACTION		query for requested output

The preconditions axioms are as follows:

$$
\begin{array}{rcl}
Poss(TurnOn(x), z) & \equiv & \neg Holds(PumpOn(x), z) \\
Poss(TurnOff(x), z) & \equiv & Holds(PumpOn(x), z) \\
Poss(SetHi(x), z) & \equiv & \neg Holds(PumpHigh(x), z) \\
Poss(SetLo(x), z) & \equiv & Holds(PumpHigh(x), z) \\
Poss(OpenValve, z) & \equiv & \neg Holds(ValveOpen, z) \\
Poss(CloseValve, z) & \equiv & Holds(ValveOpen, z) \\
Poss(CheckSteamOutput, z) & \equiv & \top \\
Poss(AskRequestedOutput, z) & \equiv & \top
\end{array}
$$

Next, we specify the effects of these actions. Having gone through the effort of defining both causal relationships for the steam boiler domain and the notion of ramifications, the update axioms for the non-sensing actions are particularly simple since in each case nothing but a single direct effect occurs:

$$
\begin{array}{l}
Poss(TurnOn(x), s) \supset [KState(Do(TurnOn(x), s), z') \equiv (\exists z)\,(KState(s, z) \wedge \\
\quad Ramify(z, PumpOn(x), \emptyset, z', Do(TurnOn(x), s)))\,] \\
Poss(TurnOff(x), s) \supset [KState(Do(TurnOff(x), s), z') \equiv (\exists z)\,(KState(s, z) \wedge \\
\quad Ramify(s, \emptyset, PumpOn(x), z', Do(TurnOff(x), s)))\,] \\
Poss(SetHi(x), s) \supset [KState(Do(SetHi(x), s), z') \equiv (\exists z)\,(KState(s, z) \wedge \\
\quad Ramify(s, PumpHigh(x), \emptyset, z', Do(SetHi(x), s)))\,] \\
Poss(SetLo(x), s) \supset [KState(Do(SetLo(x), s), z') \equiv (\exists z)\,(KState(s, z) \wedge \\
\quad Ramify(s, \emptyset, PumpHigh(x), z', Do(SetLo(x), s)))\,] \\
Poss(OpenValve, s) \supset [KState(Do(OpenValve, s), z') \equiv (\exists z)\,(KState(s, z) \wedge \\
\quad Ramify(s, ValveOpen, \emptyset, z', Do(OpenValve, s)))\,] \\
Poss(CloseValve, s) \supset [KState(Do(CloseValve, s), z') \equiv (\exists z)\,(KState(s, z) \wedge \\
\quad Ramify(s, \emptyset, ValveOpen, z', Do(CloseValve, s)))\,]
\end{array}
$$

As for the action of sensing the steam output, it is assumed that the corresponding sensor has an accuracy of ± 2 liters per second:

$$
\begin{array}{l}
Poss(CheckSteamOutput, s) \supset \\
\quad (\exists o)(\forall z')\,[KState(Do(CheckSteamOutput, s), z') \equiv KState(s, z') \wedge \\
\qquad (\exists o')\,(Holds(SteamOutput(o'), z') \wedge o - 2 \leq o' \leq o + 2)\,]
\end{array}
\tag{9.23}
$$

A pure sensing action, *CheckSteamOutput* has no physical effect. Asking for a requested steam output, on the other hand, has the nondeterministic effect of fluent *RequestedOutput* taking on any value. This value will be immediately known:

$$
\begin{array}{l}
Poss(AskRequestedOutput, s) \supset (\exists o)(\forall z') \\
\quad [\,KState(Do(AskRequestedOutput, s), z') \equiv (\exists z)\,(KState(s, z) \wedge \\
\qquad [\,(\forall o')\,\neg Holds(RequestedOutput(o'), z) \wedge \\
\qquad\quad z' = z + RequestedOutput(o)\,] \\
\qquad \vee \\
\qquad [\,(\exists o')\,(Holds(RequestedOutput(o'), z) \wedge \\
\qquad\qquad z' = z - RequestedOutput(o') + RequestedOutput(o))\,]\,)\,]
\end{array}
\tag{9.24}
$$

This completes the axiomatization of the steam boiler domain. Suppose, for instance, the agent initially knows that the requested steam output is $88\,l/\sec$ while all pumps are set to low mode and are currently switched off, the valve is closed, and the water level is $30\,l$:

$$
\begin{array}{l}
KState(S_0, z) \equiv (\exists z') \\
\quad (\,z = RequestedOutput(88) \\
\qquad\quad \circ\, WaterLevel(30) \circ SteamOutput(10) \circ z' \\
\quad \wedge\, (\forall x)\,\neg Holds(PumpOn(x), z') \\
\quad \wedge\, (\forall x)\,\neg Holds(PumpHigh(x), z') \\
\quad \wedge\, \neg Holds(ValveOpen, z')\,)
\end{array}
\tag{9.25}
$$

Let $S_2 = Do(TurnOn(2), Do(TurnOn(1), S_0))$, then the domain axioms entail

$$\begin{aligned}
&KState(S_2, z) \equiv (\exists l, o, z') \\
&\quad (z = PumpOn(1) \circ PumpOn(2) \circ RequestedOutput(88) \\
&\qquad \circ WaterLevel(l) \circ SteamOutput(o) \circ z' \\
&\quad \wedge 106 \leq l \leq 114 \wedge 86 \leq o \leq 94 \\
&\quad \wedge (\forall x) \neg Holds(PumpOn(x), z') \\
&\quad \wedge (\forall x) \neg Holds(PumpHigh(x), z') \\
&\quad \wedge \neg Holds(ValveOpen, z'))
\end{aligned} \tag{9.26}$$

Put in words, turning on pumps 1 and 2 brings about an increase of the water level and steam output by an amount between 76 and 84 liters per second according to causal relationship (9.11) and the capacity of the two pumps. Suppose the agent now checks the steam output and receives a value of 87 liters per second. Let $S_3 = Do(CheckSteamOutput, S_2)$, then with the instance $\{o/87, s/S_2\}$ of knowledge update axiom (9.23) it follows that

$$\begin{aligned}
&KState(S_3, z) \equiv (\exists l, o, z') \\
&\quad (z = PumpOn(1) \circ PumpOn(2) \circ RequestedOutput(88) \\
&\qquad \circ WaterLevel(l) \circ SteamOutput(o) \circ z' \\
&\quad \wedge 106 \leq l \leq 109 \wedge 86 \leq o \leq 89 \\
&\quad \wedge (\forall x) \neg Holds(PumpOn(x), z') \\
&\quad \wedge (\forall x) \neg Holds(PumpHigh(x), z') \\
&\quad \wedge \neg Holds(ValveOpen, z'))
\end{aligned}$$

Notice that the steam output cannot be less than $86\,l/\text{sec}$ because the water level is at least 106 liters. On the other hand, the steam output cannot be more than $89\,l/\text{sec}$ because the sensor has an accuracy of $\pm 2\,l/\text{sec}$. Suppose that in this situation the agent asks for a new requested steam output and gets the answer $60\,l/\text{sec}$, then the instance $\{o/60, s/S_3\}$ of knowledge update axiom (9.24) implies, setting $S_4 = Do(AskRequestedOutput, S_3)$,

$$\begin{aligned}
&KState(S_4, z) \equiv (\exists l, o, z') \\
&\quad (z = PumpOn(1) \circ PumpOn(2) \circ RequestedOutput(60) \\
&\qquad \circ WaterLevel(l) \circ SteamOutput(o) \circ z' \\
&\quad \wedge 106 \leq l \leq 109 \wedge 86 \leq o \leq 89 \\
&\quad \wedge (\forall x) \neg Holds(PumpOn(x), z') \\
&\quad \wedge (\forall x) \neg Holds(PumpHigh(x), z') \\
&\quad \wedge \neg Holds(ValveOpen, z'))
\end{aligned} \tag{9.27}$$

From the indirect effects that may occur in the steam boiler, it follows that it suffices to turn off pump 2 in order to produce a steam output that is close to the requested value. Let $S_5 = Do(\mathit{TurnOff}(2), S_4)$, then

$$\begin{array}{l} \mathit{KState}(S_5, z) \equiv (\exists l, o, z') \\ \quad (\, z = \mathit{PumpHigh}(1) \circ \mathit{ValveOpen} \\ \qquad\quad \circ\, \mathit{PumpOn}(1) \circ \mathit{RequestedOutput}(60) \\ \qquad\quad \circ\, \mathit{WaterLevel}(l) \circ \mathit{SteamOutput}(o) \circ z' \\ \quad \wedge\, 89 \leq l \leq 96 \wedge 59 \leq o \leq 66 \\ \quad \wedge\, (\forall x)\, \neg \mathit{Holds}(\mathit{PumpOn}(x), z') \\ \quad \wedge\, (\forall x)\, \neg \mathit{Holds}(\mathit{PumpHigh}(x), z') \\ \quad \wedge\, \neg \mathit{Holds}(\mathit{ValveOpen}, z')\,) \end{array} \tag{9.28}$$

This knowledge state is obtained via causal relationships (9.9) and (9.10), responsible for switching pump 2 to high capacity and opening the valve, respectively, along with the causal relationships by which the water level and steam output of the boiler are adjusted. □

As a general method, the use of causal relationships constitutes an extensive solution to the ramification problem. While each individual relationship specifies a local phenomenon, the induced ramification relation allows to model arbitrarily long and complex chains of indirect effects. Causal relationships are therefore a concise and modular means to describe semi-automatic environments. In addition, ramification enhances the fundamental expressiveness of standard state update axioms, as it allows to model actions with an unbounded number of effects.

Delayed Effects

Causal relationships provide means also to model an important phenomenon that is not present in our running example: In a dynamic environment, actions may have **delayed** effects, which do not materialize until after the agent has performed further actions. This aspect can be modeled with the help of a "momentum" fluent in the spirit of *LevelChange* used in the steam boiler domain. An action which has a delayed effect can be specified as causing such a fluent to hold, which eventually, under circumstances specified by a suitable causal relationship, triggers the actual delayed effect. These momentum fluents can thus be viewed as representing an internal process of the system. Having triggered a delayed effect, such a process may either terminate by another causal relationship, or continue to hold in case of repeated delayed effects. On the other hand, the agent may have at its disposal the intervening action of terminating the momentum fluent before it produces the delayed effect; see Exercise 9.5 for an example domain where effects may occur with delay. In this way, the method of causal relationships allows agents to reason about and to control various kinds of dynamic systems.

```
ramify(Z1,ThetaP,ThetaN,Z2) :- update(Z1,ThetaP,ThetaN,Z),
                               causes(Z,ThetaP,ThetaN,Z2).

causes(Z,P,N,Z2) :-
   causes(Z,P,N,Z1,P1,N1) -> causes(Z1,P1,N1,Z2)
                           ; Z2=Z.
```

Figure 9.8: FLUX kernel definition of ramification.

9.3 Causality in FLUX

Ramifications are inferred in FLUX using the kernel definition of a predicate $Ramify(z_1, p, n, z_2)$ as shown in Figure 9.8. This definition follows macro (9.19) but with the situation argument suppressed. To generalize FLUX effect specifications to update with ramification, the predicate $Update(z_1, \vartheta^+, \vartheta^-, z_2)$ used in standard update axioms just needs to be replaced by the new predicate $Ramify(z_1, \vartheta^+, \vartheta^-, z_2)$. The kernel definition of the ramification relation presupposes that the individual causal relationships of a domain are encoded by clauses which define the predicate $Causes(z_1, p_1, n_1, z_2, p_2, n_2)$. The transitive closure of the causal relation is defined in the second clause in Figure 9.8. A sink is reached if no further relationship applies. This reflects the minimality condition of the induced causal relation according to Definition 9.3.

Correctness of update with ramification in FLUX requires the domain axioms to be encoded in such a way that a unique sink is always reached after finitely many applications of causal relationships. This sink needs to be a sound FLUX representation of the resulting knowledge state according to the domain axiomatization. In practice, soundness can often be met by a straightforward encoding of the causal relationships and imposing conditions on the knowledge states which may arise in the course of an agent program.

Example 7 (continued) Figures 9.9 and 9.10 show a straightforward FLUX encoding of the causal relationships for the steam boiler domain under the assumption that the agent always knows the states of the pumps and of the valve. As an example, recall knowledge state (9.27) at the end of the preceding section. The following query shows how the ramifications of switching off pump 2 in this state are inferred:

```
?- Z1 = [pump_on(1),pump_on(2),
         water_level(L),steam_output(O) | Z],
   L :: 106..109, O #= L-20,
   not_holds_all(pump_on(_), Z),
   not_holds_all(pump_high(_), Z),
   not_holds(valve_open, Z),
   ramify(Z1, [], [pump_on(2)], Z2).
```

```
causes(Z, P, N, [pump_high(X)|Z], [pump_high(X)|P], N) :-
   holds(pump_on(Y), N), connected(X, Y),
   not_holds(pump_high(X), Z).

connected(X,Y) :- ( Y=1->X=2; Y=2->X=1; Y=3->X=4; Y=4->X=3 ).

causes(Z, P, N, [valve_open|Z], [valve_open|P], N) :-
   holds(pump_high(X), P), holds(pump_on(X), Z),
   not_holds(valve_open, Z).

causes(Z, P, N, Z, [level_change(X,25)|P], N) :-
   holds(pump_high(X), P), holds(pump_on(X), Z),
   not_holds(level_change(X,25), P).

causes(Z, P, N, Z, [level_change(X,-25)|P], N) :-
   holds(pump_high(X), N), holds(pump_on(X), Z),
   not_holds(level_change(X,-25), P).

causes(Z, P, N, Z, [level_change(X,D)|P], N) :-
   holds(pump_on(X), P), not_holds_all(level_change(X,_), P),
   capacity(X, D, Z).

causes(Z, P, N, Z, [level_change(X,D)|P], N) :-
   holds(pump_on(X), N), not_holds_all(level_change(X,_), P),
   capacity(X, D1, Z), D #= -D1.

capacity(X, D, Z) :-
   not_holds(pump_high(X),Z) -> ((X=1; X=2) -> D :: 38 .. 42 ;
                                 (X=3; X=4) -> D :: 97 .. 103)
                              ; ((X=1; X=2) -> D :: 63 .. 67 ;
                                 (X=3; X=4) -> D :: 122 .. 128).
```

Figure 9.9: FLUX specification of the causal relationships (9.9)–(9.14) of the steam boiler.

```
Z2 = [steam_output(O2),water_level(L2),
      valve_open,pump_high(1),pump_on(1) | Z]
O2 :: 59..66
L2 :: 89..96

Constraints:
not_holds_all(pump_on(_), Z)
not_holds_all(pump_high(_), Z)
not_holds(valve_open, Z)
```

Thus the (only) direct effect that *PumpOn*(2) becomes false led to the indirect effects that pump 1 switches to high capacity, the valve opens, and both water level as well as steam output acquire updated values.

Our encoding of the causal relationships would be unsound, however, if the agent had incomplete knowledge of the pumps or the valve. In the following FLUX state, for example, it is unknown whether pump 1 is running. Hence, turning off pump 2 may or may not have the indirect effect that the first pump automatically switches to high capacity. Yet our FLUX encoding implies that this indirect effect necessarily occurs:

```
?- Z1 = [pump_on(2),valve_open,
         water_level(100),steam_output(70) | Z],
   not_holds_all(pump_high(_), Z),
   ramify(Z1, [], [pump_on(2)], Z2).

Z2 = [steam_output(O),water_level(L),
      pump_high(1),valve_open,pump_on(1) | _]

Constraints:
L :: 83..87
O :: 53..57
```

The result is incorrect insofar as the domain axiomatization entails a weaker knowledge state, namely, where either *PumpHigh*(1) is true and the water level is between 83 and 87, or *PumpHigh*(1) continues to be false (because pump 1 is off) and the water level is between 68 and 72. A more general encoding of the causal relationships for the steam boiler would include cancellation of knowledge of a fluent in situations where the agent has insufficient knowledge to decide whether this fluent is subject to indirect change.

Figure 9.11 depicts FLUX definitions of both precondition and update axioms for the steam boiler agent. Along with the causal relationships, the encoding allows to infer the result of a sequence of actions including all of their ramifications. Recall, for instance, initial knowledge state (9.25), and consider the actions of turning on pumps 1 and 2:

```
init(Z0) :- Z0 = [requested_output(88),
                  water_level(30),steam_output(10)|Z],
```

```
causes(Z1, P, N, [water_level(L2)|Z], P1, [level_change(X,D)|N]) :-
   holds(level_change(X,D), P), not_holds(level_change(X,D), N),
   holds(water_level(L1), Z1, Z), L2 #= L1 + D,
   ( holds(water_level(_), P, P2)
     -> P1=[water_level(L2)|P2] ; P1=[water_level(L2)|P] ).

causes(Z1, P, N, [steam_output(O2)|Z], P1, N) :-
   holds(water_level(L), P),
   ( not_holds(valve_open, Z1) -> O2 #= L-20 ; O2 #= L-30 ),
   holds(steam_output(O1), Z1, Z), \+ O2 #= O1,
   ( holds(steam_output(O1), P, P2)
     -> P1=[steam_output(O2)|P2] ; P1=[steam_output(O2)|P] ).

causes(Z1, P, N, [steam_output(O2)|Z], P1, N) :-
   holds(valve_open, P),
   holds(water_level(L), Z1), O2 #= L-30,
   holds(steam_output(O1), Z1, Z), \+ O2 #= O1,
   ( holds(steam_output(O1), P, P2)
     -> P1=[steam_output(O2)|P2] ; P1=[steam_output(O2)|P] ).

causes(Z1, P, N, [steam_output(O2)|Z], P1, N) :-
   holds(valve_open, N),
   holds(water_level(L), Z1), O2 #= L-20,
   holds(steam_output(O1), Z1, Z), \+ O2 #= O1,
   ( holds(steam_output(O1), P, P2)
     -> P1=[steam_output(O2)|P2] ; P1=[steam_output(O2)|P] ).
```

Figure 9.10: FLUX specification of the causal relationships (9.15)–(9.18) of the steam boiler.

```
poss(turn_on(X),Z)   :- knows_not(pump_on(X),Z).
poss(turn_off(X),Z)  :- knows(pump_on(X),Z).
poss(set_hi(X),Z)    :- knows_not(pump_high(X),Z).
poss(set_lo(X),Z)    :- knows(pump_high(X),Z).
poss(open_valve,Z)   :- knows_not(valve_open,Z).
poss(close_valve,Z)  :- knows(valve_open,Z).
poss(check_steam_output,_).
poss(ask_requested_output,_).

state_update(Z1,turn_on(X),Z2,[]):- ramify(Z1,[pump_on(X)],[],Z2).

state_update(Z1,turn_off(X),Z2,[]):- ramify(Z1,[],[pump_on(X)],Z2).

state_update(Z1,set_hi(X),Z2,[]):- ramify(Z1,[pump_high(X)],[],Z2).

state_update(Z1,set_lo(X),Z2,[]):- ramify(Z1,[],[pump_high(X)],Z2).

state_update(Z1,open_valve,Z2,[]):- ramify(Z1,[valve_open],[],Z2).

state_update(Z1,close_valve,Z2,[]):- ramify(Z1,[],[valve_open],Z2).

state_update(Z,check_steam_output,Z,[O]) :-
   O1 :: O-2..O+2, holds(steam_output(O1),Z).

state_update(Z1,ask_requested_output,Z2,[O]) :-
   holds(requested_output(O1),Z1)
   -> update(Z1,[requested_output(O)],[requested_output(O1)],Z2)
    ; update(Z1,[requested_output(O)],[],Z2).
```

Figure 9.11: FLUX specification of the preconditions and effects of the actions available to our control agent.

```
            not_holds_all(pump_on(_),Z),
            not_holds_all(pump_high(_),Z),
            not_holds(valve_open,Z).

?- init(Z0),
   state_update(Z0,turn_on(1),Z1,[]),
   state_update(Z1,turn_on(2),Z2,[]),

Z1 = [steam_output(O2{48..52}),water_level(L2{68..72}),
      pump_on(1),requested_output(88)|Z]
Z2 = [steam_output(O2{86..94}),water_level(L2{106..114}),
      pump_on(2),pump_on(1),requested_output(88)|Z]
```

The computed intervals for water level and steam output correspond to the inequations in knowledge state (9.26). Suppose the agent continues with checking the steam output, asking for a new requested output, and adjusting the settings accordingly, then FLUX arrives at an incomplete state which corresponds to the result of our derivation in fluent calculus at the end of Section 9.2 (cf. knowledge state (9.28)):

```
?- init(Z0),
   state_update(Z0,turn_on(1),Z1,[]),
   state_update(Z1,turn_on(2),Z2,[]),
   state_update(Z2,check_steam_output,Z3,[87]),
   state_update(Z3,ask_requested_output,Z4,[60]),
   state_update(Z4,turn_off(2),Z5,[]).

...
Z3 = [steam_output(O2{86..89}),water_level(L2{106..109}),
      pump_on(2),pump_on(1),requested_output(88)|Z]
Z4 = [requested_output(60),steam_output(O2{86..89}),
      water_level(L2{106..109}),pump_on(2),pump_on(1)|Z]
Z5 = [steam_output(O2{59..66}),water_level(L2{89..96}),
      valve_open,pump_high(1),
      requested_output(60),pump_on(1)|Z]
```

We are now ready to write an agent program for the automatic control of the steam boiler. Figure 9.12 depicts the main loop, where the agent continually checks both the current steam output and the requested output, and modifies the setting accordingly.

The crucial part of the agent program is the definition in Figure 9.13 of the planning problem of always achieving the requested steam output. The heuristic encoding spans a plan space consisting of all possible changes in the settings of the pumps and the valve. Specifically, the first clause for *ControlPlan*/4 considers to leave the valve unchanged or to open or close it, depending on which is possible. The second clause for *ControlPlan*/4 defines the following options for each pump x.

```
main :- init(Z0), main_loop(Z0).

main_loop(Z) :- execute(check_steam_output,Z,Z),
                execute(ask_requested_output,Z,Z1),
                plan(control_plan,Z1,P), execute(P,Z1,Z2),
                main_loop(Z2).
```

Figure 9.12: Main loop for the controlbot.

---

1. The pump remains unchanged.
2. The pump is turned on. Additionally, the mode may be changed from low to high or vice versa, depending on which is possible.
3. The pump is turned off.
4. The pump is switched to high or low capacity, depending on which is possible.

For example, in a situation where the valve is closed and all pumps are off and switched to low capacity, the plans in the search space range from the empty sequence [] to

$$[\mathit{TurnOn}(1), \mathit{SetHi}(1), \mathit{TurnOn}(2), \mathit{SetHi}(2), \\ \mathit{TurnOn}(3), \mathit{SetHi}(3), \mathit{TurnOn}(4), \mathit{SetHi}(4), \mathit{OpenValve}]$$

Since the aim is to produce a steam output which is as close to the requested amount as one can get, the cost of a plan is determined by the difference between the two values. Because the actual output is usually uncertain, the auxiliary predicate $\mathit{Mean}(o, n)$ defines n to be the mean value of the interval range for domain variable o. A secondary criterion for the cost of a plan is its length.

Figure 9.14 depicts an example initial state of the steam boiler. Consistency of states is defined by three domain constraints which stipulate, respectively, that water level and steam output be unique and that there can be at most one requested output. The following shows a trace of the program wrt. a certain sequence of output requests:

```
check_steam_output   : 0

ask_requested_output : 88
turn_on(1)
turn_on(2)
close_valve
check_steam_output   : 89

ask_requested_output : 60
turn_off(2)
```

```
pumps([1,2,3,4]).

control_plan(Z,P,Zn) :-
   pumps(Pumps), control_plan(Pumps,Z,P,Zn).

control_plan([],Z,P,Zn) :-
   Zn=Z, P=[]
   ; poss(open_valve,Z) -> do(open_valve,Z,Zn,[],P)
                        ; do(close_valve,Z,Zn,[],P).

control_plan([X|Pumps],Z,P,Zn) :-
   ( Z=Z1, P=P1
     ; poss(turn_on(X),Z)
       -> do(turn_on(X),Z,Z2,P2,P),
          ( Z2=Z1, P2=P1
            ; poss(set_hi(X),Z2) -> do(set_hi(X),Z2,Z1,P1,P2)
                                 ; do(set_lo(X),Z2,Z1,P1,P2) )
     ; do(turn_off(X),Z,Z1,P1,P)
     ; poss(set_hi(X),Z1) -> do(set_hi(X),Z2,Z1,P1,P)
                          ; do(set_lo(X),Z2,Z1,P1,P) ),
   control_plan(Pumps,Z1,P1,Zn).

do(A,Z,Z1,P,[A|P]) :- state_update(Z,A,Z1,[]).

plan_cost(control_plan,P,Z,C) :-
   holds(steam_output(O1),Z), holds(requested_output(O2),Z),
   mean(O1,M), length(P,N),
   C is abs(M-O2)*100+N.

mean(O,N) :- dvar_domain(O,dom(X,_)),
             ( X=[V..W] -> N is V+(W-V)/2 ; X=[V], N=V).
```

Figure 9.13: An encoding of the planning problem for the controlbot with the aim to adjust the steam output.

```
init(Z0) :- Z0 = [water_level(30),valve_open|Z],
            not_holds_all(pump_on(_),Z),
            not_holds_all(pump_high(_),Z),
            consistent(Z0).

consistent(Z) :-
   holds(water_level(_), Z,Z1) -> not_holds_all(water_level(_),Z1),
   holds(steam_output(_),Z,Z2) -> not_holds_all(steam_output(_),Z2),
   requested_output_unique(Z), duplicate_free(Z).

requested_output_unique([F|Z]) <=>
   F=requested_output(_) -> not_holds_all(requested_output(_),Z)
                          ; requested_output_unique(Z).
```

Figure 9.14: Initial state of the steam boiler, where all pumps are in lower position and off.

```
check_steam_output   : 65

ask_requested_output : 153
set_lo(1)
turn_on(4)
close_valve
check_steam_output   : 148

ask_requested_output : 320
set_hi(1)
turn_on(3)
set_hi(3)
set_hi(4)
close_valve
check_steam_output   : 321

ask_requested_output : 320
check_steam_output   : 321

ask_requested_output : 176
set_lo(1)
turn_on(2)
turn_off(3)
set_lo(4)
open_valve
check_steam_output   : 177
```

The figures illustrate how the control agent is able to modify the settings of the steam boiler so as to produce an amount of steam which is very close to the requested value. Moreover, exploiting the causal dependencies among the components whenever these give rise to desired automatic changes, the agent behaves economically and always performs as few actions as necessary to arrive at the desired setting. □

9.4 Bibliographical Notes

The ramification problem has been uncovered by [Ginsberg and Smith, 1988], inspired by a planning method which takes into account implications of plans [Finger, 1987]. Early approaches accommodate indirect effects by exempting certain fluents from the assumption of persistence [Lifschitz, 1990; del Val and Shoham, 1993; Kartha and Lifschitz, 1994]. More elaborate methods define various categories of fluents and use a partial preference ordering among all possible changes, e.g., [Sandewall, 1995a]. In [Sandewall, 1995b], a systematic framework based on this concept has been introduced with the aim of assessing the range of applicability of different approaches that follow the same principles.

Aiming at more extensive solutions to the ramification problem, specific notions of causality have been developed as the basis for indirect effects, e.g., [Elkan, 1992; Lin, 1995] for situation calculus, or [Geffner, 1990; Brewka and Hertzberg, 1993; McCain and Turner, 1995; Łukaszewicz and Madalińska-Bugaj, 1995]. The concept of causal relationships has been introduced in [Thielscher, 1997] and compared to several other causality-based solutions to the ramification problem. Later approaches include [McIlraith, 2000] for situation calculus, [Gustafsson and Doherty, 1996; Kakas and Miller, 1997a; Shanahan, 1999] for calculi based on time points like event calculus, or [Giunchiglia *et al.*, 1997; Denecker *et al.*, 1998]. The work of [Pearl, 1993; Pearl, 1994] considers dynamic changes and their ramifications in probabilistic causal theories.

A formal justification for defining ramification as transitive closure of cause-effect relations has been given in [Sandewall, 1996]. In the series of papers [Peppas *et al.*, 1999; Prokopenko *et al.*, 1999; Prokopenko *et al.*, 2000], a unifying semantics has been defined for a variety of approaches to the ramification problem, including the concept of causal relationships. The use of ramifications to model dynamic systems has been propagated in [Watson, 1998].

9.5 Exercises

9.1. Consider the following graph:

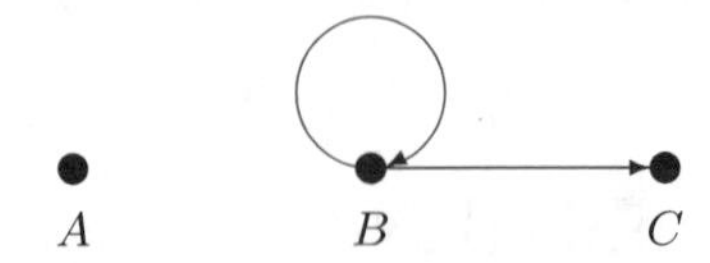

Suppose that $A \neq B \neq C$.

(a) Define the predicate $Edge(x, y)$ to be true just in case there is an edge from node x to node y in this graph.

(b) Show that the attempt to give a first-order definition of transitive closure by

$$Path(x, y) \stackrel{\text{def}}{=} x = y \vee (\exists z)\,(Path(x, z) \wedge Edge(z, y))$$

does not entail $\neg Path(A, C)$.

(c) Prove that the second-order definition

$$Path(x, y) \stackrel{\text{def}}{=} (\forall P) \left\{ \begin{array}{l} (\forall u)\, P(u, u) \wedge \\ (\forall u, v, w)\, (P(u, v) \wedge Edge(v, w) \supset P(u, w)) \\ \qquad\qquad\qquad \supset P(x, y) \end{array} \right\}$$

along with the axiom of (a) entails $\neg Path(A, C)$.
Hint: Find a λ-expression by which predicate variable P can be instantiated such that the antecedent of the implication inside the bracelets is satisfied while $P(A, C)$ is false.

(d) Prove that the axioms in (a) and (c) entail $Path(B, C)$.

9.2. Modify the FLUX specification of the steam boiler so that pump 4 has a capacity of 150 ± 5 liters in low mode and 175 ± 5 in high mode. Construct a situation (i.e., a setting of the pumps and a new requested steam output) in which the agent program of Section 9.3 generates a configuration which is sub-optimal. Modify the definition of the planning problem of Figure 9.13 so that the agent will always find the configuration by which the amount of produced steam is as close as possible to the requested output.

Hint: The problem is caused by the strict order in which the pumps are considered in a plan.

9.3. Consider the electric circuit of Figure 9.15 and the two fluents $Open(x)$, denoting that switch x is opened, and $Active(x)$, denoting that relay or light bulb x is active.

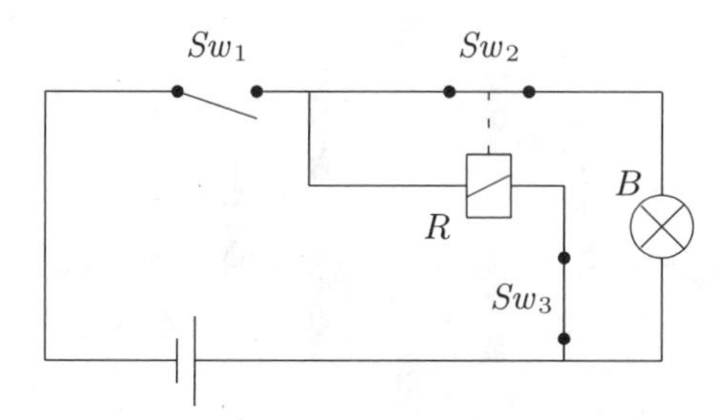

Figure 9.15: An electric circuit. Upon activation, relay R forces switch Sw_2 to jump open.

(a) Formulate domain constraints to express the dependencies among the components of the circuit.

(b) Specify a set of causal relationships for the circuit describing the various possible indirect effects.

(c) Axiomatize the state depicted in Figure 9.15, and use the knowledge update axiom for altering a switch, (9.22), along with the new causal relationships to prove that the light is still off after toggling Sw_1.

9.4. Give a FLUX specification of the circuit of Figure 9.15. The causal relationships shall be encoded under the assumption that the agent always knows the states of the three switches. Use FLUX to find a plan to switch on the light bulb in the state depicted in Figure 9.15.

9.5. This exercise is concerned with the use of ramifications for the specification of delayed effects. Consider the configuration of dominoes sketched in Figure 9.16.

(a) Give a FLUX specification of the action of waiting, which has no direct effect, and of tipping one domino in either of two directions, along with suitable causal relationships. The only effect of the action shall be that the domino starts falling down into the respective direction. As a delayed effect, in the following situation the domino will have fallen and caused the adjacent erected dominoes, if any, to start falling down. Use FLUX to show that if in the state of Figure 9.16 the domino in the center is tipped in either direction, then it takes 8 wait actions until all dominoes have fallen down. Show also that if the domino on the top is tipped in either direction, then it takes 11 wait actions until all dominoes have fallen down.

(b) Extend the specification by an update axiom for the action of stopping a domino from falling. Encode the problem of finding, for any two dominoes x, y, a plan in which only the domino in the center

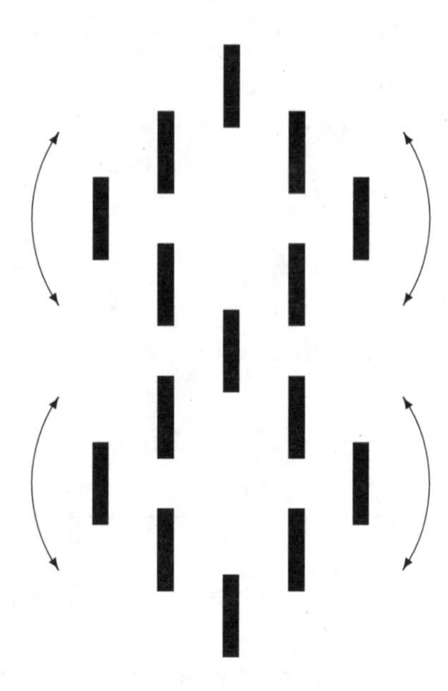

Figure 9.16: A total of 15 erected dominoes are arranged in such a way that if one of them is tipped in either direction, then eventually all other dominoes will fall down.

---

may be tipped such that in the end x has fallen down while y remains erected. Find all pairs (x, y) with $x \neq y$ for which no such plan exists.

Chapter 10

Troubleshooting: Qualification Problem

Robot programs rely on expectations regarding the effects of the actions carried out by the physical apparatus. Reality, however, does not always comply with these expectations. A complex process such as lifting an object with a gripper may fail. Pathways may be blocked. A human user may place the wrong item on the tray of the robot. The list of potential errors during the execution of a program is endless. In contrast to standard software, where a runtime error is almost always due to a programming error, a robot program can be provably correct and nonetheless fail when the robot gets stuck in the real world. This is known as the **qualification problem**.[1]

An additional problem is that errors during the execution of a program are not always immediately recognizable. For one thing, it may be too much of an effort to ascertain each and every time that an action has been successful in every aspect. Besides, the robot may be too limited in its sensing capabilities to directly realize certain errors. For example, without sophisticated image processing and object recognition algorithms, an office robot is unable to tell whether a user has placed the right item onto the tray. As a consequence, an error is detected only indirectly and much later, say, when trying to deliver the item to the wrong person.

This chapter is concerned with troubleshooting in robot control programs, that is, means of telling a robot what to do if something has gone wrong. The goal is to avoid, as best as possible, the necessity of human intervention and rebooting the system in case a discrepancy arises between the expected outcome and reality. An easy method are preprogrammed error handling routines for common problems, e.g., by having the robot repeat a failed action for a fixed number of times, then systematically working through a list of alternative ways, and finally—if still unsuccessful—notifying the user and saving the task for

[1] It has been so named because in reality the successful execution of an action may have innumerably many more qualifications than those specified in a control program.

later. A more intelligent method of troubleshooting is to have the robot search for a possible explanation once it detects an error, and then coming up with a suitable way of correcting it. For example, upon failing to deliver a package our office robot may be able to generate a plausible hypothesis as to who the right recipient should be.

Troubleshooting in general aims at enhancing the robustness of a control program. Of course, there will always be cases in which error handling fails, such as the breakdown of a crucial low-level module like navigation, or a serious problem with the mechanics of the robot. Giving the matter even the utmost considerable thought cannot help when a screwdriver is required or soldering. (Still the control program may be able to identify problems of this kind and call for help.) Dealing with the qualification problem therefore means to endow a robot with the cognitive capability of troubleshooting as far as possible within the control program. To this end, we will consider

10.1. how to model ways in which an action may go wrong;

10.2. how to distinguish plausible from less likely explanations;

10.3. how to do troubleshooting in FLUX;

10.4. how to deal with persistent problems in the environment.

10.1 Accidental Action Failure

The expectations agents and robots have concerning the effects of their actions are manifest in their internal world model. When making an observation that contradicts the model, they realize that something must have gone wrong. This may happen immediately after an action has produced an error, provided that the unexpected outcome is directly observed by closely monitoring the execution. Often, however, an error is not detected until much later by an incoherent sensing result.

The possibility of erroneous execution puts a challenge on the fundamental concept of knowledge update, which implies that agents simply *know* that actions have their specified effects. Consequently, any observation to the contrary renders the internal model inconsistent by ruling out all possible states. In Chapter 7 this problem was addressed through nondeterminism: By a nondeterministic update, world models are weakened so as to be consistent with several possible outcomes of the action in question. Nondeterministic update, however, is suitable only in case the resulting knowledge state still contains enough information for a robot to be capable of making decisions. Our mailbot of Section 7.1, for example, is programmed so as to learn immediately which of the possible outcomes of a *Pickup* or *Deliver* action has actually occurred. Otherwise, the robot would have been unable to proceed eventually, having lost all its knowledge about the contents of its bags. For a similar reason, the nondeterministic cleanbot of Section 7.2 had to be able to regain knowledge of (un-)occupied offices by returning to an adjacent location. This shows that the concept of nondeterminism should not be used to account for all possible ways in which actions may go wrong, lest the internal world model becomes too blurred to carry out a task.

The better alternative for dealing with unlikely but not impossible failures is for agents to assume that actions have their intended effects as long as the observations are in line with this assumption. Only in case they obtain a conflicting sensing result they should consider the various possible ways in which one of their previous actions might have failed. This is known as **nonmonotonic** reasoning, where **default** conclusions are adhered to until additional observations suggest the contrary. A logical entailment relation is nonmonotonic if some conclusion from a set of axioms is no longer valid after the set has been extended.

For a systematic treatment of failed execution of actions, the signature of fluent calculus is extended as follows.

Definition 10.1 A tuple $\mathcal{S} \cup \langle \mathcal{C}, Acc \rangle$ is a **fluent calculus signature for accidental qualifications** if $\mathcal{S}$ is a fluent calculus signature and

- $\mathcal{C}$ finite set of function symbols into sort ACCIDENT
- Acc : ACCIDENT $\times$ SIT.

An **accident** is a term of sort ACCIDENT. □

As part of a domain specification, the new sort for accidents can be used to define ways in which things may go wrong. An instance of $Acc(c, s)$ then indicates that accident c happens in situation s.

Example 8 Consider a very simple domain with only one fluent F, one action A, and the single accident C : ACCIDENT. Suppose that the normal effect of A is to make F true; however, this effect does not materialize if accident C happens. Ignoring preconditions for the sake of simplicity, the action is suitably described by the following update:

$$\begin{array}{l} KState(s, z') \equiv (\exists z)\,(KState(s, z) \wedge \\ \quad \neg Acc(C, s) \wedge z' = z + F \\ \quad \vee \\ \quad Acc(C, s) \wedge z' = z\,) \end{array} \tag{10.1}$$

Here, the first equation specifies the normal effect while the second one covers the occurrence of an accident, in which case the action has no effect. □

As its stands, the update specified by (10.1) is nondeterministic, because complete knowledge of the state in situation s alone does not suffice to infer whether F holds in the successor situation $Do(A, s)$—this depends on whether or not $Acc(C, s)$ happens to be true. Without taking additional measures, therefore, the occurrence of the intended effect cannot be predicted. With the aim to have agents and robots assume away accidents unless something actually went wrong, fluent calculus shall be embedded into the standard nonmonotonic formalism of **default logic**.

Given a signature S, a **default** δ is an expression of the form $\frac{:\gamma}{\gamma}$ where γ is a first-order formula over S. A default is a kind of deduction rule which allows to conclude γ provided it is consistent to do so. (We only need this special form of defaults, which are—to use the technical term—*prerequisite-free* and *normal*.) If γ contains free variables, then δ is a representative of all its ground instances. A **default theory** Δ is a pair (D, W) where D is a set of defaults and W a set of formulas. As an example, the following simple default theory says that by default $P(x)$ is false for any x while $P(A)$ or $P(B)$ is true: $\Delta = (\{\frac{:\neg P(x)}{\neg P(x)}\}, \{P(A) \vee P(B)\})$.
To define the logical consequences of a default theory $\Delta = (D, W)$ let, for any set E of closed formulas, E' be the *smallest* set of formulas such that

1. $E' = \mathcal{T}(E')$ (where $\mathcal{T}(E')$ denotes the deductive closure of E');
2. $W \subseteq E'$;
3. $\gamma \in E'$ for all $\frac{:\gamma}{\gamma} \in D$ such that $\neg\gamma \notin E$

Then E is an **extension** of Δ iff $E' = E$. Intuitively, each extension describes a particular way of adding default conclusions to the formulas W of a default theory. A default theory Δ **entails** a formula φ if $\varphi \in E$ for all extensions E of Δ. For instance, suppose that the constants A, B, C are the only function symbols of the example theory from above, then the latter has two extensions:

$$\begin{aligned} E_1 &= \mathcal{T}(\{P(A), \neg P(B), \neg P(C)\}) \\ E_2 &= \mathcal{T}(\{\neg P(A), P(B), \neg P(C)\}) \end{aligned}$$

The set $E_3 = \mathcal{T}(\{P(A), P(B), \neg P(C)\})$, say, is not an extension because $E_3' = \mathcal{T}(\{P(A) \vee P(B), \neg P(C)\}) \neq E_3$. With its two extensions, the theory entails $\neg P(C)$ and also $P(A) \equiv \neg P(B)$. This shows that $P(x)$ is false to the largest possible extent, as intended. Default logic is nonmonotonic as can be seen by the fact that if, say, $P(C)$ were added to the formulas of the example default theory, then $\neg P(C)$ would no longer be entailed.

Default reasoning allows robotic agents to assume away accidents and only use them when there is a need to explain unexpected observations. Suppose, to this end, that the observations an agent has made up to a certain point are axiomatized by a set O. Furthermore, let Σ be the underlying domain axiomatization, then the default reasoning is formalized as a default theory with axioms $\Sigma \cup O$ and which contains a single, universal default on the non-occurrence of accidents.

Definition 10.2 Consider a fluent calculus axiomatization Σ based on a signature with accidents. A **fluent calculus default theory for accidental**

qualifications is of the form

$$\Delta = \left(\left\{ \frac{: \neg Acc(c,s)}{\neg Acc(c,s)} \right\}, \Sigma \cup O \right) \tag{10.2}$$

where O is a (possibly empty) set of situation formulas. □

Example 8 (continued) Let Σ be knowledge update axiom (10.1) along with the foundational axioms of fluent calculus. Without any prior observation, that is, $O = \{\}$, the fluent calculus default theory has a unique extension in which $\neg Acc(C,s)$ holds for any s. Hence, by (10.1), $Holds(F, Do(A, S_0))$. On the other hand, suppose that it is unexpectedly observed that F has not been achieved by executing A, that is, $O = \{\neg Holds(F, Do(A, S_0))\}$. By (10.1), $Acc(C, S_0)$. Hence, the default theory admits a unique extension which contains this accident along with $\neg Acc(C,s)$ for any $s \neq S_0$. With this extension, the theory entails, for instance, $KState(Do(A, S_0), z) \equiv KState(S_0, z)$; hence, $\neg Holds(F, S_0)$. □

A crucial property of fluent calculus default theories is to propose accidents only as far as this is necessary. That is to say, whenever an extension of a theory suggests one or more accidents, then no strict subset of these accidents is consistent with the given observations.

Theorem 10.3 *For a set E of closed formulas let $\overline{Acc}(E)$ denote the set of ground literals $\neg Acc(\chi, \sigma)$ such that $E \not\models Acc(\chi, \sigma)$.*

Let Δ be a fluent calculus default theory with formulas $\Sigma \cup O$, then for any extension E of Δ and any $Acc(\chi, \sigma) \in E$,

$$\Sigma \cup O \cup \overline{Acc}(E) \cup \{\neg Acc(\chi, \sigma)\} \models \bot \tag{10.3}$$

Proof: Let E be an extension of Δ. The minimality requirement for extensions of default theories implies that

$$E = \mathcal{T}(\Sigma \cup O \cup \{\neg Acc(\chi, \sigma) : Acc(\chi, \sigma) \notin E\})$$

Since E is deductively closed, this is equivalent to

$$E = \mathcal{T}(\Sigma \cup O \cup \overline{Acc}(E))$$

Hence, for any $Acc(\chi, \sigma) \in E$,

$$\Sigma \cup O \cup \overline{Acc}(E) \models Acc(\chi, \sigma)$$

which proves (10.3). ■

As a corollary, a fluent calculus default theory entails that no accident has happend as long as the observations are in line with the normal effects of actions.

Corollary 10.4 *Let* $\overline{Acc}$ *be the set of all ground literals* $\neg Acc(\chi, \sigma)$. *Consider a fluent calculus default theory* Δ *with formulas* $\Sigma \cup O$ *such that*

$$\Sigma \cup O \cup \overline{Acc} \text{ is consistent} \tag{10.4}$$

Then Δ *has* $\mathcal{T}(\Sigma \cup O \cup \overline{Acc})$ *as its unique extension.*

Proof: Let $E = \mathcal{T}(\Sigma \cup O \cup \overline{Acc})$. From (10.4) it follows that $E \subseteq E'$ (where E' is as in the definition of extensions); hence, since E' is minimal, $E = E'$. Consequently, E is an extension of Δ. From (10.4) and Theorem 10.3 it follows that there can be no other extension. ■

In the following, we extend the mail delivery domain so as to account for several accidents that may happen when the robot interacts with its human users. For the sake of simplicity, we take the axiomatization of Chapter 1 as the starting point.

Example 1 (continued) Consider the two accidents

$$\begin{aligned} WrongPackage &: \mathbb{N} \mapsto \text{ACCIDENT} \\ WrongBag &: \mathbb{N} \mapsto \text{ACCIDENT} \end{aligned}$$

An instance $WrongPackage(r')$ shall denote the accident of receiving a package addressed to room r' instead of the correct one. An instance $WrongBag(b')$ denotes the accident of a user either putting a package into the wrong mail bag b' (when the robot performs a *Pickup* action) or taking a package out of the wrong mail bag (when the robot performs a *Deliver* action). Accounting for a possible accident, the original state update axiom for *Deliver* (cf. (1.12), page 17) is extended as follows:

$$\begin{array}{l}
Poss(Deliver(b), s) \supset [KState(Do(Deliver(b), s), z) \equiv (\exists z)\,(KState(s, z) \wedge \\
\quad (\exists r)\,(\,Holds(At(r), z) \wedge \\
\qquad [\,(\forall c)\,\neg Acc(c, s) \wedge \\
\qquad\quad z' = z - Carries(b, r) + Empty(b) \\
\qquad \vee \\
\qquad (\exists b')\,(Acc(\,WrongBag(b'), s) \wedge Holds(Carries(b', r), z) \wedge \\
\qquad\quad z' = z - Carries(b', r) + Empty(b')) \\
\qquad \vee \\
\qquad (\exists b', r')\,(Acc(\,WrongBag(b'), s) \wedge Holds(Carries(b', r'), z) \wedge r' \neq r \wedge \\
\qquad\quad z' = z - Carries(b', r') + Empty(b') \circ Request(r, r'))]\,)\,)]
\end{array}$$

Put in words, if no accident happens then the usual effects materialize. If, on the other hand, a package is taken out of the wrong bag b', then this bag is emptied. If, moreover, the package was addressed to a different room r', then a side-effect of the accident is to automatically issue a new request of bringing the package from the current room r to the right address r'.

The extended update axiom requires a generalization of the precondition axiom, too. As this needs to refer to the occurrence of an accident and since accidents are situation-dependent, the precondition has to be formulated wrt. a situation rather than a state:

$$
\begin{array}{l}
Poss(Deliver(b), s) \equiv \\
\quad [(\forall c)\, \neg Acc(c, s) \supset (\exists r)\, (Holds(At(r), s) \wedge Holds(Carries(b, r), s))] \\
\quad \wedge \\
\quad [(\forall b')\, (Acc(WrongBag(b'), s) \supset \neg Holds(Empty(b'), s)]
\end{array}
$$

That is to say, if no accident happens then the preconditions are as before, while a package taken out of the wrong bag requires the latter not to be empty.

In a similar fashion, the following extended update axiom accounts for possible accidents of picking up a package:

$$
\begin{array}{l}
Poss(Pickup(b, r), s) \supset [KState(Do(Pickup(b, r), s), z') \equiv (\exists z)\, (KState(s, z) \wedge \\
\quad (\exists r_1)\, (\, Holds(At(r_1), z) \wedge \\
\qquad [\, (\forall c)\, \neg Acc(c, s) \wedge \\
\qquad\quad z' = z - Empty(b) \circ Request(r_1, r) + Carries(b, r) \\
\qquad \vee \\
\qquad (\exists b')\, (Acc(WrongBag(b'), s) \wedge Holds(Empty(b'), z) \wedge \\
\qquad\qquad (\forall r')\, \neg Acc(WrongPackage(r'), s) \wedge \\
\qquad\quad z' = z - Empty(b') \circ Request(r_1, r) + Carries(b', r)) \\
\qquad \vee \\
\qquad (\exists r')\, (Acc(WrongPackage(r'), s) \wedge Holds(Request(r_1, r'), z) \wedge \\
\qquad\qquad (\forall b')\, \neg Acc(WrongBag(b'), s) \wedge \\
\qquad\quad z' = z - Empty(b) \circ Request(r_1, r') + Carries(b, r')) \\
\qquad \vee \\
\qquad (\exists b', r')\, (Acc(WrongBag(b'), s) \wedge Holds(Empty(b'), z) \wedge \\
\qquad\qquad Acc(WrongPackage(r'), s) \wedge Holds(Request(r_1, r'), z) \wedge \\
\qquad\quad z' = z - Empty(b') \circ Request(r_1, r') + Carries(b', r'))]\,))\,]
\end{array}
$$

Put in words, if no accident happens then again the action has its usual effects. The other three disjuncts account for the user putting, respectively, the right package into the wrong bag, the wrong package into the right bag, and the wrong package into the wrong bag. The precondition axiom is adapted accordingly:

$$
\begin{array}{l}
Poss(Pickup(b, r), s) \equiv \\
\quad [(\forall b')\, \neg Acc(WrongBag(b'), s) \supset Holds(Empty(b), s)] \\
\quad \wedge \\
\quad [(\forall b')\, (Acc(WrongBag(b'), s) \supset Holds(Empty(b'), s))] \\
\quad \wedge \\
\quad (\exists r_1)\, (Holds(At(r_1), s) \wedge \\
\qquad [(\forall r')\, \neg Acc(WrongPackage(r'), s) \supset Holds(Request(r_1, r), s)] \\
\qquad \wedge \\
\qquad [(\forall r')\, (Acc(WrongPackage(r'), s) \supset Holds(Request(r_1, r'), s))]\,)
\end{array}
$$

That is to say, if the intended bag is chosen it must be empty, otherwise any wrongly chosen one must be so. Likewise, if the right package is chosen there must be a corresponding request, otherwise there must be one for any wrongly chosen package.

The extended update axioms help the mailbot to explain and to recover after having made unexpected observations. Consider, for example, the following scenario. Initially, the robot is at room 4, carries in bag 1 a package for 5, and there are requests from room 4 to both 1 and 6, and from room 6 to 2 (see also Figure 10.1(a)):

$$\begin{array}{l} KState(S_0, z) \equiv z = At(4) \circ Carries(1,5) \circ Empty(2) \circ Empty(3) \circ \\ \qquad Request(4,1) \circ Request(4,6) \circ Request(6,2) \end{array}$$

Suppose that the robot considers to execute the following sequence of actions:

$$\begin{array}{l} S_1 = Do(Pickup(3,6), S_0) \\ S_2 = Do(Go(Up), S_1) \\ S_3 = Do(Deliver(1), S_2) \\ S_4 = Do(Go(Up), S_3) \\ S_5 = Do(Deliver(3), S_4) \end{array}$$

Without further observation, the default theory admits a unique extension, in which there are no accidents and the update axioms imply the following expectation concerning the state in situation S_4 (cf. Figure 10.1(b)):

$$\begin{array}{l} KState(S_4, z) \equiv z = At(6) \circ Empty(1) \circ Empty(2) \circ Carries(3,6) \circ \\ \qquad Request(4,1) \circ Request(6,2) \end{array}$$

But suppose that when action $Deliver(3)$ is executed in situation S_4, the user in room 6 informs the robot that mail bag 3 actually contains a package with the wrong address, as illustrated in Figure 10.1(c). This observation can be formally expressed as $(\exists r)\,(Holds(Carries(3,r), S_4) \wedge r \neq 6)$. Consider the fluent calculus default theory which consists of this single observation along with the extended mailbot domain axiomatization, including the initial knowledge state from above. Of the foregoing actions, only $Pickup(3,6)$ in situation S_0 could have had the effect of introducing a package into bag 3. From the knowledge update axiom and the observation it follows that $Acc(WrongPackage(r), S_0)$ for some $r \neq 6$. The precondition axiom then implies that $Holds(Request(4,r), S_0)$. According to the given initial state, $r = 1$. Since it is consistent to assume away any other accident, the unique extension therefore contains the sole accident $Acc(WrongPackage(1), S_0)$. With this, the default theory with the troublesome observation entails

$$\begin{array}{l} KState(S_4, z) \equiv z = At(6) \circ Empty(1) \circ Empty(2) \circ Carries(3,1) \circ \\ \qquad Request(4,6) \circ Request(6,2) \end{array}$$

This shows that as a side-effect the explanation generates reasonable expectations concerning the state in situation S_4 in which the unexpected observation has been made: Bag 3 is assumed to contain the package for room 1 while the package for 6 is still waiting in room 4 (Figure 10.1(d)). This is an example

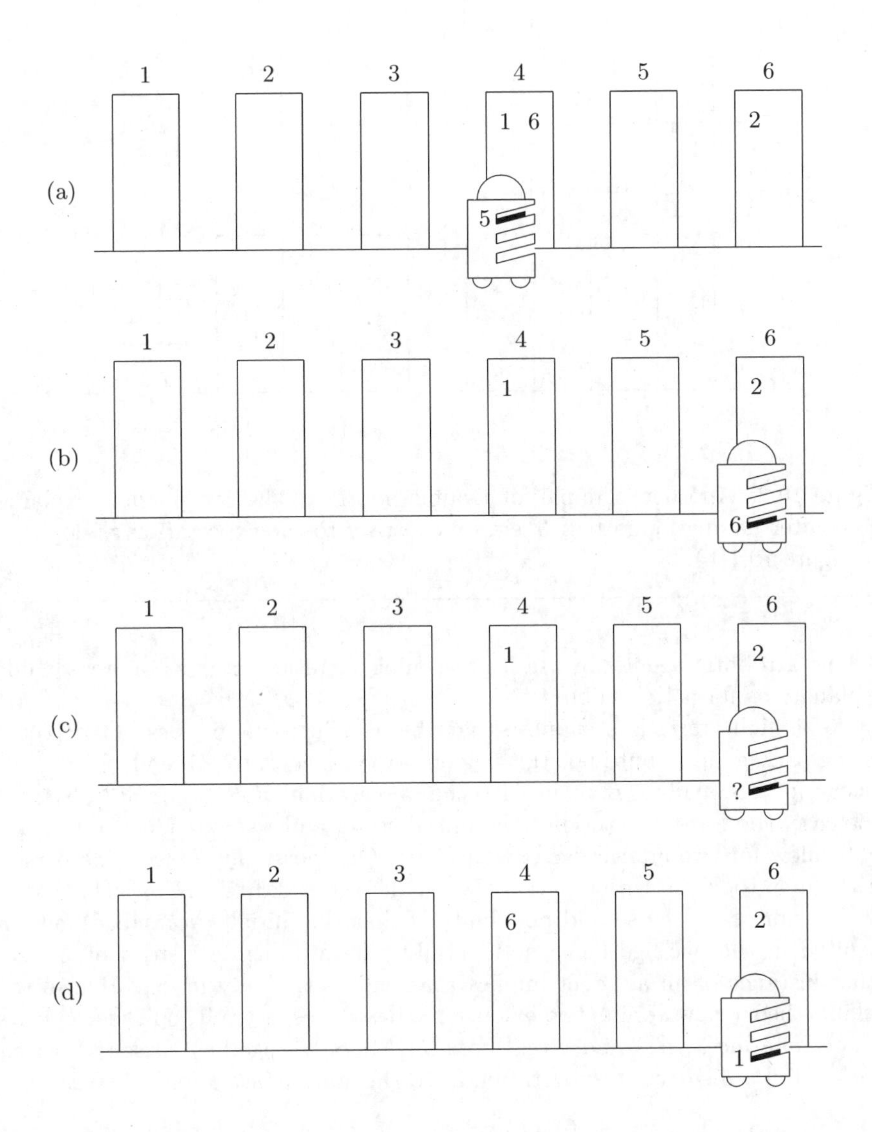

Figure 10.1: Example for troubleshooting: (a) initial state, (b) expected state four actions later, (c) package in bag 3 found not to be for room 6, (d) expected state after explaining.

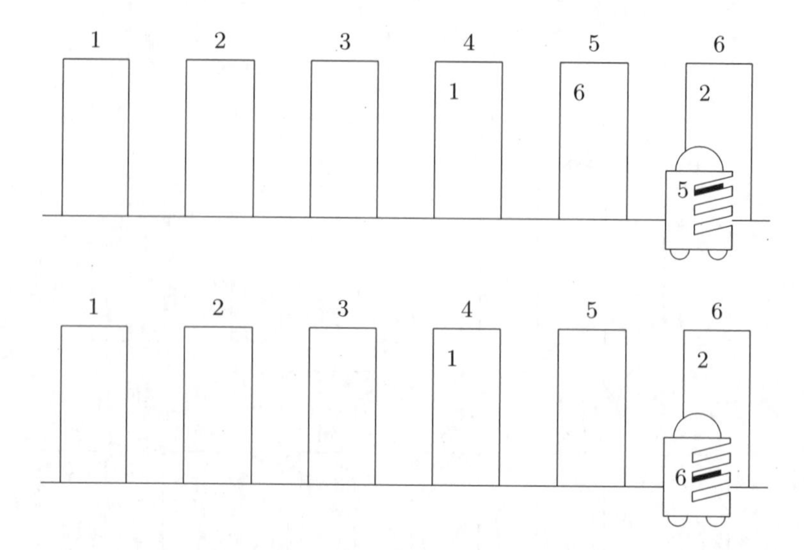

Figure 10.2: Another example for troubleshooting: The two possible explanations after having found bag 3 empty contrary to the expectation as depicted in Figure 10.1(b).

---

of how a robotic agent can use its reasoning faculties to arrive at reasonable explanations for action failure.

As a slight variation, suppose that the user in room 6 does not detect a wrong package in 3 but finds this bag empty. Consider, to this end, the default theory for the mailbot domain with the observation $Holds(Empty(3), S_4)$. In this case, the foregoing actions, the initial state, and the extended update axioms allow for two alternative explanations: One possibility is that the person in room 5 took the wrong package when the robot did the $Deliver(1)$ action in situation S_2. The second possibility is that the initial $Pickup(3, 6)$ action resulted in the package being accidentally put into bag 2 instead of 3. No other instance of an accident implies that bag 3 is empty in S_4. Hence, the default theory now admits two extensions, based, respectively, on the accidents $Acc(WrongBag(3), S_2)$ and $Acc(WrongBag(2), S_0)$. Together, these two extensions entail a disjunctive expectation as to the state in situation S_4, viz.

$$\begin{aligned} KState(S_4, z) \equiv\ & z = At(6) \circ Carries(1,5) \circ Empty(2) \circ Empty(3) \circ \\ & \quad Request(4,1) \circ Request(6,2) \circ Request(5,6) \\ & \vee \\ & z = At(6) \circ Empty(1) \circ Carries(2,6) \circ Empty(3) \circ \\ & \quad Request(4,1) \circ Request(6,2) \end{aligned} \tag{10.5}$$

According to the first explanation, the package for room 5 is still in bag 1 and the package for 6 is now in room 5, while the second explanation implies that the package for 6 should be in bag 2 (cf. Figure 10.2). □

10.2 Preferred Explanations

The more potential accidents are included in a specification, the less erroneous situations will leave the robot without explanation. On the other hand, accounting for even the most unlikely of errors has the disadvantage that the same problem may allow for many different explanations. Explanations come in form of extensions of the underlying fluent calculus default theory, and in most cases each extension implies different assumptions about the current state. By definition, a default theory entails only what can be derived in all extensions. Hence, in case of multiple extensions the robot can derive a mere disjunction covering all possible explanations, as in (10.5) in our last example. With too many explanations, the information that can thus be extracted may be too little for the robot to continue.

In order to get around this problem, agents should select among possible extensions and continue under the hypothesis of one particular explanation. This selection could be based on experience concerning the relative likelihood of possible errors, but it may also be done arbitrarily if no such information is available. The agent should in any case be ready to revise its hypothesis should subsequent observations still suggest a discrepancy between the expectations and reality. For example, if the mail delivery robot of Figure 10.2 selects the first explanation, then it would eventually go back to room 5 to claim the wrongly delivered package. If this action fails, too, then there is still the second explanation according to which the lost package is in bag 2.

The selection among different explanations, i.e., extensions, is based on a preference ordering among the underlying defaults.

A **prioritized default theory** Δ is a triple $(D, W, <)$ where (D, W) is a standard default theory and $<$ a total ordering over the elements of D. (This suffices for our purpose; in a more general definition the ordering can be partial.) A **preferred extension** E of Δ is an extension of (D, W) which satisfies the following: For every extension E' of (D, W) and every default $\frac{:\gamma'}{\gamma'}$ in D, if $\gamma' \in E'$ but $\gamma' \notin E$, then there is a default $\frac{:\gamma}{\gamma} < \frac{:\gamma'}{\gamma'}$ in D such that $\gamma \in E$ but $\gamma \notin E'$. A prioritized default theory **entails** a formula if the latter holds in all preferred extensions.

Consider, e.g., the prioritized default theory $(\{\frac{:\neg P(x)}{\neg P(x)}\}, \{P(A) \vee P(B)\}, <)$ such that $\frac{:\neg P(A)}{\neg P(A)} < \frac{:\neg P(B)}{\neg P(B)} < \frac{:\neg P(C)}{\neg P(C)}$. The unique preferred extension is

$$E_2 = \mathcal{T}(\{\neg P(A), P(B), \neg P(C)\})$$

The other extension, $E_1 = \mathcal{T}(\{P(A), \neg P(B), \neg P(C)\})$, of the non-prioritized default theory is not preferred because in comparison with E_2 there is no 'compensation' for not having applied the default $\frac{:\neg P(A)}{\neg P(A)}$.

In order to use the concept of preferred extensions for troubleshooting, the ground instances of the generic default for accidents need to be ordered. If some errors are known to be more likely than others, this could be reflected in the ordering. If no such information is available, the most efficient way is to specify the ordering such that the robot always takes the first explanation it finds.

Definition 10.5 Consider a fluent calculus axiomatization Σ based on a signature with accidents. A **prioritized fluent calculus default theory for accidental qualifications** is of the form

$$\Delta = \left(\left\{ \frac{: \neg Acc(c,s)}{\neg Acc(c,s)} \right\}, \Sigma \cup O, < \right)$$

where O is a (possibly empty) set of situation formulas. □

By computing preferred extensions, agents make a selection among the possible explanations for an observed inconsistency. If additional observations suggest that the chosen hypothesis turns out to be incorrect, then again the notion of preferred extensions helps to select the now most plausible explanation.

Example 8 (continued) Suppose the simple domain of Section 10.1 is augmented by the fluent G. Suppose further that under rare circumstances, action A not only fails to achieve effect F, but instead produces the anomalous effect that G becomes true. Let this accident be denoted by C' : ACCIDENT, then knowledge update axiom (10.1) is extended as follows:

$$\begin{array}{l} KState(Do(A,s),z') \equiv (\exists z)\,(KState(s,z) \wedge \\ \quad (\forall c)\neg Acc(c,s) \wedge z' = z + F \\ \quad \vee \\ \quad Acc(C,s) \wedge \neg Acc(C',s) \wedge z' = z \\ \quad \vee \\ \quad Acc(C',s) \wedge \neg Acc(C,s) \wedge z' = z + G\,) \end{array} \tag{10.6}$$

Let $<$ be any total ordering which satisfies

$$\frac{: \neg Acc(C',\sigma)}{\neg Acc(C',\sigma)} < \frac{: \neg Acc(C,\sigma)}{\neg Acc(C,\sigma)} \tag{10.7}$$

for all ground situations σ, and suppose that G is false initially:

$$KState(S_0, z) \equiv \neg Holds(G, z) \tag{10.8}$$

Consider the case where after executing A, the agent observes that unexpectedly F does not hold. The fluent calculus default theory with initial knowledge state (10.8) and observation $\neg Holds(F, Do(A, S_0))$ admits two extensions E_1 and E_2 such that

$$Acc(C,S_0), \neg Acc(C',S_0) \in E_1 \quad \text{and} \quad Acc(C',S_0), \neg Acc(C,S_0) \in E_2$$

From (10.7) it follows that E_1 but not E_2 is a preferred extension. Consequently, the prioritized default theory entails $Knows(\neg F \wedge \neg G, Do(A, S_0))$. Should the agent additionally observe, however, that G holds after A, that is, $Holds(G, Do(A, S_0))$, then E_1 is no longer an extension. Now being the only extension, E_2 becomes preferred, and hence the default theory entails that $Knows(\neg F \wedge G, Do(A, S_0))$. □

Defining a preference ordering will also help the mail delivery robot making a selection among alternative explanations.

Example 1 (continued) Suppose that a user of the delivery robot accidentally inserting the right package into the wrong bag happens more frequently than selecting the wrong package. This is stipulated by

$$\frac{: \neg Acc(WrongPackage(\rho), \sigma)}{\neg Acc(WrongPackage(\rho), \sigma)} < \frac{: \neg Acc(WrongBag(\beta), \sigma)}{\neg Acc(WrongBag(\beta), \sigma)} \tag{10.9}$$

for all $\rho, \beta \in \mathbb{N}$ and all ground situations σ. Moreover, let more recent abnormalities be preferred in general. (This will allow the FLUX control program to choose the first explanation that is computed.) Formally,

$$\frac{: \neg Acc(\chi, \sigma)}{\neg Acc(\chi, \sigma)} < \frac{: \neg Acc(\chi', Do(\alpha, \sigma))}{\neg Acc(\chi', Do(\alpha, \sigma))} \tag{10.10}$$

for all ground accidents χ, χ', situations σ, and actions α.

Recall, for instance, the initial situation depicted in Figure 10.1(a) and the action sequence that leads to the expected state shown in Figure 10.1(b). As in the preceding section, suppose that the robot makes the unexpected observation that $Holds(Empty(3), S_4)$. We have seen that the corresponding non-prioritized fluent calculus default theory admits two extensions, one of which is obtained by applying all defaults but

$$\frac{: \neg Acc(WrongBag(3), S_2)}{\neg Acc(WrongBag(3), S_2)}$$

and one of which by applying all defaults but

$$\frac{: \neg Acc(WrongBag(2), S_0)}{\neg Acc(WrongBag(2), S_0)} \tag{10.11}$$

(see again Figure 10.2). Under the preference ordering according to (10.10), the latter default is preferred over the former due to the different situation arguments. Hence, the only preferred extension is the one where the conclusion of (10.11) holds, that is, the explanation illustrated in the top of Figure 10.2.

Suppose, for the sake of argument, that having chosen this explanation, the robot tries to proceed with the action $Pickup(2, 2)$. It is then told that actually bag 2 is not empty. This observation is inconsistent with the hypothesis. Formally speaking, after adding the observation $\neg Holds(Empty(2), S_5)$, the default theory admits only the second explanation—where $Acc(WrongBag(2), S_0)$ holds—as extension, which now becomes the preferred one.

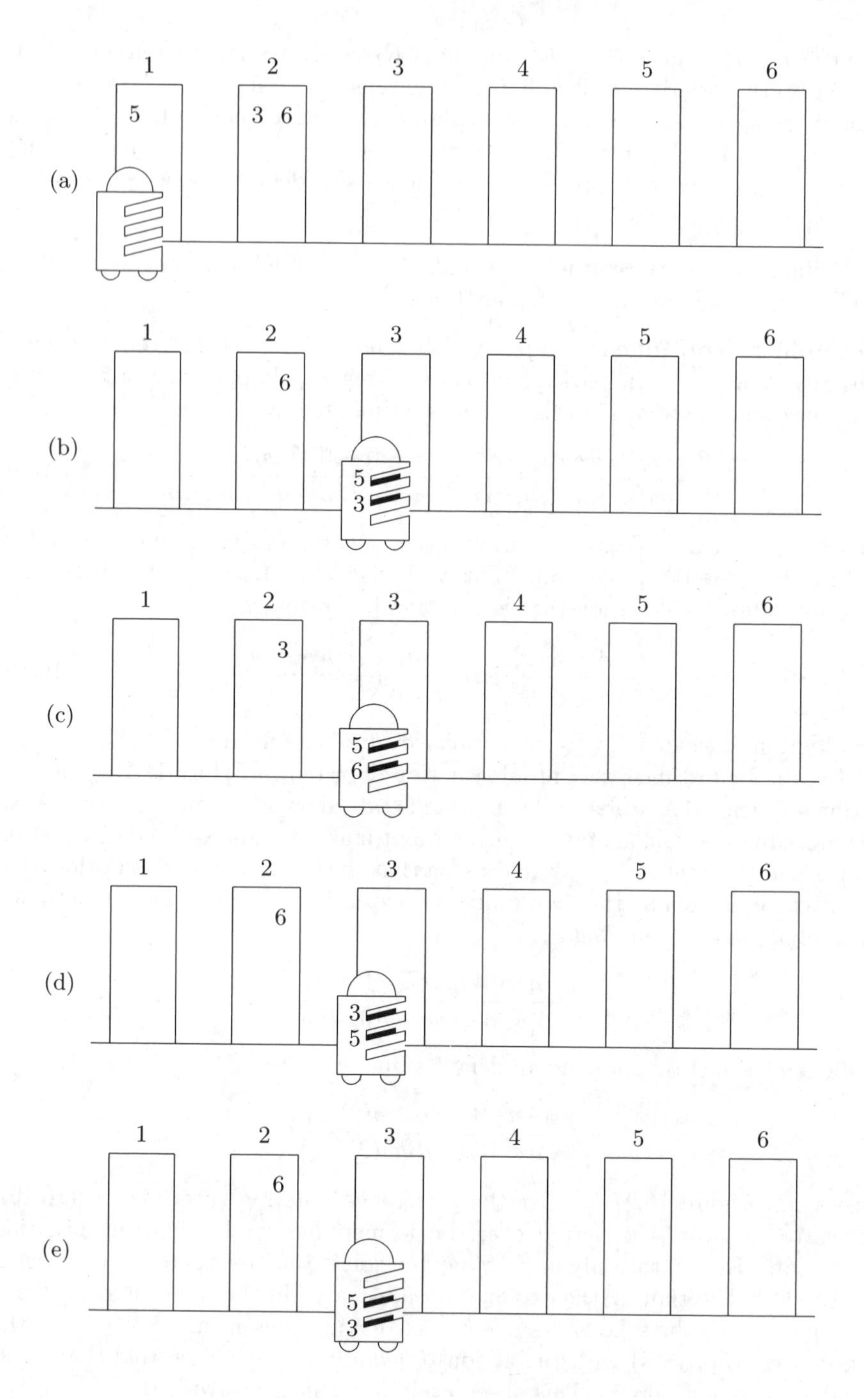

Figure 10.3: (a) initial state, (b) expected state four actions later, (c)–(e) three different explanations for having found the wrong package in bag 2.

As another interesting example, consider the following initial situation (see also Figure 10.3(a)):

$$KState(S_0, z) \equiv z = At(1) \circ Empty(1) \circ Empty(2) \circ Empty(3) \circ Request(1,5) \circ Request(2,3) \circ Request(2,6) \tag{10.12}$$

Take the situation where the robot arrives at room 3 after having put the package for room 5 into mail bag 1 and the package for 3 into bag 2, that is,

$$S_4 = Do(Go(Up), Do(Pickup(2,3), Do(Go(Up), Do(Pickup(1,5), S_0))))$$

Suppose the user in room 3 tells the robot that the package in bag 2 is not his (contrary to the expectations, cf. Figure 10.3(b)), that is,

$$(\exists r)\,(Holds(Carries(2,r), S_4) \wedge r \neq 3) \tag{10.13}$$

The underlying default theory admits three extensions as explanations, viz.

$$\begin{array}{r} Acc(WrongPackage(6), S_2) \in E_1 \\ Acc(WrongBag(2), S_0) \wedge Acc(WrongBag(1), S_2) \in E_2 \\ Acc(WrongBag(2), S_0) \wedge Acc(WrongBag(3), S_2) \in E_3 \end{array} \tag{10.14}$$

where $S_2 = Do(Go(Up), Do(Pickup(1,5), S_0))$. Put in words, in the first case the user in room 2 handed over the wrong package (Figure 10.3(c)), while in the two other cases the user in room 1 has put the package into bag 2 and the user in room 2 his package into either bag 1 or bag 3 (Figure 10.3(d) and (e)). According to the preference specification (cf. (10.9)–(10.10)), the fact that the default on $\neg Acc(WrongPackage(6), S_2)$ is not applied in E_1 is compensated for by having applied the (stronger) default on $\neg Acc(WrongBag(2), S_0)$. Similarly for E_3. Consequently, E_1 is preferred. There is no such compensation in E_2 or E_3 for not applying the default on $\neg Acc(WrongBag(2), S_0)$. Consequently, neither E_2 nor E_3 is preferred. □

10.3 Troubleshooting in FLUX

The need for troubleshooting arises when during the computation of a FLUX program an observation, i.e., sensing result, is inconsistent with the world model. Observations are evaluated against the internal model whenever this clause (cf. Figure 5.9, page 131) is applied:

$$Execute(a, z_1, z_2) \leftarrow Perform(a, \vec{y}),\ StateUpdate(z_1, a, z_2, \vec{y})$$

The sensing result $\vec{y}$ obtained while performing action a is incorporated into the world model when resolving $StateUpdate(z_1, a, z_2, \vec{y})$. If the observation contradicts the model, then this predicate fails. This is the point where troubleshooting is called for.

Troubleshooting means to postulate accidents that may have happened in one or more of the foregoing actions. For an efficient encoding of state update axioms with accidents, the standard predicate *StateUpdate* is accompanied by a new predicate *AbStateUpdate*, by which the accidental effects of an action are specified. Exploiting the fixed computation strategy of Prolog, the order in which possibly multiple accidents are encoded should reflect the preference ordering among the corresponding defaults. In this way, a less likely accident is considered (via backtracking) only if all other, preferred accidents fail to explain the troublesome observation. Writing state update axioms in this fashion avoids the explicit encoding of the predicate *Acc* and the underlying default theory. The construction of (preferred) extensions as explanations is then implicit in the Prolog computation.

Example 8 (continued) The simple, abstract knowledge update axiom of (10.6) can be encoded in FLUX as follows:

```
state_update(Z1,a,Z2,[]) :-
   update(Z1,[f],[],Z2).

ab_state_update(Z1,a,Z2,[]) :-
   Z2=Z1 ;
   update(Z1,[g],[],Z2).
```

The first clause describes the normal effect of the action. In the second clause, the first one of the two alternative updates describes the effect under accident C (that is, nothing changes) while the second possibility accounts for accident C' (that is, effect G materializes instead of F). The order in which the two accidental updates are specified reflects the preference of assuming $\neg Acc(C', \sigma)$ over $\neg Acc(C, \sigma)$. If there is no observation to the contrary, the encoding allows to predict the usual effects of action A:

```
?- not_holds(g,Z0),
   state_update(Z0,a,Z1,[]).

Z1=[f|Z0]
not_holds(g,Z0)
```

If, on the other hand, the agent makes the surprising observation that F is false after A, then the normal update fails. When considering the abnormal update, the explanation is preferred by which the state remains unchanged:

```
?- not_holds(g,Z0),
   ( state_update(Z0,a,Z1,[]) ; ab_state_update(Z0,a,Z1,[]) ),
   not_holds(f,Z1).

Z0=Z1
```

The less likely explanation, that is, the accidental arising of G, is granted only if the observation suggests that there is no other possibility:

```
ab_res([],Z) :- init(Z).
ab_res([[A,Y]|S],Z) :- ab_res(S,Z1),
                       ( state_update(Z1,A,Z,Y)
                         ;
                         ab_state_update(Z1,A,Z,Y) ).

execute(A,Z1,Z2) :-
   perform(A,Y)   -> ( nonvar(Z1), Z1=[sit(S)|Z], ! ; S=[], Z=Z1 ),
                     ( state_update(Z,A,Z3,Y) ; ab_res([[A,Y]|S],Z3) ),
                     !, Z2=[sit([[A,Y]|S])|Z3] ;
   A=[A1|A2]      -> execute(A1,Z1,Z), execute(A2,Z,Z2) ;
   A=if(F,A1,A2) -> (holds(F,Z1) -> execute(A1,Z1,Z2)
                                  ; execute(A2,Z1,Z2)) ;
   A=[]           -> Z1=Z2 ;
   complex_action(A,Z1,Z2).
```

Figure 10.4: A universal definition of troubleshooting.

```
?- not_holds(g,Z0),
   ( state_update(Z0,a,Z1,[]) ; ab_state_update(Z0,a,Z1,[]) ),
   holds(g,Z1).

Z1=[g|Z0]
```

□

Troubleshooting requires to find a sequence of updates which is consistent with an unexpected observation. Of course, any explanation must also be consistent with any state information the agent has acquired prior to the troublesome observation. The leading two clauses in Figure 10.4 constitute a universal definition of troubleshooting in FLUX by a new predicate called *AbRes*. Its first argument is a sequence of pairs $(a_n, \vec{y}_n), \ldots, (a_1, \vec{y}_1)$ such that the situation $Do(a_n, \ldots, Do(a_1, S_0) \ldots)$ is the one that requires troubleshooting, and $\vec{y}_1, \ldots, \vec{y}_n$ encodes the sensing results obtained in the course of this action history. An instance $AbRes(\sigma, z)$ is true if z is a FLUX state reachable from the initial state by a sequence of updates according to the actions and observations in σ. Because of the fixed computation rule in Prolog, the first computed answer substitution for z depends on the order in which the alternative updates are specified in the encodings of the update axioms. In particular, the very first attempt to compute an answer for $AbRes(\sigma, z)$ is by assuming no accident at all, using just regular updates via *StateUpdate*. Only if this is inconsistent with the observations in σ will the first computed answer postulate the occurrence of one or more accidental updates via *AbStateUpdate*. The order in which the answers are computed then defines a particular preference ordering among the accidents: Later accidents are generally preferred, while the preferences among alternative updates for the same action are according to the order of their oc-

```
state_update(Z1,pickup(B,R),Z2,[NotEmpty]) :-
   NotEmpty=false, holds(empty(B),Z1),
   holds(at(R1),Z1),
   update(Z1,[carries(B,R)],[empty(B),request(R1,R)],Z2).

state_update(Z1,deliver(B),Z2,[Empty,WrongAddress]) :-
   Empty=false, WrongAddress=false, holds(carries(B,R),Z1),
   holds(at(R),Z1),
   update(Z1,[empty(B)],[carries(B,R)],Z2).
```

Figure 10.5: Encoding of the normal effects of the mailbot actions.

currence in the respective update axiom.

The second part of Figure 10.4 shows how troubleshooting is integrated into the definition of execution in FLUX (taken from Chapter 6; Figure 6.12, page 163). Troubleshooting requires having recorded the foregoing actions and sensing results. These are added to the world model using the special fluent *Sit*. Its argument is initialized with the empty list. This situation fluent is always subtracted prior to inferring the update, and added again afterwards, augmented by the latest action and observation. If the update fails, then troubleshooting is carried out via *AbRes*.

Example 1 (continued) Figure 10.5 and 10.6 show the FLUX encoding of the extended update axioms for *Pickup* and *Deliver*, distinguishing between the normal and the accidental effects. As a means of informing the mailbot that something went wrong, the two axioms have been extended by a sensing component. In case of *Pickup*, the robot can be told that the offered bag is not empty. Likewise, when the robot attempts a *Deliver* action it is informed whether the bag in question is empty or contains a package for a different person. Under these circumstances, the actions do not have any physical effect. As this indicates that some accident must have happended, the standard update axiom only allows the values *False* for the sensed properties. The encodings of precondition axioms are not adapted to accidents for the sake of efficiency, because they are used by FLUX agents solely for predicting the executability of actions, for which accidents are assumed away anyway.

As an example, recall the scenario depicted in Figure 10.1(c), where the latest delivery fails because bag 3 contains a package for a different user. The computed answer for z_5 in the following query corresponds to the explanation depicted in Figure 10.1(d):

```
init(Z0) :- Z0 = [at(4),carries(1,5),empty(2),empty(3),
                  request(4,1),request(4,6),request(6,2) | Z],
            not_holds_all(request(_,_), Z),
            consistent(Z0).
```

```
ab_state_update(Z1,pickup(B,R),Z2,[NotEmpty]) :-
   NotEmpty=false,
   holds(at(R1),Z1),
   ( holds(empty(BP),Z1), BP #\= B, holds(request(R1,R),Z1),
        update(Z1,[carries(BP,R)],[empty(BP),request(R1,R)],Z2) ;
     holds(empty(B),Z1), holds(request(R1,RP),Z1), RP #\= R,
        update(Z1,[carries(B,RP)],[empty(B),request(R1,RP)],Z2) ;
     holds(empty(BP),Z1), holds(request(R1,RP),Z1), BP #\= B, RP #\= R,
        update(Z1,[carries(BP,RP)],[empty(BP),request(R1,RP)],Z2) ) ;
   NotEmpty=true, not_holds(empty(B),Z1), Z2 = Z1.

ab_state_update(Z1,deliver(B),Z2,[Empty,WrongAddress]) :-
   Empty=false, WrongAddress=false,
   holds(at(R),Z1),
   ( holds(carries(BP,R),Z1), BP #\= B,
        update(Z1,[empty(BP)],[carries(BP,R)],Z2) ;
     holds(carries(BP,RP),Z1), RP #\= R,
        update(Z1,[empty(BP),request(R,RP)],[carries(BP,RP)],Z2) ) ;
   Empty=true, holds(empty(B),Z1), Z2 = Z1 ;
   WrongAddress=true, holds(at(R),Z1), holds(carries(B,RP),Z1),
   RP #\= R, Z2 = Z1.
```

Figure 10.6: Encoding of the accidental effects of the mailbot actions.

```
?- ab_res([[deliver(3), [false,true]],  [go(up),[]],
           [deliver(1), [false,false]], [go(up),[]],
           [pickup(3,6),[false]]],      Z5).

Z5=[at(6),empty(1),carries(3,1),empty(2),
   request(4,6),request(6,2)|Z]
```

Next, consider the case where the bag is observed to be empty when attempting to execute *Deliver*(3). The first computed explanation is the one shown in the top of Figure 10.2. The one shown in the bottom, which is less preferred, is computed as the secondary answer:

```
?- ab_res([[deliver(3), [true,false]],  [go(up),[]],
           [deliver(1), [false,false]], [go(up),[]],
           [pickup(3,6),[false]]],      Z5).

Z5=[at(6),request(5,6),empty(3),carries(1,5),empty(2),
   request(4,1),request(6,2)|Z]                          More?

Z5=[at(6),empty(1),carries(2,6),empty(3),
   request(4,1),request(6,2)|Z]
```

In fact, there are four more answers, which postulate super-sets of the accidents granted in the given two answers, hence which do not even constitute extensions of the default theory, let alone preferred ones.

Finally, recall the scenario with initial situation (10.12) and troublesome observation (10.13) discussed at the end of Section 10.2:

```
init(Z0) :- Z0 = [at(1),empty(1),empty(2),empty(3),
                  request(1,5),request(2,3),request(2,6) | Z],
            not_holds_all(request(_,_), Z),
            consistent(Z0).

?- ab_res([[deliver(2), [false,true]], [go(up),[]],
           [pickup(2,3),[false]],       [go(up),[]],
           [pickup(1,5),[false]]],      Z5).

Z5=[at(3),carries(2,6),carries(1,5),empty(3),request(2,3)|Z]
```

The inferred FLUX state corresponds to the preferred explanation which we have derived above (viz. E_1 in (10.14)). □

10.4 Persistent Qualifications

Accidents are only postulated in retrospect and never predicted: Without observation about future situations, it is always consistent to assume that all actions will have their normal effects; hence, no extension will make the prediction that there be a particular accident in the future. This is independent of any previous accident. If, for example, the mailbot has concluded that it must have got a wrong package, then it will nonetheless assume, by default, that it will get the right one next time it asks the same person.

Robots may, however, encounter problems during the execution of a program of which we desire that they are predicted to persist. Like accidents, these **abnormalities** are to be assumed away by default at the beginning. Yet once it is observed, an abnormality is assumed to stay on as the program continues, unless explicit measures are taken to overcome the problem.

A typical example from the mail delivery world is if at some point some room is not accessible, or the person in that room is not reachable. If the mailbot makes an observation to this extent, it should not assume that this is without effect on the continuation of the program. If being unreachable were postulated as a mere accident, then the robot would simply try the same action again and again, a rather unfruitful behavior. Instead of continually interpreting every unsusccessful attempt as an accident of the past, the mailbot should be programmed so as to record the problem as a more serious abnormality. Thus we can prepare the robot for taking measures, such as automatically sending an E-Mail to the person in question, asking him or her for a confirmation when back in office. The robot will then assume that a person is unreachable until this exogenous action (that is, the confirmation) happens, with the explicit effect of causing the abnormality to disappear.

Abnormalities are of sort FLUENT, because they represent properties of states that, once there, do not change unless by actions which effect such a change. Still, abnormalities represent exceptional circumstances and should be assumed away if possible. It is tempting to introduce a generic default rule in the spirit of the one used for accidents, that is,

$$\frac{: \neg Holds(Ab(x), s)}{\neg Holds(Ab(x), s)} \tag{10.15}$$

where the fluent $Ab(x)$ represents an abnormality. The conclusion of this default would be made whenever it is consistent to assume that $Ab(x)$ is false in a situation s. If, on the other hand, the observations suggest that $Holds(Ab(x), \sigma)$ in some situation σ, then the conclusion of (10.15) with $\{s/\sigma\}$ would not be made. The default conclusion would then also not be made with a successor situation $\{s/Do(\alpha, \sigma)\}$ unless the update axiom for α implies that $Ab(x)$ is subtracted from $State(\sigma)$. In this way, the update axioms ensure that (10.15) does not apply until a situation is reached with an action whose effect is that $Ab(x)$ becomes false.

Despite being intuitive, default rule (10.15) is unsuited because it gives rise to **anomalous extensions**.

Example 9 Consider a simple domain signature with just the two fluents $Ab(1)$ and $Ab(2)$ and one action A. The action shall be possible unless $Ab(1)$ holds. The effect of A is to bring about $Ab(2)$. To get an intuition, suppose A is the (exogenous) action of informing the mailbot that a user leaves his office. This causes him to be unreachable, represented by $Ab(2)$. The action is possible unless the wireless connection to the robot is down, represented by $Ab(1)$.

Without any observation, the intuitively expected conclusion would be that action A is possible (by default, the connection is working) in situation S_0. Consequently, $Ab(2)$ should hold in situation $Do(A, S_0)$ (that is, the user is expected not to be reachable after having received the information). Formally:

1. $\neg Holds(Ab(1), S_0)$;
2. $Poss(A, S_0)$;
3. $Holds(Ab(2), Do(A, S_0))$.

However, none of these statements is entailed if default rule (10.15) were used. To see why, consider the precondition and knowledge update axiom for A,

$$\begin{aligned} & Poss(A, s) \equiv \neg Holds(Ab(1), s) \\ & Poss(A, s) \supset \\ & \quad [KState(Do(A, s), z') \equiv (\exists z)\,(KState(s, z) \wedge z' = z + Ab(2))] \end{aligned} \tag{10.16}$$

along with the foundational axioms of fluent calculus. The default theory with these axioms and with default rule (10.15) admits two extensions. The first one, call it E_1, contains

$$\neg Holds(Ab(1), S_0),\ Holds(Ab(2), Do(A, S_0))$$

as expected. There is, however, a second extension E_2 containing

$$Holds(Ab(1), S_0),\ \neg Holds(Ab(2), Do(A, S_0))$$

This extension is obtained by applying default rule (10.15) with the default conclusion $\neg Holds(Ab(2), Do(A, S_0))$, which implies $Holds(Ab(1), S_0)$ by way of the domain axioms in (10.16). In this second extension, none of the intended conclusions is true, hence they are not entailed by the default theory.[2] □

The reason for anomalous models to arise is that default rule (10.15) does not allow to tell apart abnormalities that are explicitly effected by an action. Conditions may be abnormal *a priori*, but once they are the effect of an action, it is inappropriate to still assume them away by default. The problem of anomalous models therefore requires to assume away not the abnormalities themselves but their unexpected arising. This makes it necessary to adopt the notion of causal relationships and ramifications of the previous chapter.

Definition 10.6 A tuple $\mathcal{S} \cup \langle \mathcal{B}, Ab, AbCaused \rangle$ is a **fluent calculus signature for qualifications** if $\mathcal{S}$ is a fluent calculus signature for both ramifications and accidental qualifications, and

- $\mathcal{B}$ finite set of function symbols into sort ABNORMALITY
- Ab : ABNORMALITY $\mapsto$ FLUENT
- $AbCaused$: ABNORMALITY $\times$ SIT.

An **abnormality** is a term of sort ABNORMALITY. □

The set $\mathcal{B}$ contains the abnormalities of a domain. The fluents $Ab(x)$ then indicate whether abnormality x is true in a state. Usually, abnormalities do not hold unless they are effected by an action that has occurred. However, just like in case of accidents the occurrence of an abnormality can be postulated in order to explain observations that contradict the expectations of a robot. This is the purpose of predicate $AbCaused(x, s)$. It is false by default, but if postulated it brings about effect $Ab(x)$ in situation s. As this would always be a side-effect of the action that leads to s, it requires a causal relationship as given in following definition. The default assumption $\neg AbCaused(x, s)$ is accompanied by the assumption that abnormalities do not hold initially.

[2] To better understand why E_2 is indeed unwanted, it may be helpful to give a probabilistic account of the scenario. Let $P(Ab(1) = False) = P(Ab(2) = False) = 0.99$ be the *a priori* likelihood of, respectively, the connection being down and the user being unreachable. By the effect of action A, the *conditional* probability $P(Ab(2) = True \mid Ab(1) = False)$ is 1 (because if the connection is not down, the action does have the usual effect). It follows that the joint probability distribution satisfies $P(Ab(1) = False, Ab(2) = True) = 0.99$. Due to the high probability, this truth assignment (which corresponds to extension E_1) should be the default conclusion. The assignment which corresponds to the anomalous extension, E_2, satisfies $P(Ab(1) = True, Ab(2) = False) \leq 0.01$. In other words, although the *a priori* likelihood of the two abnormalities is the same, the conditional dependency causes the asymmetric distribution in the joint probability.

Definition 10.7 Consider a fluent calculus axiomatization Σ based on a signature for qualifications such that the causal relationships in Σ include

$$\begin{array}{l} AbCaused(x,s) \wedge \neg Holds(Ab(x),z) \supset \\ \quad Causes(z,p,n,z+Ab(x),p+Ab(x),n,s) \end{array} \tag{10.17}$$

A **fluent calculus prioritized default theory** is of the form

$$\Delta = \left(\left\{ \frac{:\neg Acc(c,s)}{\neg Acc(c,s)}, \frac{:\neg AbCaused(x,s)}{\neg AbCaused(x,s)}, \frac{:\neg Holds(Ab(x),S_0)}{\neg Holds(Ab(x),S_0)} \right\}, \Sigma \cup O, < \right)$$

where O is a (possibly empty) set of situation formulas. □

Example 9 (continued) Recall from above the domain with the two abnormalities $Ab(1)$ and $Ab(2)$ along with action A. Consider the precondition and update axioms

$$\begin{array}{l} Poss(A,s) \equiv \neg Holds(Ab(1),s) \\ Poss(A,s) \supset \\ \quad [KState(Do(A,s),z') \equiv (\exists z)\,(KState(s,z) \wedge \\ \qquad Ramify(z,Ab(2),\emptyset,z',Do(A,s)))] \end{array} \tag{10.18}$$

There are no other causal relationships besides (10.17) in this domain. The fluent calculus default theory with the empty set of observations admits a unique extension E, in which all default conclusions hold: From $\neg Holds(Ab(1),S_0)$ it follows that $Poss(Ab(1),S_0)$. Since $\neg AbCaused(x,s)$ for all abnormalities x and situations s, the causal relation induced by (10.17) is empty, and ramification reduces to the identity relation; hence, by the knowledge update axiom in (10.18), $Knows(Ab(2),Do(A,S_0))$. This abnormality follows without assuming $AbCaused(Ab(2),Do(A,S_0))$. In contrast, the anomalous extension from above would still require not to apply a default conclusion, namely, $\neg Holds(Ab(1),S_0)$. □

In the following, we extend the mailbot domain by an account of the abnormality that some user is not reachable.

Example 1 (continued) Consider $Reachable : \mathbb{N} \mapsto$ ABNORMALITY, where an instance of the fluent $Ab(Reachable(r))$ shall denote that the person in room r is currently not reachable. For the sake of simplicity, let us ignore accidents for the moment. The two actions of picking up and delivering packages cannot be performed if the person in question is not reachable:

$$\begin{array}{l} Poss(Pickup(b,r),z) \equiv \\ \quad Holds(Empty(b),z) \wedge \\ \quad (\exists r_1)\,(Holds(At(r_1),z) \wedge Holds(Request(r_1,r),z) \\ \qquad \wedge \neg Holds(Ab(Reachable(r_1)),z)) \\ \\ Poss(Deliver(b),z) \equiv \\ \quad (\exists r)\,(Holds(At(r),z) \wedge Holds(Carries(b,r),z) \\ \qquad \wedge \neg Holds(Ab(Reachable(r)),z)) \end{array} \tag{10.19}$$

```
poss(pickup(B,R),Z) :- knows_val([B1],empty(B1),Z), B=B1,
                       knows_val([R1,R2],request(R1,R2),Z), R=R2,
                       knows(at(R1),Z),
                       \+ knows(ab(reachable,R1),Z) .

poss(deliver(B),Z) :- knows_val([R],at(R),Z),
                      knows_val([B1],carries(B1,R),Z), B=B1,
                      \+ knows(ab(reachable,R),Z).
```

Figure 10.7: The precondition axioms for the mailbot with abnormalities.

The actions have their usual effects; however, the update axioms need to employ the expression *Ramify* to account for possible ramifications. Recall, for instance, the scenario depicted in the top of Figure 10.1, where

$$\begin{aligned} KState(S_0, z) \equiv\ & Holds(At(4), z) \wedge Holds(Empty(2), z) \wedge Holds(Empty(3), z) \\ & \wedge Holds(Request(4, 1), z) \wedge Holds(Request(4, 6), z) \wedge \\ & \wedge Consistent(z) \end{aligned}$$

Without additional observations, the corresponding fluent calculus default theory admits a unique extension, which includes $\neg Holds(Ab(Reachable(4)), S_0)$ and $\neg AbCaused(Reachable(4), Do(Pickup(2,1), S_0))$ as default consequences. Hence, by (10.19) and the update axiom for *Pickup* it follows that it is possible to first pick up in bag 2 the package for 1 and, then, to pick up in bag 3 the package for 6. Suppose, however, the first action is possible but not the second one:

$$Poss(Pickup(2,1), S_0) \wedge \neg Poss(Pickup(3,6), Do(Pickup(2,1), S_0))$$

The fact that the first action is possible implies $\neg Holds(Ab(Reachable(4)), S_0)$. According to precondition axioms (10.19) and the update axiom for *Pickup*, the only consistent explanation for failing to pick up the second package is $Holds(Ab(Reachable(4)), Do(Pickup(2,1), S_0))$. This, in turn, makes it necessary that $AbCaused(Reachable(4), Do(Pickup(2,1), S_0))$ holds, that is, in between the two actions the user must have become unavailable. A persistent qualification, this continues to be the case as the robot proceeds; consequently, the update axioms entail, e.g.,

$$Knows(Ab(Reachable(4)), Do(Go(Up), Do(Pickup(2,1), S_0)))$$

□

For an efficient encoding of abnormalities in FLUX, the explicit introduction of default rules should again be avoided. Instead of ensuring the absence of an abnormality by assuming it away if possible, the agent simply verifies that an abnormality is not known to be true. Figure 10.7 shows how this can

```
state_update(Z1,pickup(B,R),Z2,[NotEmpty,NotReachable]) :-
   NotReachable=false, ...

state_update(Z1,deliver(B),Z2,[Empty,WrongAddress,NotReachable]) :-
   NotReachable=false, ...

ab_state_update(Z1,pickup(B,R),Z2,[NotEmpty,NotReachable]) :-
   NotEmpty=false, NotReachable=false,
      ... ;
   NotEmpty=true, NotReachable=false, ... ;
   NotReachable=true,
      holds(at(R1),Z1), holds(ab(reachable,R1),Z1), Z2 = Z1.

ab_state_update(Z1,deliver(B),Z2,[Empty,WrongAddress,NotReachable]) :-
   Empty=false, WrongAddress=false, NotReachable=false,
      ... ;
   Empty=true, NotReachable=false, ... ;
   WrongAddress=true, NotReachable=false, ... ;
   NotReachable=true,
      holds(at(R),Z1), holds(ab(reachable,R),Z1), Z2 = Z1.
```

Figure 10.8: The extended update axioms for the mailbot with abnormalities.

be achieved using the principle of negation-as-failure in the encoding of the extended precondition axioms for the mailbot (cf. (10.19)). It is necessary to **flatten** the argument of an abnormality, so that, e.g., $Ab(Reachable(r))$ is encoded as $Ab(Reachable, r)$. This is required because the arguments of fluents must not be compound terms in order that they are correctly treated by the FLUX constraint solver.

To account for the fact that users may not be present, the updates for the actions *Pickup* and *Deliver* are both extended by yet another sensing component, which indicates whether or not the person in question is responding at all. The encoding of the normal update does not change besides requiring that the sensed value for the additional parameter *NotReachable* be *False*. The same holds for accidental effects of these actions, which, too, are only possible if the user in question is reachable. If, however, the robot receives the value *True* for *NotReachable*, then the abnormality is asserted, and the state remains unchanged. Figure 10.8 depicts the corresponding additions to the state update axioms of Figures 10.5 and 10.6, respectively.

For the sake of efficiency, the arising of an abnormality is not actually inferred by ramification in FLUX but simply asserted. This is possible because the progression of the world model ensures that this assertion does not have an effect on the past, where this abnormality has been assumed away by default when checking preconditions. Consider, for example, the following query:

```
init(Z0) :- Z0 = [at(4),carries(1,5),empty(2),empty(3),
```

```
continue_delivery(D, Z) :-
   ( knows_val([B],empty(B),Z), knows_val([R1,R2],request(R1,R2),Z),
     \+ knows(ab(reachable,R1),Z), \+ knows(ab(reachable,R2),Z)
     ;
     knows_val([R1],carries(_,R1),Z), \+ knows(ab(reachable,R1),Z) ),
   knows_val([R],at(R),Z),
   ( R < R1 -> D = up ; D = down ).
```

Figure 10.9: New definition of continuation for the mailbot.

```
                     request(4,1),request(4,6),request(6,2) | Z],
             not_holds_all(request(_,_), Z),
             consistent(Z0).

?- init(Z0), poss(pickup(2,1),Z0),
   state_update(Z0,pickup(2,1),Z1,[false,false]),
   ab_state_update(Z1,pickup(3,6),Z1,[false,true]).

Z1=[carries(2,1),at(4),carries(1,5),empty(3),
    request(4,6),request(6,2),ab(reachable,4) | Z]
Z0=[at(4),carries(1,5),empty(2),empty(3),
    request(4,1),request(4,6),request(6,2),ab(reachable,4) | Z]
```

Although the last, abnormal update introduces the fluent $Ab(Reachable(4))$ into z_1 and, thus, into z_0, this fluent is not present at the stage of the computation when $Poss(Pickup(2,1), z_0)$ is derived.

As long as a user is known not to be reachable, the mailbot will no longer attempt to pick up or deliver an item to the respective office. This is ensured by the extended precondition axioms. In a similar fashion, the definition of the auxiliary predicate *ContinueDelivery* can be extended so as to consider moving to an office only if it is not known to be unreachable; see Figure 10.9.

To complete the error handling in case a user is not reachable, consider the exogenous action $EMail(r)$ by which the person in room r informs the mailbot that he or she is back again. The effect of this action is to cancel the corresponding abnormality:

```
state_update(Z1,e_mail(R),Z2) :- cancel(ab(reachable,R),Z1,Z2).
```

In this way, the abnormality will no longer be known, but it is not explicitly false so that later on it can be asserted again if the observations suggest so.

The combination of qualifications with exogenous actions requires to extend the kernel clauses for troubleshooting. Besides the observations made in the course of its actions, an agent also needs to record the exogenous actions that have happened in the course of the program. Their effects, too, need to be inferred when searching for an explanation. Figure 10.10 depicts an additional

```
ab_res([],Z) :- init(Z).
ab_res([S1|S],Z) :-
   ab_res(S,Z1),
   ( S1=[A,Y] -> ( state_update(Z1,A,Z,Y)
                 ; ab_state_update(Z1,A,Z,Y) )
     ;
     S1=[A,Y,E], ( state_update(Z1,A,Z2,Y)
                 ; ab_state_update(Z1,A,Z2,Y) ),
                state_updates(Z2, E, Z) ).

state_updates(Z,[],Z).
state_updates(Z1,[A|S],Z2) :-
   state_update(Z1,A,Z), state_updates(Z,S,Z2).
```

Figure 10.10: A universal definition of troubleshooting including exogenous actions.

---

clause for the predicate *AbRes* for troubleshooting in order to account for exogenous actions. Figure 10.11 shows how exogenous actions are incorporated into the definition of execution. To this end, the ternary variant of the standard predicate $Perform(a, y, e)$ indicates that upon performing action a the agent perceives sensor information y, and a (possibly empty) sequence e of exogenous actions happens during the performance of the action.

10.5 Bibliographical Notes

The qualification problem has been introduced in [McCarthy, 1977] as the challenge to develop a formalism by which the successful execution of an action is predicted by default only and so to account for the inherent uncertainty of natural environments. The first proposed solution [McCarthy, 1986] has used the nonmonotonic formalism of circumscription [McCarthy, 1980] in order to minimize unlikely qualifications, that is, to assume these away unless the observations imply that something must have gone wrong. However, in [Lifschitz, 1987] this approach has been shown to suffer from the problem of anomalous extensions (cf. Section 10.4). As a consequence, an improved minimization technique based on a chronological preference relation has been developed [Shoham, 1987; Shoham, 1988]. In [Kautz, 1986; Sandewall, 1993b; Stein and Morgenstern, 1994], it has been shown that the applicability of chronological minimization is intrinsically restricted to reasoning problems which do not involve indeterminate information, that is, nondeterministic actions or incomplete state knowledge (see also Exercise 10.4). The refined method of prioritized chronological minimization [White *et al.*, 1998] has been aimed at overcoming these restrictions by means of a special three-valued logic and tailor-made semantics.

The use of default logic to address the qualification problem in fluent calcu-

```
execute(A,Z1,Z2) :-
   is_predicate(perform/2),
   perform(A,Y)     ->
                        ( nonvar(Z1), Z1=[sit(S)|Z], ! ; S=[], Z=Z1 ),
                        ( state_update(Z,A,Z3,Y)
                          ; ab_res([[A,Y]|S],Z3) ),
                        !, Z2=[sit([[A,Y]|S])|Z3] ;

   is_predicate(perform/3),
   perform(A,Y,E)   ->
                        ( nonvar(Z1), Z1=[sit(S)|Z], ! ; S=[], Z=Z1 ),
                        ( state_update(Z,A,Z3,Y), state_updates(Z3,E,Z4)
                          ; ab_res([[A,Y,E]|S],Z4) ),
                        !, Z2=[sit([[A,Y,E]|S])|Z4] ;

   A = [A1|A2]      ->
                        execute(A1,Z1,Z), execute(A2,Z,Z2) ;

   A = if(F,A1,A2) ->
                        (holds(F,Z1) -> execute(A1,Z1,Z2)
                                      ; execute(A2,Z1,Z2)) ;

   A = []           ->
                        Z1=Z2 ;

   complex_action(A,Z1,Z2).
```

Figure 10.11: Adding exogenous actions to the definition of execution with troubleshooting.

---

lus has been introduced in [Thielscher, 2001c] along with the use of causality to overcome anomalous extensions. Similar approaches to the qualification problem have been taken in [Kakas and Michael, 2003; Giunchiglia *et al.*, 2004]. Minimization of uncaused abnormalities has also been a key element in the action theories of [Stein and Morgenstern, 1994; Amsterdam, 1991]. Some restrictions with the latter approaches have been pointed out and overcome in [Ortiz, 1999] based on a special Kripke-style semantics. Some authors, e.g., [Ginsberg and Smith, 1988; Lin and Reiter, 1994; Shanahan, 1997] have solved a variant of the qualification problem where implicit preconditions of actions are derived from domain constraints. The computational aspect to the qualification problem has been stressed in [Elkan, 1995], where it has been argued that agents must be able to predict the success of a plan without even considering all possible qualifying causes for the actions, unless some piece of knowledge hints at their presence. A first computational approach to the qualification problem in FLUX has been developed in [Martin and Thielscher, 2001].

10.6 Exercises

10.1. Consider a scenario in the mail delivery domain where initially the robot is at room 1 with all bags empty and three packages which are sitting in rooms 1, 2, and 3 and are addressed to, respectively, 4, 5, and 6. Suppose the mailbot consecutively picks up and delivers the packages while moving all the way up to room 6. Find all extensions which would explain a failure of delivering, in room 6, the contents of bag 3 because the latter is empty. What are the possible explanations if the bag were not found to be empty but to contain the wrong package? Verify your answers with FLUX.

10.2. Consider a domain with two toggle switches 1 and 2 which together are used to control a light bulb according to the specification

$$Light(s) \stackrel{\text{def}}{=} Holds(Closed(1), s) \wedge Holds(Closed(2), s)$$

(See also Figure 9.3, page 215.)

(a) Give a knowledge update axiom for the action $Alter(x)$ of altering the position of switch x which accounts for the possible accident, denoted by $C(x)$ of sort ACCIDENT, that the state of x does actually not change.

Hint: For the sake of simplicity, let $Closed(x)$ be the only fluent in this domain. The status of the light is simply determined according to the definition above.

(b) Find all extensions which explain the consecutive observations

$$\begin{array}{l} \neg Light(S_0) \\ \neg Light(Do(Alter(1), S_0)) \\ \neg Light(Do(Alter(2), Do(Alter(1), S_0))) \\ \neg Light(Do(Alter(1), Do(Alter(2), Do(Alter(1), S_0)))) \end{array}$$

Hint: The initial state of the switches is not known.

(c) Formulate the update axiom of (a) in FLUX and use the encoding to compute the answer for (b).

(d) Suppose that altering switch 2 is generally more prone to an accident than performing this action with switch 1. Formulate this by a prioritized fluent calculus default theory and solve Exercise (b) in the extended setting. What can be concluded about the initial state of the two switches?

10.3. Extend the mail delivery domain by

$$NoPackage : \text{ACCIDENT}$$

indicating the accident that no package at all is received or taken, respectively, when the robot performs a *Pickup* or *Deliver* action. Reformulate the precondition and effect axioms accordingly. Solve Exercise 10.1 in the light of this additional accident.

10.4. One known attempt to overcome the problem of anomalous extensions is to use the chronological preference ordering

$$\frac{: \neg Holds(Ab(x), s)}{\neg Holds(Ab(x), s)} < \frac{: \neg Holds(Ab(y), Do(a, s))}{\neg Holds(Ab(y), Do(a, s))}$$

(a) Show that this indeed gives the intended solution in the motivating example of Section 10.4, which is given by precondition and update axioms (10.16).

(b) Consider the example of leaving a car overnight in a parking lot—action *ParkOvernight*—which may have the abnormal effect that the car gets stolen, denoted by $Ab(Stolen)$. Suppose that the car is observed to having been stolen after two consecutive *ParkOvernight* actions. Formulate this scenario using chronological preference and show that this gives rise to the intuitively unjustified conclusion that the car must have been stolen during the second night.

(c) Reformulate the scenario of Exercise (b) using the axiomatization technique for persistent qualifications of Section 10.4. Show that the resulting default theory admits two extensions, one of which implies that the car has been stolen during the first *ParkOvernight* action while the other one postulates the theft happening in the second night.

10.5. Extend the domain of the cleaning robot (Chapter 3) to account for the persistent qualification that the passageway between two squares may be blocked. If so, the *Go* action shall have no effect, that is, the robot remains at its location. Reformulate the update axiom for *Go* accordingly, and modify the control program for the cleanbot of Chapter 5 to account for possibly blocked passageways.

Chapter 11

Robotics

The focus in this book has been on programming high-level strategies for reasoning robotic agents. To this end, tasks like locomotion, recognizing and picking up objects, or even emptying waste bins have been considered elementary actions. This abstraction has allowed us to develop a general framework for endowing robots with the ability of deductive reasoning in order that they can carry out complex missions such as following long-term strategies, generating and executing composite plans, and performing troubleshooting. The aim of this final chapter is to show briefly how high-level robot programs can be integrated with systems that bring inanimate matter to actually carry out actions. Naturally, it would be impossible within the scope of this book to give in-depth introductions to the whole spectrum of robotics, ranging from the elaborate kinematics of two-legged robots to the interpretation of complex stereo images. We therefore restrict our attention to the most basic task of an autonomous robot, namely, to move around. To this end, we consider

11.1. the integration of basic and high-level control of autonomous robots;

11.2. a method for localization;

11.3. a method for navigation.

11.1 Control Architectures

Figure 11.1 depicts a mobile robot with typical features of autonomous systems for indoor environments. The robot is equipped with a camera, a ring of sonar sensors, and a laser range finder. The latter two devices are used to measure the distance to the nearest obstacle which reflects the sonar signal or laser beam, respectively. Furthermore, an odometry sensor is connected to the wheels, providing information about the path the robot takes. All of these devices are typical **sensors**, to be used to acquire information about the current state.

The robot manipulates its environment with the help of its **effectors**. The two independent, motor-controlled wheels, together with a supporting wheel,

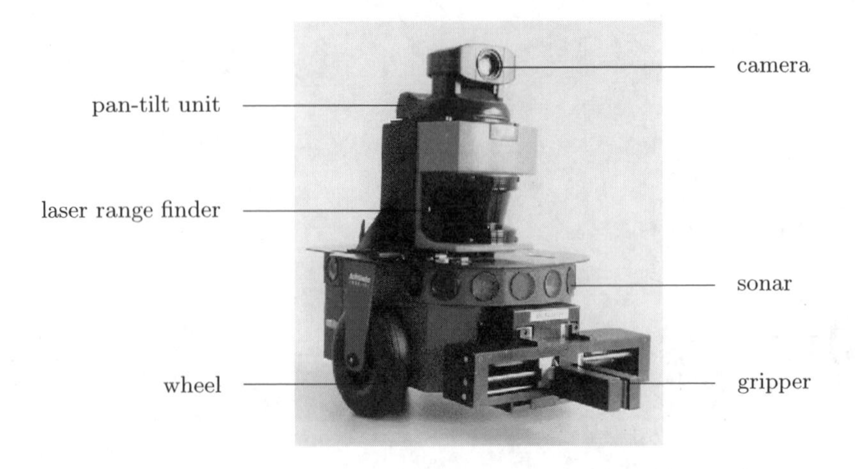

Figure 11.1: A typical indoor mobile robot.

allow the robot to move on the ground on a straight or curved trajectory with variable speed. The robot of Figure 11.1 is furthermore equipped with a two-armed gripper, which can be opened or closed and moved upward, so that it is possible to grab and lift small objects. A third effector is the pan-tilt unit on which the camera is mounted; with its help the robot can enlarge its visual field.

Running a robotic agent on a real robot calls for an intermediate control layer which fills the gap between the high-level actions and the plain motor control and raw sensor data. A schematic architecture for integrated robot control is depicted in Figure 11.2, including a **user interface** and a **low-level control** layer.

The user interface is evoked when executing an action which requires to communicate with a user. Communication can be carried out locally via an on-board notebook or a speech interface, as well as remotely via a wireless connection to personal computers. If a communicative action asks for input, then the latter is returned as sensor information to the agent program. Exogenous actions are a second form of input which may come from the user interface, e.g., when a user adds or cancels a request in a delivery scenario.

The low-level control layer endows the robot with the basic capabilities needed to carry out the high-level actions. Depending on the kind of task (navigation, object manipulation), the environment (office building, rough terrain), and the available sensors and effectors, low-level control may require a variety of elaborate methods, which often need to be tailored to specific problems. We will exemplify low-level control with the basic operation of moving around in a known environment. This task of **navigation** is of course rather elementary if the robot moves on a fixed path, like for example on rails. If the robot can

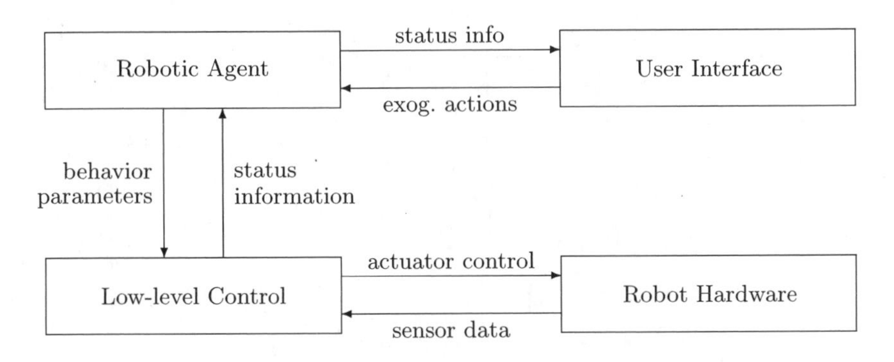

Figure 11.2: A schematic architecture for robot control.

---

move freely, however, then it needs to be able to decide which way to go without bumping into obstacles. Goal-oriented navigation furthermore presupposes that the robot is able to find out where it is. This task of **localization** can be easy if the environment is suitably prepared, e.g., furnished with a number of primitive transmitters. If the robot cannot employ such aids, however, then it has to use its own sensors to determine its position.

11.2 Localization

Localization means to determine one's position relative to a map of the environment. A typical example of a map of an office floor is depicted in Figure 11.3. The map indicates walls between rooms as well as a few stationary obstacles in the offices and the hallway. A mobile indoor robot can use the information provided by such a map to determine its position from sensor data.

To begin with, localization requires to define the space of possible locations. This can be done in two different ways. A **topological** space is composed of qualitative descriptions of locations, such as "in the hallway at room 3." This representation has the advantage that it can be directly mapped onto corresponding fluents used in a high-level control program. A further advantage is that topological state spaces can be comparably small even if the environment is of large absolute size. An obvious disadvantage, then, is the inaccuracy caused by such a coarse definition of locations. The alternative of a **grid-based** representation is obtained by embedding a map into a discretized coordinate system. The position of a robot is then described, for example, by a three-dimensional vector $\langle x, y, \theta \rangle$, where $\langle x, y \rangle$ denotes the square which the robot inhabits and θ the (discretized) direction it faces. A suitably fine-grained discretization allows for a very accurate description, but then this is usually traded for efficiency due to a larger state space.

Uncertainty is a general challenge when it comes to low-level robot control. In particular, real sensor data are always inaccurate to a certain extent, and

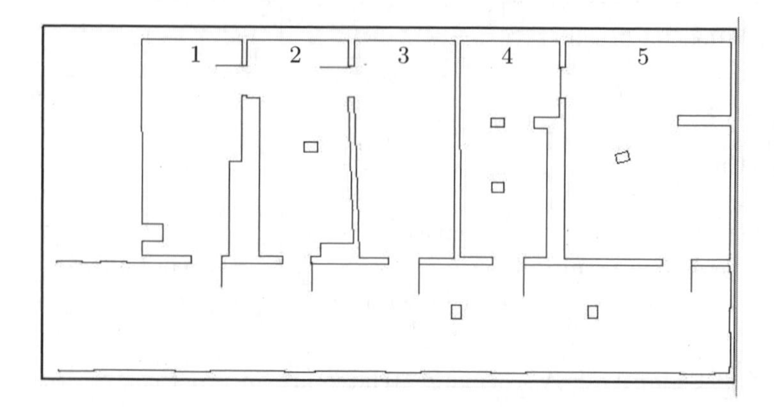

Figure 11.3: Section of an office floor with five rooms and a hallway.

sometimes even completely corrupted. Many standard methods for low-level control are therefore probabilistic in nature. For the task of localization this means to estimate the position of the robot by a **probability distribution** over all possible locations. Consider, to this end, a random variable L_t so that $P(L_t = l)$ denotes the probability that the robot is at location l at some time t. If, for example, the robot knows with absolute certainty that it is now at l_0, then $P(L_{now} = l_0) = 1$ while $P(L_{now} = l) = 0$ for all $l \neq l_0$. The other extreme is a probability distribution which is uniform over all locations, thus indicating that the robot is completely uncertain about where it is. A sufficiently reliable position estimation is characterized by high probabilities near the actual location and negligible probabilities elsewhere.

As the robot gathers more sensor information and moves along, it has to adjust its position estimation. This requires both a **perceptual model** and a **motion model**. The former is needed to interpret new sensor data while the latter is used to reestimate the position in accordance with a movement.

Perceptual Models

A perceptual model describes which sensory input the robot can expect at each location. This obviously depends on the type of sensor and the structure of the environment. The standard method for localization is to use a proximity sensor such as sonar or laser, which determines the distance to the nearest visible object. The perceptual model is then given by conditional probabilities $P(d \mid l)$, which state the likelihood of measuring distance d at location l. The distance from a location to a known obstacle can be read from the map. The probability of detecting such an obstacle depends on the accuracy of the sensors.

The perceptual model is used to adjust the position estimation upon receiving new sensor input. Suppose the robot's original belief about its location is described by the random variable L_{old}, when the proximity sensor measures the distance d. The updated position estimation, described by the random variable L_{new}, can then be calculated via the following equation:

$$P(L_{new} = l) \; = \; P(L_{old} = l) \cdot P(d \,|\, l) \cdot \alpha \qquad (11.1)$$

Here, α is the normalization factor which ensures that the resulting probabilities add up to 1. For the purpose of sensor fusion, (11.1) can be applied iteratively to incorporate an entire stream of sensor data so as to improve the position estimation.

The precision of localization depends on the accuracy of the perceptual model. Environments can be inhabited by unknown obstacles, both stationary and dynamic (such as humans). The likelihood of encountering an object that is not in the map needs to be incorporated into the model (see also Exercise 11.2). The iterative application of (11.1) furthermore relies on the **Markov assumption** that the sensory input at any given location is independent of previous measurements. In cluttered environments, however, the sonar or laser signal is frequently reflected by an unknown object. It is then likely that this object is still there when the next measurement is taken just an instant later. This violates the assumption of independence of the data in a stream of sensor data. A robot can thus get easily lost because more often than not it senses distances closer than expected according to the map. This may require the use of additional **filtering** techniques with the aim to detect and ignore sensor data coming from unknown obstacles, or to dynamically modify the map by previously unknown but static obstacles.

Motion Models

A motion model describes how the position of a robot is expected to change upon moving. The model depends on the accuracy of the effectors and odometry reading as well as on the structure of the ground. To account for the inherent uncertainty, the motion model is given by conditional probabilities $P(l \,|\, a, l')$, which state the likelihood of arriving at location l when the move action a is performed starting in location l'. The uncertainty is usually proportional to the length of the movement.

Let again random variable L_{old} describe the robot's original belief about its position prior to taking a particular action a. After the move, the updated position estimation, represented by L_{new}, can be calculated with the help of the motion model via this equation:

$$P(L_{new} = l) \; = \; \sum_{l'} P(L_{old} = l') \cdot P(l \,|\, a, l') \qquad (11.2)$$

Under the Markov assumption that given the starting location the effect of any action is independent of previous actions, (11.2) can be iteratively applied to account for a sequence of moves. Since effectors and odometry are usually far less reliable than laser range finders, a robot should interleave the use of the motion model with position estimation via (11.1) and the perceptual model.

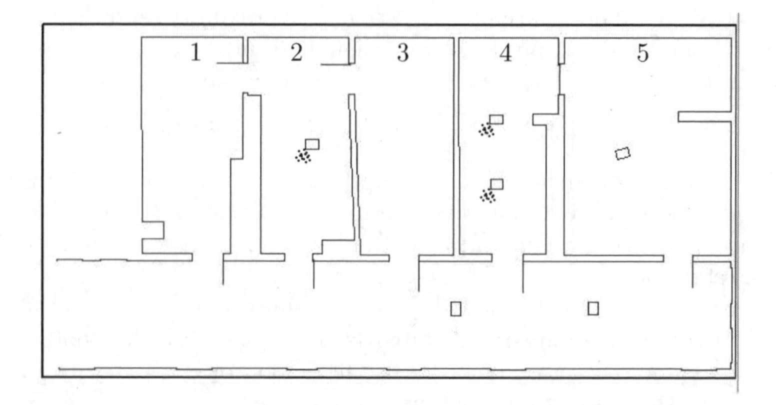

Figure 11.4: A multi-modal position estimation with three different areas, illustrated by point clouds, of locations of non-negligible probabilities in offices 2 and 4.

Mapping Position Information between Low and High Level

While a probabilistic representation captures best the inherent uncertainty on the low level of control, high-level programs usually abstract from probability values for the sake of efficiency. For the same reason, the location space is usually topological and coarse on this level. If a grid-based representation is used for localization, then this needs to be transformed into a qualitative description of the position. To this end, the map can be divided into (possibly overlapping) regions, represented by fluents like *InRoom*(1) and *AtDoor*(2). A given position estimation can then be mapped onto a high-level representation as follows. The probability values for all locations belonging to a region are summed up. If the result reaches a threshold, say 1%, then the corresponding fluent possibly holds in the current situation. In this way, uncertainty about the position is preserved, but only as far as it is relevant for high-level decision making.

Consider, as an example, the probability distribution illustrated in Figure 11.4. The distribution is multi-modal as it contains more than one group of locations with non-negligible probability values. Mapped onto the high-level, the incomplete knowledge of the position of the robot is described by the formula $[Holds(InRoom(2), S) \vee Holds(InRoom(4), S)] \wedge (\forall d)\, \neg Holds(AtDoor(d), S)$, with S being the current situation. This reflects the amount of uncertainty about the location which is relevant for the high-level control. The concrete probabilities have been abstracted away just like the uncertainty regarding the two possible positions of the robot in room 4.

The information flow between the two levels of control need not be unidirectional. Knowledge that has been acquired while executing a high-level program may help improve the position estimation on the low level. Suppose, for example, in the scenario just mentioned high-level reasoning tells the robot that it cannot be in office 4, hence that it must be in room 2. The robot may have come to this conclusion by successfully executing an action that would not have been possible in room 4, or by having observed an exogenous action which implies that it must be in room 2, etc. High-level information saying that the robot must be in some region(s) R can be incorporated as additional sensor information into the position estimation on the low level. Let, to this end, $P(d \mid l)$ be a probability distribution which is uniform over all locations l that belong to the regions R. Applying (11.1) with this probability distribution has the effect to filter out high probability values at wrong locations and thus to possibly transform a multi-modal distribution into a unimodal one. In the aforementioned scenario, for example, the multi-modal probability distribution of Figure 11.4 will thus be updated to a distribution with far less uncertainty, where the high probability values all center around a location in room 2.

11.3 Navigation

Navigation means to have the robot move safely to a given target location. The problem can be divided into two sub-tasks. For **path planning**, the robot uses knowledge of the structure of its environment to find the optimal path from its current position to the goal. The route is then executed by **local navigation**, where the robot follows the chosen path while avoiding collision with stationary or moving obstacles.

Path Planning

A simple method for path planning is to use a graph whose edges correspond to crucial passageways in the environment. A possible graph for our example office floor is shown in Figure 11.5. The task, then, is to find a path from a starting node to some goal node. In some situations, the robot may know that certain passageways are blocked, e.g., due to a closed door. This information can be exploited by pruning away corresponding edges before planning a path.

If several paths to the goal location exist, an optimal one should be chosen. To this end, the edges in the graph are labeled with the estimated time it would take the robot to complete the passage. Time is more accurate a criterion than the mere length of a segment, because narrow or cluttered passages require the robot to move with slow speed or to take a curved route. If knowledge is given as to the likelihood with which a passageway may be blocked, then this, too, can be incorporated into the weight of an edge in order to have the robot choose a path that is likely to be unobstructed. Finally, it may be desirable that the robot avoids certain segments altogether, such as entering an office with the sole purpose of passing through it. Soft constraints of this kind can also be reflected by giving suitable weights to the edges, which may be done dynamically depending on the path planning problem at hand. Given a graph with weighted edges, an optimal path is one whose accumulated cost is minimal.

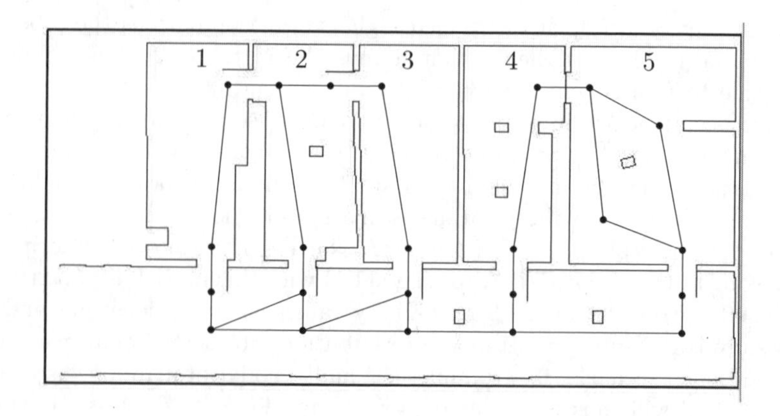

Figure 11.5: A graph representing the passageways in the map.

When a suitable path has been found, the robot starts executing it using local navigation as will be explained below. In the course of the execution it may turn out that a chosen passage is blocked or otherwise impossible to negotiate for the robot. This requires to plan an alternative route, starting from the current location. If (re-)planning a path fails, then this implies failure of the high-level action of moving to the goal position, calling for suitable measures on the high level of control.

Local Navigation

Once an optimal path has been planned, it must be executed. To this end, the robot attempts to negotiate, one after the other, the passageways on the path. Each such step is performed locally, that is, a new target location is given relative to the current position. To calculate orientation and distance to the next goal, the robot uses the global coordinates of the nodes and its ability to localize itself.

During the actual journey, the robot must keep a safe distance to obstacles. Besides walls and other stationary barriers, all kinds of unknown obstacles may be encountered, including dynamic ones like humans or fellow robots. Local navigation therefore cannot rely solely on knowledge of the structure of the environment. Rather, the robot has to continually monitor its surroundings and to correct its movement whenever necessary. One way of representing the surroundings is by a discretized **polar map** centered around the current position of the robot as shown in Figure 11.6. This map can then be used to locally plan a movement as a sequence of segments that lead from the center to the border. If the sensors detect an obstacle within the range of the polar map, then all segments are pruned on which the robot would risk collision. Consider, for example, the scenario illustrated in Figure 11.7(a). Having detected the

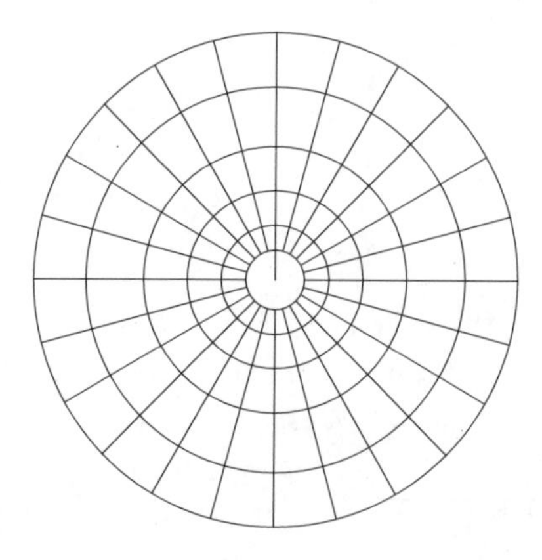

Figure 11.6: A polar map with the robot in the center.

wall with its proximity sensors, the robot employs the reduced map shown in Figure 11.7(b) for planning a local path.

In order to find an optimal path, the remaining nodes in the polar map are valued according to their distance to the target location. A second, possibly conflicting criterion is the clearance of a node, that is, its proximity to an obstacle. Finally, a smoother motion is generally obtained by assigning higher values to nodes which lie on the current trajectory of the robot in motion. Figure 11.8 depicts an example valuation given the target location as shown in Figure 11.7(b). After having decided how to proceed, the motors which control the wheels of the robot are set accordingly, taking into account the clearance of the path when choosing the velocity.

Local navigation allows a robot to take detours when executing a global path in order to avoid collisions with obstacles. A timeout should apply in cases where the target location is unreachable or all local paths are blocked. The global path then needs to be replanned, or troubleshooting is called for on the high level of control.

Mapping Navigation Information between Low and High Level

Navigation is used to execute the high-level action of moving to a specific location. High-level knowledge of temporarily obstructed passageways, such as closed doors, is passed on to the path planning module, where it is used to prune the graph representing the structure of the environment. Conversely, when local navigation fails to reach a target, this information can be passed on to the high level and cause an update of the knowledge of the robotic agent. Depending on the situation and the high-level strategy, this may trigger a communicative action by which a human user is informed about the problem and

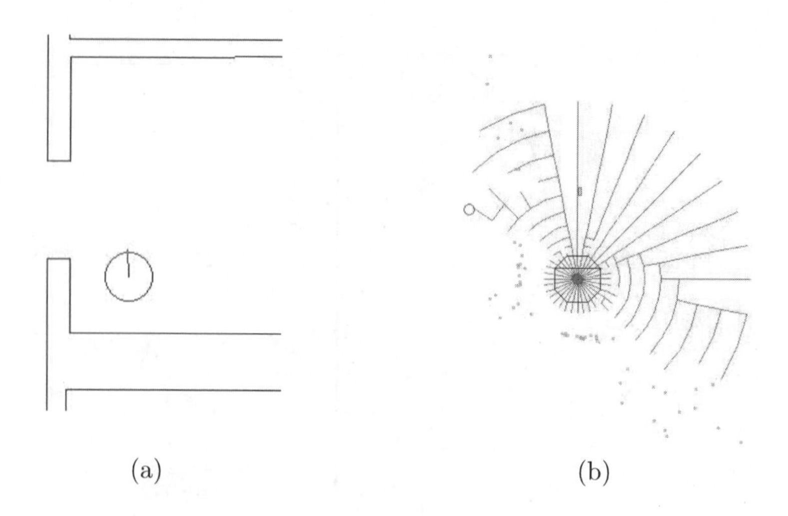

Figure 11.7: (a) Being in the corner of a room, the robot intends to move outside. (b) The sonar readings, shown as points, indicate obstacles to the left and behind the robot. This information is used to prune the local polar map so as to contain safe paths only. The small circle to the upper left of the robot indicates the target position.

possibly asked for help—opening a door, for instance.

11.4 Bibliographical Notes

The first case study on high-level robot control has been described in [Nilsson, 1969; Nilsson, 1984], proving it feasible in principle to build robots that plan ahead and use their plans to actually pursue their goals. A layered architecture which combines high-level, symbolic planning with execution on a robot platform has been developed in [Haigh and Veloso, 1997]. The combination of action programming languages with low-level robot control has been described, e.g., in [Hähnel *et al.*, 1998; Shanahan and Witkowski, 2000; Fichtner *et al.*, 2003]. One such integrated system has been successfully employed to control a tour guide robot in a museum [Burgard *et al.*, 1999]. An overview of probabilistic methods for low-level robot control in uncertain environments can be found in [Thrun, 2000]. The use of perceptual and motion models for localization has been introduced in [Fox *et al.*, 1999]. A number of successful robot systems focusing on many different low-level control tasks have been described in the collection [Kortenkamp *et al.*, 1998].

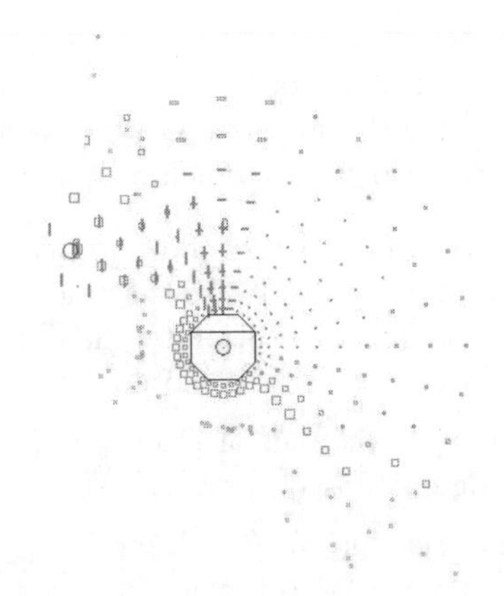

Figure 11.8: Example evaluation of the nodes in the pruned polar map wrt. a goal location which is in front and to the left of the robot. The closer a node is to this target, the higher it is valued.

11.5 Exercises

11.1. Implement a user interface for the mail delivery domain which allows to issue new requests and to cancel existing ones. The input needs to be transformed into exogenous actions for the FLUX mailbot program. The interface should also allow a user to inquire about the status of a request at any time, and to accept or refuse a package that is being delivered.

11.2. Consider a discretization $d_1, \ldots, d_n$ of the distances that can be measured by a proximity sensor with maximal range d_n. Given a map m, let $P_m(d_i \,|\, l)$ denote the probability of detecting a known object at distance d_i from location l. Let $P_u(d_i)$ denote the (location-independent) probability that an unknown object is encountered at distance d_i. Calculate the perceptual model $P(d_i \,|\, l)$ in which these two probabilities are combined.

Hint: A proximity sensor can detect an object at a distance d_i only if there is no object at a distance d_j with $j < i$.

11.3. To obtain a high-level representation of the locations in the map of Figure 11.3, consider the fluents $\mathit{InRoom}(r)$ with $r \in \{0, \ldots, 5\}$ (where 0 represents the hallway) and $\mathit{AtDoor}(d)$ with $d \in \{1, \ldots, 8\}$ (representing the eight doors in this environment). Using a suitably fine grid, write a program that transforms a FLUX knowledge state based on these fluents into a probabilistic position estimation that conveys the same information about the location of the robot.

11.4. Encode the passageway graph of Figure 11.5 in Prolog, including an assignment of estimated times for the robot to travel along each edge. Write a program for planning the shortest path between any two nodes and given a (possibly empty) list of edges which are known to be blocked.

11.5. (a) Use the fluents of Exercise 11.3 to modify the FLUX program for the mailbot so as to be applicable to the office environment of Figure 11.3. Extend the state representation by the fluent *Closed*(d) representing that a door $d \in \{1, \ldots, 8\}$ is closed.

(b) Integrate the path planner of Exercise 11.4 in such a way that high-level knowledge of closed doors is respected.

(c) Integrate the user interface of Exercise 11.1 and extend the control program by having the robot ask a user via the interface to open a particular door in case path planning fails.

Appendix A

FLUX Manual

This glossary contains concise descriptions of all FLUX predicates that are meant to be used in agent programs. Predicates provided by the FLUX kernel are explained in Section A.1. Section A.2 covers the predicates which need to be defined by the user as part of the background theory for an agent. All predicates are listed in alphabetical order.

A.1 Kernel Predicates

The reference at the end of each description points to the page with the most general clausal definition of the predicate.

A.1.1 `cancel/3`

Predicate `cancel(F,Z1,Z2)` means that knowledge state `Z2` is knowledge state `Z1` without any knowledge of fluent `F`. It is used in state update axioms to cancel possible knowledge of an indeterminate effect.

Examples:

```
state_update(Z1,toss(C),Z2,[]) :- cancel(heads(C),Z1,Z2).

?- holds(heads(coin),Z1), state_update(Z1,toss(coin),Z2,_).

Z1 = [heads(coin) | Z2]

?- or_holds([heads(c1),heads(c2),heads(c3)],Z1),
   state_update(Z1,toss(c2),Z2,_).

Z1 = Z2
```

See also: `state_update/4`

Clausal definition: Figure 4.8, page 94

A.1.2 duplicate_free/1

Constraint `duplicate_free(Z)` means that list `Z` contains no two occurrences of the same element. It is used to ensure that a list with variable tail encodes a state.

Example:

```
?- Z0 = [open(X),open(Y) | Z], duplicate_free(Z0).

Z0 = [open(X),open(Y)|Z]

Constraints:
not_holds(open(X), Z)
not_holds(open(Y), Z)
duplicate_free(Z)

X \= Y
```

Constraint handling rules: Figure 4.2, page 80

A.1.3 execute/3

Predicate `execute(A, Z1, Z2)` means that the agent performs `A` and, simultaneously, that state `Z1` is updated to state `Z2` according to the effects of `A`. The latter should always be ground and can be an elementary or a complex action, a conditional `if(F, A1, A2)`, or a list `[A1, ..., An]`. The predicate is used in agent programs to trigger the performance of an action and to update the current world model accordingly.

The performance of an elementary action follows the user-defined predicate `perform/2` or `perform/3`, respectively. A complex action is executed according to its definition using the predicate `complex_action/3`. Performing a conditional means to check whether the condition, fluent `F`, holds, and if so then `A1` is executed, else `A2`. A list is executed by serial execution of its elements.

The update is inferred on the basis of the user-defined state update axioms for the elementary actions using the predicate `state_update/4`, possibly followed by the update according to the observed exogenous actions based on their effect specification by the predicate `state_update/3`. If the update fails and alternative, abnormal effects have been specified with the help of the predicate `ab_state_update/4`, then these are used to update the state.

Examples:

```
perform(go(D),[]) :-
   write('Performing '), write(go(D)), nl.

state_update(Z1,go(D),Z2,[]) :-
   D = out,  update(Z1,[],[in],Z2) ;
   D = home, update(Z1,[in],[],Z2).

complex_action(go,Z1,Z2) :-
   knows(in,Z1) -> execute(go(out),Z1,Z2)
                 ; execute(go(home),Z1,Z2).

?- duplicate_free(Z1),
   execute([go,if(in,go(out),[])],Z1,Z2).

Performing go(home)
Performing go(out)

Constraints:
not_holds(in,Z2)
```

Clausal definition: Figure 10.11, page 270

A.1.4 holds/2

Predicate `holds(F,Z)` is used to assert that fluent `F` holds in state `Z`.

Example:

```
?- holds(open(X), [open(1),open(Y) | Z]).

X = 1  More?

X = Y  More?

Z = [open(X)|_]
```

See also: `holds/3`, `not_holds/2`, `knows/2`

Clausal definition: Figure 4.5, page 90

A.1.5 holds/3

Predicate `holds(F,Z,Zp)` means that fluent `F` holds in state `Z` and that state `Zp` is `Z` without `F`. It is used to assert that a fluent holds in a state and to access the residue state. The latter can then be used to assert additional information, e.g., that no further instance of the fluent holds.

Example:

```
?- holds(open(1), Z, Z1), not_holds_all(open(_), Z1)

Z = [open(1) | Z1]

Constraints:
not_holds_all(open(_), Z1)
```

See also: holds/2, not_holds/2, knows/2

Clausal definition: Figure 4.5, page 90

A.1.6 knows/2

Predicate knows(F,Z) means that fluent F is known to hold in state Z. It is used to verify knowledge of a particular fluent instance. In order to compute known instances of a non-ground fluent expression, the predicate knows_val/3 should be used instead.

Examples:

```
?- Z = [f(1)|_], not_holds(f(2), Z), knows(f(1), Z).

Yes

?- Z = [f(1)|_], not_holds(f(2), Z), knows(f(2), Z).

No

?- Z = [f(1)|_], not_holds(f(2), Z), knows(f(3), Z).

No
```

See also: knows/3, knows_val/3, holds/2, knows_not/2

Clausal definition: Figure 5.2, page 109

A.1.7 knows/3

Predicate knows(F,S,Z) means that fluent F will be known to hold after performing the actions in S starting in state Z. When solving planning problems, this predicate is used to verify knowledge of a fluent in a future situation. Argument S is of the form do(ak,...,do(a1, [])...) and may include the predefined actions if_true(F) and if_false(F), respectively, which are used to indicate which of the different branches is currently being searched in case of conditional planning. Much like `knows/2`, predicate `knows/3` is used to verify knowledge of a particular fluent instance rather than to derive known instances of a non-ground fluent expression, for which the predicate `knows_val/4` should be used.

Examples:

```
state_update(Z1, open_and_check(X), Z2, [E]) :-
   update(Z1, [open(X)], [], Z2),
   ( E = true, holds(empty(X), Z2)
     ; E = false, not_holds(empty(X), Z2) ).

?- duplicate_free(Z),
   knows(open(1), do(open_and_check(1),[]), Z).

Yes

?- duplicate_free(Z),
   knows(empty(1), do(open_and_check(1),[]), Z).

No

?- duplicate_free(Z),
   knows(empty(1), do(if_true(empty(1)),
                      do(open_and_check(1),[])), Z).

Yes
```

See also: `knows/2`, `knows_val/4`, `knows_not/3`
Clausal definition: Figure 6.12, page 163

A.1.8 knows_not/2

Predicate `knows_not(F,Z)` means that fluent `F` is known not to hold in state `Z`. Like `knows/2` it is used to verify knowledge of a particular fluent instance.

Examples:

```
?- Z = [f(1)|_], not_holds(f(2), Z), knows_not(f(1), Z).

No

?- Z = [f(1)|_], not_holds(f(2), Z), knows_not(f(2), Z).

Yes
```

```
?- Z = [f(1)|_], not_holds(f(2), Z), knows_not(f(3), Z).

No
```

See also: knows_not/3, not_holds/2, knows/2

Clausal definition: Figure 5.2, page 109

A.1.9 knows_not/3

Predicate knows_not(F, S, Z) means that fluent F will be known not to hold after performing the actions in S starting in state Z. When solving planning problems, this predicate is used to verify knowledge of a fluent in a future situation. Argument S is of the form do(ak, ..., do(a1, [])...) and may include the pre-defined actions if_true(F) and if_false(F), respectively, which are used to indicate different search branches in case of conditional planning. Much like knows_not/2, predicate knows_not/3 is used to verify knowledge of a particular fluent instance.

Examples:

```
state_update(Z1, open_and_check(X), Z2, [E]) :-
   update(Z1, [open(X)], [], Z2),
   ( E = true, holds(empty(X), Z2)
     ; E = false, not_holds(empty(X), Z2) ).

?- not_holds(open(1), Z),
   knows_not(open(1), [], Z).

Yes

?- not_holds(open(1), Z),
   knows_not(empty(1), do(open_and_check(1),[]), Z).

No

?- not_holds(open(1), Z),
   knows_not(empty(1), do(if_false(empty(1)),
                          do(open_and_check(1),[])), Z).

Yes
```

See also: knows_not/2, knows/3

Clausal definition: Figure 6.12, page 163

A.1.10 knows_val/3

Predicate knows_val(X,F,Z) computes ground instances for the variables in list X for which fluent F is known to hold in state Z. List X should always be non-empty and only contain variables that occur in F. The predicate is re-satisfiable if multiple values are known.

Examples:

```
?- Y :: 2..3,
   knows_val([X], open(X), [open(1),open(Y),open(4) | _]).

X = 1  More?

X = 4  More?

No

?- R#=1 #\/ R#=2, Z = [at(2,R) | _],
   knows_val([X,Y], at(X,Y), Z).

No

?- R#=1 #\/ R#=2, Z = [at(2,R) | _],
   knows_val([X], at(X,_), Z).

X = 2  More?

No
```

See also: knows/2, knows_val/4

Clausal definition: Figure 5.2, page 109

A.1.11 knows_val/4

Predicate knows_val(X,F,S,Z) generates ground instances for the variables in list X such that fluent F will be known to hold after performing the actions in S starting in state Z. When solving planning problems, this predicate is used to compute known values for a fluent in a future situation. Argument S is of the form do(ak,...,do(a1,[])...) and may include the pre-defined actions if_true(F) and if_false(F), respectively, which are used to indicate different search branches in case of conditional planning.

Examples:

```
state_update(Z1, open_and_check(X), Z2, [E]) :-
```

```
   update(Z1, [open(X)], [], Z2),
   ( E = true, holds(empty(X), Z2)
     ; E = false, not_holds(empty(X), Z2) ).

init(Z0) :- or_holds([empty(1),empty(2)], Z0),
            duplicate_free(Z0).

?- init(Z0),
   knows_val([X], open(X), do(open_and_check(1),[]), Z0).

X = 1  More?

No

?- init(Z0),
   knows_val([X], empty(X), do(open_and_check(1),[]), Z0).

No

?- init(Z0),
   knows_val([X], empty(X), do(if_false(empty(1)),
                               do(open_and_check(1),[])), Z0).

X = 2  More?

No
```

See also: knows/3, knows_val/3

Clausal definition: Figure 6.12, page 163

A.1.12 not_holds/2

Predicate not_holds(F,Z) is used to assert that fluent F does not hold in state Z.

Example:

```
?- not_holds(open(X), [open(1),open(2) | Z]), X :: 1..3.

X = 3

Constraints:
not_holds(open(3), Z)
```

See also: holds/2, knows_not/2, not_holds_all/2

Constraint handling rules: Figure 4.2, page 80

A.1.13 not_holds_all/2

Constraint not_holds_all(F, Z) is used to assert that no instance of fluent F holds in state Z. None of the variables in F should occur elsewhere, but the fluent may contain multiple occurrences of variables.

Examples:

```
?- not_holds_all(cleaned(_,5), [cleaned(1,Y),cleaned(X,2)|Z]).

Constraints:
Y #\= 5
not_holds_all(cleaned(_,5), Z)

?- not_holds_all(at(_,_), [at(_,_) | _]).

No

?- or_holds([occupied(2,4),occupied(3,3)], Z),
   not_holds_all(occupied(X,X), Z).

Z = [occupied(2,4) | Z1]

Constraints:
not_holds_all(occupied(X,X), Z1)
```

See also: holds/2, not_holds/2

Constraint handling rules: Figure 4.2, page 80

A.1.14 or_holds/2

Constraint or_holds(D, Z) is used to assert that at least one of the fluents in list D holds in state Z.

Example:

```
?- not_holds_all(occupied(0,_), Z),
   not_holds(occupied(2,3), Z),
   or_holds([occupied(0,3),occupied(1,3),occupied(2,3)], Z).

Z = [occupied(1,3) | Z1]

Constraints:
not_holds_all(occupied(0,_), Z1)
not_holds(occupied(2,3), Z1)
```

See also: holds/2, not_holds/2

Constraint handling rules: Figure 4.3, page 83

A.1.15 plan/3

Predicate plan(N, Z, P) means that plan P is an optimal solution to the planning problem named N with initial state Z. The predicate fails if no solution can be found.

It is assumed that a finite space of possible solutions is described by a domain-dependent definition of a predicate N(Z, P, Zn), where N is the name of the problem, in such a way that P is an executable action sequence that leads from state Z to state Zn and Zn satisfies the goal. Alternatively, the programmer can define a predicate N(Z, P) in such a way that P is a conditional plan which can be executed in state Z and which will always lead to a state that satisfies the goal. Conditional plans are sequences of both actions and the construct if(F, P1, P2) for branching, where F is a fluent and P1, P2 are conditional plans.

It is moreover assumed that for planning problem N a domain-dependent definition of the predicate plan_cost(N, P, Zn, C) (respectively, plan_cost(N, P, C) in case of a conditional planning problem) is given such that positive rational number C is the cost of plan P, with resulting state Zn in case of unconditional planning. If two or more solutions with the same minimal cost exist, then only the one that occurs first in the space of solutions is returned.

Example:

```
% One action only: deliver a package from room R1 to R2

poss(deliver(R1,R2),Z) :- knows_val([R1,R2],request(R1,R2),Z).

state_update(Z1,deliver(R1,R2),Z2,[]) :-
   update(Z1,[],[request(R1,R2)],Z2).

% Plan space contains all executable action sequences

delivery_plan(Z,[],Z) :-
   \+ knows_val([R1,R2],request(R1,R2),Z).

delivery_plan(Z,[A|P],Zn) :-
   poss(A,Z), state_update(Z,A,Z1,[]), delivery_plan(Z1,P,Zn).

% Plan cost: total walking distance, starting in room 1

plan_cost(delivery_plan,P,_,C) :- walking_distance(P,1,C).
```

```
walking_distance([],_,0).
walking_distance([deliver(R1,R2)|P],R,C1) :-
   walking_distance(P,R2,C2),
   C1 is C2 + abs(R1-R) + abs(R2-R1).

% Example planning problem; the inferred solution has cost 11

?- Z = [request(1,6),request(2,4),request(3,1) | _],
   plan(delivery_plan,Z,P).

P = [deliver(2,4),deliver(3,1),deliver(1,6)]
```

Clausal definition: Figure 6.12, page 163

A.1.16 `ramify/4`

Predicate `ramify(Z1,P,N,Z2)` means that state `Z2` is the overall result after effecting the direct positive and negative effects `P` and `N`, respectively, followed by the automatic state transitions caused by indirect effects. It is assumed that the causal relationships, which determine the indirect effects, are specified by a domain-dependent definition of the predicate `causes(Z,P,N,Z1,P1,N1)` in such a way that state `Z` in conjunction with the positive and negative effects (both direct and indirect) `P` and `N` leads to an automatic transition to state `Z1` along with updated positive and negative effects `P1` and `N1`. The order in which the causal relationships are applied depends on the order of their specification. The overall resulting state is reached when no further causal relationship applies.

Example:

```
state_update(Z1,alter_switch,Z2,[]) :-
   knows(closed,Z1)      -> ramify(Z1,[],[closed],Z2) ;
   knows_not(closed,Z1) -> ramify(Z1,[closed],[],Z2).

% Switch controls a light bulb

causes(Z,P,N,[light_on|Z],[light_on|P],N) :-
   holds(closed,P), not_holds(light_on,Z).
causes(Z,P,N,Z1,P,[light_on|N]) :-
   holds(closed,N), holds(light_on,Z,Z1).

% Light permanently registered by light detecting device

causes(Z,P,N,[photo|Z],[photo|P],N) :-
   holds(light_on,P), not_holds(photo,Z).
?- not_holds(closed,Z1),
```

```
    not_holds(light_on,Z1),
    not_holds(photo,Z1),
    duplicate_free(Z1),
    state_update(Z1,alter_switch,Z2,[]),
    state_update(Z2,alter_switch,Z3,[]).

Z2 = [photo,light_on,closed | Z1]
Z3 = [photo | Z1]
```

Clausal definition: Figure 9.8, page 230

A.1.17 update/4

Predicate update(Z1, P, N, Z2) means that state Z2 is the result of updating state Z1 by a list of positive effects P and a list of negative effects N. The negative effects are inferred first, so that if a fluent occurs in both P and N, then it will hold in the resulting state, Z2. Positive effects which are known to be true in Z1 cause no change just like negative effects which are known to be false in Z1.

Examples:

```
state_update(Z1, close(X), Z2, []) :-
   update(Z1, [closed(X)], [], Z2).
state_update(Z1, open(X), Z2, []) :-
   update(Z1, [], [closed(X)], Z2).

?- not_holds(closed(1), Z1),
   state_update(Z1, open(1), Z2, []),
   state_update(Z2, close(1), Z3, []).

Z2 = Z1
Z3 = [closed(1) | Z1]

Constraints:
not_holds(closed(1), Z1)

?- or_holds([closed(1),closed(2)], Z1),
   state_update(Z1, close(1), Z2, []).

Z2 = [closed(1) | Z]
```

Clausal definition: Figure 4.7, page 93

A.2 User-Defined Predicates

A.2.1 `ab_state_update/4`

Predicate `ab_state_update(Z1, A, Z2, Y)` means that action `A` can cause an abnormal update from state `Z1` to state `Z2` while observing `Y`. It can be used to specify the ways in which an action may go wrong. The sensed information `Y` is a (possibly empty) list of values which the agent is assumed to acquire during the actual execution of the action.

The abnormal updates are invoked if the agent makes an observation which contradicts its world model, indicating that one of its past actions must have failed to produce the usual effect. The various ways in which an action may go wrong should be specified in the order of their relative likelihood, since the agent automatically searches through the possible abnormalities until it finds the first sequence of updates that explains all observations of the past. To this end, the executed sequence of actions along with the acquired sensing information is stored in the state using the special fluent `sit`.

Each individual update is usually specified with the help of the kernel predicate `update/4`, possibly accompanied by `cancel/3` to account for uncertainty due to incomplete state knowledge or a nondeterministic action. In the presence of indirect effects, the kernel predicate `ramify/4` should be used instead of `update/4`.

Example:

```
% Normally, picking up an object succeeds

state_update(Z1,pickup(X),Z2,[Ok]) :-
   Ok = true, update(Z1,[carries(X)],[onfloor(X)],Z2).

% In some cases, the action fails and nothing changes

ab_state_update(Z,pickup(_),Z,[Ok]) :-
   Ok = false.

% In exceptional cases, the object is out of sight after
% unsuccessfully trying to pick it up

ab_state_update(Z1,pickup(X),Z2,[Ok]) :-
   Ok = false, update(Z1,[],[onfloor(X)],Z2).

% A sensing action and initial state

state_update(Z,find_object(X),Z,[F]) :-
   F = true,       holds(onfloor(X),Z) ;
```

```
   F = false, not_holds(onfloor(X),Z).

init(Z0) :- Z0 = [onfloor(cup)|Z], duplicate_free(Z0),
            not_holds_all(carries(_),Z).

?- init(Z0), execute(pickup(cup),Z0,Z1).

Perfoming pickup(cup)       Ok = true

Z0 = [onfloor(cup) | Z]
Z1 = [sit([[pickup(cup),[true]]]), carries(cup) | Z]

?- init(Z0), execute(pickup(cup),Z0,Z1).

Perfoming pickup(cup)       Ok = false

Z0 = [onfloor(cup) | Z]
Z1 = [sit([[pickup(cup),[false]]]), onfloor(cup) | Z]

?- init(Z0),
   execute(pickup(cup),Z0,Z1),
   execute(find_object(cup),Z1,Z2).

Perfoming pickup(cup)       Ok = false
Perfoming find_object(cup)  F  = false

...
Z2 = [sit([[find_object(cup),[false]],
           [pickup(cup),[false]]]) | Z3]

Constraints:
not_holds(onfloor(cup),Z3)
...
```

See also: `state_update/4`

A.2.2 `causes/6`

Predicate `causes(Z,P,N,Z1,P1,N1)` means that state `Z` in conjunction with the positive and negative effects (both direct and indirect) `P` and `N` leads to an automatic transition to state `Z1` along with updated positive and negative effects `P1` and `N1`. The predicate is used to specify the causal relationships of a domain in order to account for indirect effects of actions. The order in which the causal relationships are given determines the order of their application after the direct effects of an action have been inferred. The set of relationships should be defined in such a way that their successive application terminates after finitely many steps. In the presence of causal relationships, the kernel predicate `ramify/4` should be used when defining `state_update/4`, in order to account for indirect effects after each action execution.

Example:

```
% f(X) causes not f(X-1), if X>1

causes(Z, P, N, Z1, P, [f(X1)|N]) :-
   holds(f(X), P), X > 1,
   X1 is X-1, not_holds(f(X1), N),
   update(Z, [], [f(X1)], Z1).

% not f(X) causes f(X-1), if X>1

causes(Z, P, N, Z1, [f(X1)|P], N) :-
   holds(f(X), N), X > 1,
   X1 is X-1, not_holds(f(X1), P),
   update(Z, [f(X1)], [], Z1).

state_update(Z1, trigger(X), Z2, []) :-
   ramify(Z1, [f(X)], [], Z2).

?- duplicate_free(Z1),
   state_update(Z1, trigger(7), Z2, []),
   state_update(Z2, trigger(8), Z3, []).

Z2 = [f(1),f(3),f(5),f(7) | Z1]
Z3 = [f(2),f(4),f(6),f(8) | Z1]

Constraints:
not_holds(f(1), Z1)
...
not_holds(f(8), Z1)
```

See also: `state_update/4`, `ab_state_update/4`

A.2.3 `complex_action/3`

Predicate `complex_action(A,Z1,Z2)` means that state `Z2` is the result of executing in state `Z1` the complex action named `A`. Complex actions are usually employed for planning. The predicate `complex_action/3` is then used to define the actual execution of a complex action in terms of primitive actions.

Example:

```
complex_action(bring(X,R),Z1,Z5) :-
   knows_val([R1],loc(X,R1),Z1), execute(goto(R1),Z1,Z2),
   execute(pickup(X),Z2,Z3), execute(goto(R),Z3,Z4),
   execute(drop(X),Z4,Z5).

complex_action(goto(R),Z1,Z2) :-
   knows_val([R1],at(R1),Z1),
   ( R1=R -> Z2=Z1 ;
     (R1<R -> execute(go(up),Z1,Z)
            ; execute(go(down),Z1,Z)),
     complex_action(goto(R),Z,Z2) ).

state_update(Z1,go(D),Z2,[]) :-
   holds(at(R),Z1), ( D=up -> R1#=R+1 ; R1#=R-1 ),
   update(Z1,[at(R1)],[at(R)],Z2).

state_update(Z1,pickup(X),Z2,[]) :-
   holds(loc(X,R),Z1),
   update(Z1,[carries(X)],[loc(X,R)],Z2).

state_update(Z1,drop(X),Z2,[]) :-
   holds(at(R),Z1),
   update(Z1,[loc(X,R)],[carries(X)],Z2).

?- Z1=[at(1),loc(letter,5)|_], execute(bring(letter,2),Z1,Z2).

Perfoming go(up)
Perfoming go(up)
Perfoming go(up)
Perfoming go(up)
Perfoming pickup(letter)
Perfoming go(down)
Perfoming go(down)
Perfoming go(down)
Perfoming drop(letter)

Z1 = [at(1),loc(letter,5) | Z]
Z2 = [loc(letter,2),at(2) | Z]
```

A.2.4 init/1

Predicate init(Z) is used to specify the initial state knowledge. The kernel refers to this predicate when inferring a sequence of abnormal state updates in order to explain observations that are inconsistent with the normal effects of the past actions.

Example:

```
init(Z0) :- Z0=[at(1),loc(cup,1) | Z],
            not_holds_all(carries(_),Z),
            consistent(Z0).

% Consistent states have unique value for "at"
% Consistent states have unique locations for all objects

consistent(Z) :- holds(at(_),Z,Z1),
                 not_holds_all(at(_),Z1),
                 location_consistent(Z),
                 duplicate_free(Z).

% Constraint handling rule for domain-dependent constraint

location_consistent([F|Z]) <=>
   ( F=loc(X,_) -> not_holds_all(loc(X,_),Z) ; true ),
   location_consistent(Z).

% Usual effect of picking up an object

state_update(Z1,pickup(X),Z2,[]) :-
   holds(loc(X,Y),Z1),
   update(Z1,[carries(X)],[loc(X,Y)],Z2).

% Abnormal effect of picking up an object

ab_state_update(Z,pickup(_),Z,[]) :-
   holds(gripper_blocked,Z).

% A sensing action

state_update(Z,check_gripper,Z,[Empty]) :-
   Empty=true,  not_holds_all(carries(_),Z) ;
   Empty=false, holds(carries(_),Z).

?- init(Z0).

Z0 = [at(1),loc(cup,1) | Z]

Constraints:
not_holds_all(carries(_),Z)
```

```
not_holds_all(at(_),Z)
not_holds_all(loc(cup,_),Z)
location_consistent(Z)
duplicate_free(Z)

?- init(Z0),
   execute(pickup(cup),Z0,Z1),
   execute(check_gripper,Z1,Z2).

Perfoming pickup(cup)
Perfoming check_gripper  Empty = true

Z2 = [sit(...),at(1),loc(cup,1),gripper_blocked | Z]
```

See also: `ab_state_update/4`

A.2.5 perform/2

Predicate `perform(A, Y)` means that action `A` is actually performed and sensor information `Y` is returned. The latter is a (possibly empty) list of sensed values acquired as a result of executing the action. The structure of `Y` should be identical to the one used when defining the update of `A` via `state_update/4` and possibly `ab_state_update/4`. Predicate `perform/2` is used by the kernel predicate `execute/3` in order to trigger the actual execution of primitive actions.

Example:

```
% Simulating the roll of a dice

perform(roll(_), [X]) :- random(N), X is N mod 6+1.

% Updating the state accordingly

state_update(Z1, roll(D), Z2, [X]) :-
   holds(dice(D,X1), Z1),
   update(Z1, [dice(D,X)], [dice(D,X1)], Z2).

?- Z1 = [dice(1,1),dice(2,1) | Z],
   not_holds_all(dice(_,_), Z),
   execute(roll(1), Z1, Z2),
   execute(roll(2), Z2, Z3).

Z1 = [dice(1,1),dice(2,1) | Z]
Z2 = [dice(1,6),dice(2,1) | Z]
Z3 = [dice(2,3),dice(1,6) | Z]
```

See also: `state_update/4`

A.2.6 `perform/3`

Predicate `perform(A,Y,E)` means that action `A` is actually performed and sensor information `Y` is returned. Moreover, `E` is a (possibly empty) list of exogenous actions that are observed during the execution. Sensor information `Y` is a (possibly empty) list of sensed values acquired as a result of executing the action. The structure of `Y` should be identical to the one used when defining the update of `A` via `state_update/4` and possibly `ab_state_update/4`. Predicate `perform/3` is used by the kernel predicate `execute/3` in order to trigger the actual execution of primitive actions. On the basis of their update definition via `state_update/3`, the effects of the exogenous actions are inferred in the order in which they occur in `E`.

Example:

```
% Simulating the 10% likely loss of carried object

perform(goto(R),[],E) :-
   random(N),
   ( N mod 10 =:= 1 -> E=[loose_object] ; E=[] ),
   write('Exogenous actions: '), write(E), nl.

% State update axioms

state_update(Z1,loose_object,Z2) :-
   holds(carries(X),Z1) -> update(Z1,[],[carries(X)],Z2)
                         ; Z2 = Z1.

state_update(Z1,goto(R),Z2,[]) :-
   holds(at(R1),Z1),
   update(Z1,[at(R)],[at(R1)],Z2).

?- Z1=[at(1),carries(cup)|Z], not_holds_all(at(_),Z),
   execute(goto(5),Z1,Z2),
   execute(goto(2),Z2,Z3).

Exogenous actions: []
Exogenous actions: [loose_object]

Z2 = [at(5),carries(cup) | Z]
Z3 = [at(2) | Z]
```

See also: `state_update/3`, `state_update/4`

A.2.7 plan_cost/3

Predicate plan_cost(PlanningProblem, P, C) means that conditional plan P for the planning problem named PlanningProblem has cost C. The latter should always be a positive rational number. The predicate is used by the kernel predicate plan/3 for evaluating and comparing different solutions to a conditional planning problem. The predicate accompanies the definition of the domain-dependent predicate PlanningProblem/2 by which conditional planning problems are specified.

Example:

```
% Cost is total number of actions

plan_cost(search_plan,[],0).

plan_cost(search_plan,[A|P],C) :-
   A = if(_,P1,P2) -> plan_cost(search_plan,P1,C1),
                      plan_cost(search_plan,P2,C2),
                      plan_cost(search_plan,P,C3),
                      C is C1+C2+C3
                    ;
                      action_cost(A,C1),
                      plan_cost(search_plan,P,C2),
                      C is C1+C2.

action_cost(_,1).

?- plan_cost(search_plan,
             [search_room(2),
              if(key(2),[get_key(2)],
                        [search_room(3),
                         if(key(3),[get_key(3)],
                                   [get_key(5)])])],C).

C = 5.
```

See also: PlanningProblem/2, plan_cost/4

A.2.8 plan_cost/4

Predicate plan_cost(PlanningProblem, P, Z, C) means that action sequence P leading to goal state Z for the planning problem named PlanningProblem has cost C. The latter should always be a positive rational number. The predicate is used by the kernel predicate plan/3 for evaluating and comparing different solutions to a planning problem. The predicate accompanies the definition of the domain-dependent predicate PlanningProblem/3 by which (unconditional) planning problems are specified.

Example:

```
% Cost is total amount paid plus travel cost

plan_cost(shopping_plan,P,Z,C) :-
   holds(total_amount(C1),Z),
   travel_cost(P,C2),
   C is C1+C2.

% Travel cost is 1 per "go" action

travel_cost([],0).
travel_cost([A|P],C) :-
   travel_cost(P,C1),
   ( A = go(_) -> C is C1+1 ; C is C1 ).

?- plan_cost(shopping_plan,[go(1),buy(1),go(2),buy(2)],
                           [total_amount(17.5),at(2)],C).

C = 19.5

?- plan_cost(shopping_plan,[go(2),buy(1),buy(2)],
                           [total_amount(18),at(2)],C).

C = 19
```

(Another example is given in the description for plan/3 in Section A.1.15.)

See also: PlanningProblem/3, plan_cost/3

A.2.9 PlanningProblem/2

Any conditional planning problem is given a unique, domain-dependent name, here called PlanningProblem, the clauses for which determine the solution space. Specifically, PlanningProblem(Z,P) needs to be defined in such a way that in state Z action sequence P is executable and always leads to a state which satisfies the goal of the planning problem at hand.

The solution space is usually defined by iteratively expanding an initially empty action sequence by both ordinary actions and branching. To continue the search on a particular branch, the special functions if_true(F) and if_false(F) can be used to continue planning under the assumption that fluent F is true or false, respectively. Branching should only be effected if the truth-value of the fluent in question has been sensed (directly or indirectly) earlier in the plan.

Since the result of sensing actions cannot be predicted at planning time, in general there is no unique resulting knowledge state for future situations. Hence, knowledge should always be verified against the initial state and the actions planned thus far, by using the kernel predicates `knows/3`, `knows_not/3`, and `knows_val/4` with the additional situation argument.

Example:

```
% Systematically search for the key, starting in room 1

search_plan(Z,P) :- search_plan(Z,[],1,P).

% Planning goal is to get the key at last,
% which requires to know where it is

search_plan(Z,S,_,[get_key(R)]) :- knows_val([R],key(R),S,Z).

% Search room R unless it is known whether the key is there

search_plan(Z,S,R,[search_room(R)|P]) :-
   \+ knows(key(R),S,Z), \+ knows_not(key(R),S,Z),
   branch_plan(R,Z,do(search_room(R),S),P).

% Either get the key or continue if key not found

branch_plan(R,Z,S,[if(key(R),[get_key(R)],P)]) :-
   search_plan(Z,do(if_false(key(R)),S),R,P).

% Skip searching room R

search_plan(Z,S,R,P) :- R < 6, R1 is R+1,
                         search_plan(Z,S,R1,P).

% Update axiom

state_update(Z,search_room(R),Z,[K]) :-
   K = true,  holds(key(R),Z) ;
   K = false, not_holds(key(R),Z).

% Generate all plans

?- or_holds([key(2),key(3),key(5)],Z), not_holds(key(4),Z),
   search_plan(Z,P).
```

```
P = [search_room(1)
     if(key(1),[get_key(1)],
               [search_room(2),
                if(key(2),[get_key(2)],
                          [search_room(3),
                           if(key(3),[get_key(3)],
                                     [get_key(5)])])])]  More?

...

% Find optimal plan

?- or_holds([key(2),key(3),key(5)],Z), not_holds(key(4),Z),
   plan(search_plan,Z,P).

P = [search_room(2),
     if(key(2),[get_key(2)],
               [search_room(3),
                if(key(3),[get_key(3)],[get_key(5)])])]
```

(For the definition of the planning cost in this scenario see the example given in the description for `plan_cost/3` in Section A.2.7.)

See also: `plan_cost/3`, `state_update/4`, `PlanningProblem/3`

A.2.10 PlanningProblem/3

Any (unconditional) planning problem is given a unique, domain-dependent name, here called `PlanningProblem`, the clauses for which determine the solution space. Specifically, `PlanningProblem(Z,P,Zn)` needs to be defined in such a way that in state `Z` action sequence `P` is executable and leads to state `Zn` which satisfies the goal of the planning problem at hand. The solution space is usually defined by successively expanding an initially empty action sequence and using the domain-dependent definition of `state_update/4` to infer the effects of each added action.

Example:

```
% Planning goal is to have no requests left

shopping_plan(Z,[],Z) :- not_holds_all(request(_),Z).

% Plans are sequences of sub-plans to go and get
% a single requested item

shopping_plan(Z1,P,Zn) :- holds(request(X),Z1),
```

```
                              go_and_get(X,Z1,Z2,P1),
                              shopping_plan(Z2,P2,Zn),
                              append(P1,P2,P).

% Get an item or go elsewhere and get it there

go_and_get(X,Z1,Z2,P) :- get(X,Z1,Z2,P).

go_and_get(X,Z1,Z2,[go(Y)|P]) :-
   poss(go(Y),Z1), state_update(Z1,go(Y),Z,[]),
   get(X,Z,Z2,P).

% To get an item, buy it

get(X,Z1,Z2,[buy(X)]) :-
   poss(buy(X),Z1), state_update(Z1,buy(X),Z2,[]).

% Precondition and update axioms

poss(go(Y), Z) :- (Y=1 ; Y=2), not_holds(at(Y),Z).
poss(buy(_),Z) :- holds(at(1),Z) ; holds(at(2),Z).

state_update(Z1,go(Y),Z2,[]) :-
   holds(at(Y1),Z1), update(Z1,[at(Y)],[at(Y1)],Z2).

price(1,1,9.5). price(1,2,10).  % price(item,store,price)
price(2,1,10).  price(2,2,8).

state_update(Z1,buy(X),Z2,[]) :-
   holds(at(Y),Z1), holds(total_amount(C1),Z1),
   price(X,Y,C), C2 is C1+C,
   update(Z1,[total_amount(C2)],
             [total_amount(C1),request(X)],Z2).

% Generate all plans

?- Z = [at(0),request(1),request(2),total_amount(0)],
   shopping_plan(Z,P,Zn).

P  = [go(1),buy(1),buy(2)]
Zn = [total_amount(19.5),at(1)]  More?

P  = [go(1),buy(1),go(2),buy(2)]
Zn = [total_amount(17.5),at(2)]  More?

P  = [go(2),buy(1),buy(2)]
```

```
Zn = [total_amount(18),at(2)]     More?

...

% Find optimal plan

?- Z = [at(0),request(1),request(2),total_amount(0)],
   plan(shopping_plan,Z,P).

P = [go(2),buy(1),buy(2)]
```

(For the definition of the planning cost in this scenario see the example given in the description for `plan_cost/4` in Section A.2.8. Another example is given in the description for `plan/3` in Section A.1.15.)

See also: `plan_cost/3`, `state_update/4`, `PlanningProblem/2`

A.2.11 state_update/3

Predicate `state_update(Z1, A, Z2)` means that the effect of exogenous action `A` is to update state `Z1` to state `Z2`. The effects are usually defined with the help of the kernel predicate `update/4`, possible accompanied by `cancel/3` to account for uncertainty due to incomplete state knowledge or a nondeterministic action. In the presence of indirect effects, the kernel predicate `ramify/4` should be used instead of `update/4`. The update should be specified in such a way that there is always a unique result for ground instances of `A`.

Examples:

```
state_update(Z1, issue_request(R1,R2), Z2) :-
   update(Z1, [request(R1,R2)], [], Z2).

?- state_update(Z1, issue_request(3,6), Z2).

Z2 = [request(3,6) | Z1]

?- holds(request(3,6), Z1),
   state_update(Z1, issue_request(3,6), Z2).

Z1 = [request(3,6) | Z]
Z2 = [request(3,6) | Z]

state_update(Z, observe_door_closed(D), Z) :-
   not_holds(open(D), Z).

?- or_holds([open(1),open(2),open(3)], Z1),
```

```
    state_update(Z1, observe_door_closed(1), Z2),
    state_update(Z2, observe_door_closed(3), Z3).

Z3 = [open(2) | Z]

Constraints:
not_holds(open(1), Z)
not_holds(open(2), Z)

state_update(Z1, lost_gripper_control, Z2) :-
   knows_not(carries(_), Z1) -> Z2 = Z1
                              ; cancel(carries(_), Z1, Z2).

?- not_holds_all(carries(_), Z1),
   state_update(Z1, lost_gripper_control, Z2).

Z2 = Z1

Constraints:
not_holds_all(carries(_), Z1)

?- Z1 = [at(3),carries(X) | _],
   state_update(Z1, lost_gripper_control, Z2).

Z2 = [at(3) | _]
```

See also: `state_update/4`

A.2.12 `state_update/4`

Predicate `state_update(Z1, A, Z2, Y)` means that the effect of primitive action `A` is to update state `Z1` to state `Z2` in the light of sensor information `Y`. The latter is a (possibly empty) list of values which the agent is assumed to acquire during the actual execution of the action. The structure of `Y` should be identical to the one used when defining the actual performance of `A` via `perform/2` or `perform/3`.

The effects are usually defined with the help of the kernel predicate `update/4`, possible accompanied by `cancel/3` to account for uncertainty due to incomplete state knowledge or a nondeterministic action. In the presence of indirect effects, the kernel predicate `ramify/4` should be used instead of `update/4`. The update should be specified in such a way that there is always a unique result for ground instances of `A` and `Y`.

Examples:

```
state_update(Z1, alter(X), Z2, []) :-
   knows(open(X), Z1)     -> update(Z1, [], [open(X)], Z2) ;
   knows_not(open(X), Z1) -> update(Z1, [open(X)], [], Z2) ;
   cancel(open(X), Z1, Z2).

?- not_holds(open(t1), Z0),
   or_holds([open(t2),open(t3)], Z0),
   duplicate_free(Z0),
   state_update(Z0, alter(t1), Z1),
   state_update(Z1, alter(t2), Z2).

Z2 = [open(t1) | Z0]

Constraints:
not_holds(open(t1), Z0)

state_update(Z1, enter(R), Z2, [Ok]) :-
   Ok = true,  update(Z1, [in(R)], [in(0)], Z2) ;
   Ok = false, holds(entry_blocked(R), Z1), Z2 = Z1.

?- Z1 = [in(0) | Z], not_holds_all(in(_), Z),
   state_update(Z1, enter(1), Z2, [true]).

Z1 = [in(0) | Z]
Z2 = [in(1) | Z]

?- state_update(Z1, enter(1), Z2, [Ok]).

Ok = true
Z2 = [in(1) | Z1]

Constraints:
not_holds(in(0), Z1)
not_holds(in(1), Z1)  More?

Ok = false
Z2 = [entry_blocked(1) | _]  More?

No
```

See also: ab_state_update/4, state_update/3

Bibliography

[Amsterdam, 1991] J. B. Amsterdam. Temporal reasoning and narrative conventions. In J. F. Allen, R. Fikes, and E. Sandewall, editors, *Proceedings of the International Conference on Principles of Knowledge Representation and Reasoning (KR)*, pages 15–21, Cambridge, MA, 1991.

[Apt and Bol, 1994] Krzysztof R. Apt and Roland Bol. Logic programming and negation: A survey. *Journal of Logic Programming*, 19/20:9–71, 1994.

[Apt, 1997] Krzysztof R. Apt. *From Logic Programming to Prolog.* Prentice-Hall, 1997.

[Bacchus and Kabanza, 2000] Fahiem Bacchus and Froduald Kabanza. Using temporal logic to express search control knowledge for planning. *Artificial Intelligence*, 116(1–2):123–191, 2000.

[Bacchus *et al.*, 1999] Fahiem Bacchus, Joseph Halpern, and Hector Levesque. Reasoning about noisy sensors and effectors in the situation calculus. *Artificial Intelligence*, 111(1–2):171–208, 1999.

[Bäckström and Klein, 1991] C. Bäckström and I. Klein. Planning in polynomial time: The SAS-PUBS class. *Journal of Computational Intelligence*, 7(3):181–197, 1991.

[Bäckström and Nebel, 1993] C. Bäckström and B. Nebel. Complexity Results for SAS$^+$ Planning. In R. Bajcsy, editor, *Proceedings of the International Joint Conference on Artificial Intelligence (IJCAI)*, pages 1430–1435, Chambéry, France, August 1993. Morgan Kaufmann.

[Baker, 1991] Andrew B. Baker. Nonmonotonic reasoning in the framework of situation calculus. *Artificial Intelligence*, 49:5–23, 1991.

[Baral and Son, 1997] Chitta Baral and Tran Cao Son. Approximate reasoning about actions in presence of sensing and incomplete information. In J. Maluszynski, editor, *Proceedings of the International Logic Programming Symposium (ILPS)*, pages 387–401, Port Jefferson, NY, October 1997. MIT Press.

[Baral, 1995] Chitta Baral. Reasoning about actions: Non-deterministic effects, constraints and qualification. In C. S. Mellish, editor, *Proceedings of the International Joint Conference on Artificial Intelligence (IJCAI)*, pages 2017–2023, Montreal, Canada, August 1995. Morgan Kaufmann.

[Bibel, 1986] Wolfgang Bibel. A deductive solution for plan generation. *New Generation Computing*, 4:115–132, 1986.

[Bibel, 1998] Wolfgang Bibel. Let's plan it deductively! *Artificial Intelligence*, 103(1–2):183–208, 1998.

[Blum and Furst, 1997] Avrim L. Blum and Merrick L. Furst. Fast planning through planning graph analysis. *Artificial Intelligence*, 90(1–2):281–300, 1997.

[Bobrow, 1980] Daniel G. Bobrow, editor. *Artificial Intelligence 13: Special Issue on Non-Monotonic Reasoning.* Elsevier, 1980.

[Boutilier and Friedmann, 1995] Craig Boutilier and Neil Friedmann. Nondeterministic actions and the frame problem. In C. Boutilier and M. Goldszmidt, editors, *Extending Theories of Actions: Formal Theory and Practical Applications*, volume SS–95–07 of *AAAI Spring Symposia*, pages 39–44, Stanford University, March 1995. AAAI Press.

[Boutilier *et al.*, 2000] Craig Boutilier, Ray Reiter, Mikhail Soutchanski, and Sebastian Thrun. Decision-theoretic, high-level agent programming in the situation calculus. In H. Kautz and B. Porter, editors, *Proceedings of the AAAI National Conference on Artificial Intelligence*, pages 355–362, Austin, TX, July 2000.

[Brewka and Hertzberg, 1993] Gerhard Brewka and Joachim Hertzberg. How to do things with worlds: On formalizing actions and plans. *Journal of Logic and Computation*, 3(5):517–532, 1993.

[Burgard *et al.*, 1999] Wolfram Burgard, Armin B. Cremers, Dieter Fox, Dirk Hähnel, Gerhard Lakemeyer, Dieter Schulz, Walter Steiner, and Sebastian Thrun. Experiences with an interactive museum tour-guide robot. *Artificial Intelligence*, 114(1–2):3–55, 1999.

[Bylander, 1994] Tom Bylander. The computational complexity of propositional STRIPS planning. *Artificial Intelligence*, 69:165–204, 1994.

[Clark, 1978] Keith L. Clark. Negation as failure. In H. Gallaire and J. Minker, editors, *Logic and Data Bases*, pages 293–322. Plenum Press, 1978.

[Clocksin and Mellish, 1994] William F. Clocksin and Chris S. Mellish. *Programming in PROLOG.* Springer, 1994.

[Colmerauer *et al.*, 1972] Alain Colmerauer, Henry Kanoui, Robert Pasero, and Philippe Roussel. *Un système de communication homme-machine en français.* Technical report, University of Marseille, 1972.

[Davis, 1993] Martin Davis. First order logic. In D. Gabbay, C. J. Hogger, and J. A. Robinson, editors, *Handbook of Logic in Artificial Intelligence and Logic Programming*, chapter 2, pages 31–65. Oxford University Press, 1993.

[del Val and Shoham, 1993] Alvaro del Val and Yoav Shoham. Deriving properties of belief update from theories of action (II). In R. Bajcsy, editor, *Proceedings of the International Joint Conference on Artificial Intelligence (IJCAI)*, pages 732–737, Chambéry, France, August 1993. Morgan Kaufmann.

[Demolombe and del Parra, 2000] Robert Demolombe and Maria del Parra. A simple and tractable extension of situation calculus to epsitemic logic. In Z. W. Ras and S. Ohsuga, editors, *International Symposium on Methodologies for Intelligent Systems (ISMIS)*, volume 1932 of *LNCS*, pages 515–524. Springer, 2000.

[Denecker *et al.*, 1998] Marc Denecker, Daniele Theseider Dupré, and Kristof Van Belleghem. An inductive definition approach to ramifications. *Electronic Transactions on Artificial Intelligence*, 2(1–2):25–67, 1998.

[Dennet, 1984] Daniel C. Dennet. Cognitive wheels: The frame problem of AI. In C. Hookway, editor, *Minds, Machines, and Evolution: Philosophical Studies*, pages 129–151. Cambridge University Press, 1984.

[Doherty *et al.*, 1998] Patrick Doherty, Joakim Gustafsson, Lars Karlsson, and Jonas Kvarnström. Temporal action logics (TAL): Language specification and tutorial. *Electronic Transactions on Artificial Intelligence*, 2(3–4):273–306, 1998.

[Elkan, 1992] Charles Elkan. Reasoning about action in first-order logic. In *Proceedings of the Conference of the Canadian Society for Computational Studies of Intelligence (CSCSI)*, pages 221–227, Vancouver, Canada, May 1992. Morgan Kaufmann.

[Elkan, 1995] Charles Elkan. On solving the qualification problem. In C. Boutilier and M. Goldszmidt, editors, *Extending Theories of Actions: Formal Theory and Practical Applications*, volume SS–95–07 of *AAAI Spring Symposia*, Stanford University, March 1995. AAAI Press.

[Ernst *et al.*, 1997] Michael D. Ernst, Todd D. Millstein, and Daniel S. Weld. Automatic SAT-compilation of planning problems. In M. E. Pollack, editor, *Proceedings of the International Joint Conference on Artificial Intelligence (IJCAI)*, pages 1169–1176, Nagoya, Japan, August 1997. Morgan Kaufmann.

[Etzoni *et al.*, 1997] Oren Etzoni, Keith Golden, and Daniel Weld. Sound and efficient closed-world reasoning for planning. *Artificial Intelligence*, 89(1–2):113–148, 1997.

[Fichtner *et al.*, 2003] Matthias Fichtner, Axel Großmann, and Michael Thielscher. Intelligent execution monitoring in dynamic environments. *Fundamenta Informaticae*, 57(2–4):371–392, 2003.

[Fikes and Nilsson, 1971] Richard E. Fikes and Nils J. Nilsson. STRIPS: A new approach to the application of theorem proving to problem solving. *Artificial Intelligence*, 2:189–208, 1971.

[Finger, 1987] Joseph J. Finger. *Exploiting Constraints in Design Synthesis.* PhD thesis, Stanford University, CA, 1987.

[Fox *et al.*, 1999] Dieter Fox, Wolfram Burgard, and Sebastian Thrun. Markov localization for mobile robots in dynamic environments. *Journal of Artificial Intelligence Research*, 11:391–427, 1999.

[Frege, 1879] Gottlob Frege. *Begriffsschrift.* Louis Nebert, Halle, 1879.

[Frühwirth, 1998] Thom Frühwirth. Theory and practice of constraint handling rules. *Journal of Logic Programming*, 37(1–3):95–138, 1998.

[Geffner, 1990] Hector Geffner. Causal theories for nonmonotonic reasoning. In *Proceedings of the AAAI National Conference on Artificial Intelligence*, pages 524–530, Boston, MA, 1990.

[Gelfond and Lifschitz, 1993] Michael Gelfond and Vladimir Lifschitz. Representing action and change by logic programs. *Journal of Logic Programming*, 17:301–321, 1993.

[Giacomo and Levesque, 1999] Giuseppe De Giacomo and Hector Levesque. An incremental interpreter for high-level programs with sensing. In H. Levesque and F. Pirri, editors, *Logical Foundations for Cognitive Agents*, pages 86–102. Springer, 1999.

[Giacomo and Levesque, 2000] Giuseppe De Giacomo and Hector Levesque. ConGolog, a concurrent programming language based on the situation calculus. *Artificial Intelligence*, 121(1–2):109–169, 2000.

[Giacomo *et al.*, 1997] Giuseppe De Giacomo, Luca Iocchi, Daniele Nardi, and Riccardo Rosati. Planning with sensing for a mobile robot. In *Proceedings of the European Conference on Planning (ECP)*, volume 1348 of *LNAI*, pages 158–170. Springer, 1997.

[Giacomo *et al.*, 2002] Giuseppe De Giacomo, Yves Lespérance, Hector Levesque, and Sebastian Sardiña. On the semantics of deliberation in IndiGolog—from theory to practice. In D. Fensel, D. McGuinness, and M.-A. Williams, editors, *Proceedings of the International Conference on Principles of Knowledge Representation and Reasoning (KR)*, pages 603–614, Toulouse, France, April 2002. Morgan Kaufmann.

[Ginsberg and Smith, 1988] Matthew L. Ginsberg and David E. Smith. Reasoning about action II: The qualification problem. *Artificial Intelligence*, 35:311–342, 1988.

[Giunchiglia *et al.*, 1997] Enrico Giunchiglia, G. Neelakantan Kartha, and Vladimir Lifschitz. Representing action: Indeterminacy and ramifications. *Artificial Intelligence*, 95:409–443, 1997.

[Giunchiglia *et al.*, 2004] Enrico Giunchiglia, Joohyung Lee, Vladimir Lifschitz, Norman McCain, and Hudson Turner. Nonmonotonic causal theories. *Artificial Intelligence*, 153(1–2):49–104, 2004.

[Golden and Weld, 1996] Keith Golden and Daniel Weld. Representing sensing actions: The middle ground revisited. In L. C. Aiello, J. Doyle, and S. Shapiro, editors, *Proceedings of the International Conference on Principles of Knowledge Representation and Reasoning (KR)*, pages 174–185, Cambridge, MA, November 1996. Morgan Kaufmann.

[Green, 1969] Cordell Green. Theorem proving by resolution as a basis for question-answering systems. *Machine Intelligence*, 4:183–205, 1969.

[Große *et al.*, 1996] Gerd Große, Steffen Hölldobler, and Josef Schneeberger. Linear deductive planning. *Journal of Logic and Computation*, 6(2):233–262, 1996.

[Grosskreutz and Lakemeyer, 2000] Henrik Grosskreutz and Gerhard Lakemeyer. cc-Golog: Towards more realistic logic-based robot control. In H. Kautz and B. Porter, editors, *Proceedings of the AAAI National Conference on Artificial Intelligence*, pages 476–482, Austin, TX, July 2000.

[Gustafsson and Doherty, 1996] Joakim Gustafsson and Patrick Doherty. Embracing occlusion in specifying the indirect effects of actions. In L. C. Aiello, J. Doyle, and S. Shapiro, editors, *Proceedings of the International Conference on Principles of Knowledge Representation and Reasoning (KR)*, pages 87–98, Cambridge, MA, November 1996. Morgan Kaufmann.

[Haas, 1987] Andrew R. Haas. The case for domain-specific frame axioms. In F. M. Brown, editor, *The Frame Problem in Artificial Intelligence*, pages 343–348, Los Altos, CA, 1987. Morgan Kaufmann.

[Hähnel *et al.*, 1998] Dirk Hähnel, Wolfram Burgard, and Gerhard Lakemeyer. GOLEX: Bridging the gap between logic (GOLOG) and a real robot. In O. Herzog and A. Günter, editors, *Proceedings of the German Annual Conference on Artificial Intelligence (KI)*, volume 1504 of *LNAI*, pages 165–176, Bremen, Germany, September 1998. Springer.

[Haigh and Veloso, 1997] Karen Z. Haigh and Manuela M. Veloso. High-level planning and low-level execution: Towards a complete robotic agent. In *International Conference on Autonomous Agents*, pages 363–370, Menlo Park, CA, February 1997.

[Hanks and McDermott, 1987] Steve Hanks and Drew McDermott. Nonmonotonic logic and temporal projection. *Artificial Intelligence*, 33(3):379–412, 1987.

[Herzig *et al.*, 2000] Andreas Herzig, Jérôme Lang, Dominique Longin, and Thomas Polascek. A logic for planning under partial observability. In H. Kautz and B. Porter, editors, *Proceedings of the AAAI National Conference on Artificial Intelligence*, pages 768–773, Austin, TX, July 2000.

[Hölldobler and Kuske, 2000] Steffen Hölldobler and Dietrich Kuske. Decidable and undecidable fragments of the fluent calculus. In M. Parigot and A. Voronkov, editors, *Proceedings of the International Conference on Logic Programming and Automated Reasoning (LPAR)*, volume 1955 of *LNAI*, pages 436–450, Reunion Island, France, November 2000. Springer.

[Hölldobler and Schneeberger, 1990] Steffen Hölldobler and Josef Schneeberger. A new deductive approach to planning. *New Generation Computing*, 8:225–244, 1990.

[Holzbaur and Frühwirth, 2000] Christian Holzbaur and Thom Frühwirth, editors. *Journal of Applied Artificial Intelligence: Special Issue on Constraint Handling Rules*, volume 14(4). Taylor & Francis, April 2000.

[Jaffar and Maher, 1994] Joxan Jaffar and Michael J. Maher. Constraint logic programming: A survey. *Journal of Logic Programming*, 19/20:503–581, 1994.

[Jaffar *et al.*, 1992] Joxan Jaffar, S. Michaylov, Peter Stuckey, and R. Yap. The $CLP(\mathcal{R})$ language and system. *ACM Transactions on Programming Languages*, 14(3):339–395, 1992.

[Kakas and Michael, 2003] Antonis Kakas and Loizos Michael. On the qualification problem and elaboration tolerance. In *Logical Formalizations of Commonsense Reasoning*, AAAI Spring Symposia, Stanford, CA, March 2003. AAAI Press.

[Kakas and Miller, 1997a] Antonis Kakas and Rob Miller. Reasoning about actions, narratives, and ramifications. *Electronic Transactions on Artificial Intelligence*, 1(4):39–72, 1997.

[Kakas and Miller, 1997b] Antonis Kakas and Rob Miller. A simple declarative language for describing narratives with actions. *Journal of Logic Programming*, 31(1–3):157–200, 1997.

[Kakas *et al.*, 2001] Antonis Kakas, Rob Miller, and Francesca Toni. E-res: Reasoning about actions, narratives, and ramifications. In T. Eiter, W. Faber, and M. Trusczynski, editors, *Proceedings of the International Conference on Logic Programming and Nonmonotonic Reasoning (LPNMR)*, volume 2173 of *LNCS*, pages 254–266, Vienna, Austria, September 2001. Springer.

[Kartha and Lifschitz, 1994] G. Neelakantan Kartha and Vladimir Lifschitz. Actions with indirect effects. In J. Doyle, E. Sandewall, and P. Torasso, editors, *Proceedings of the International Conference on Principles of Knowledge Representation and Reasoning (KR)*, pages 341–350, Bonn, Germany, May 1994. Morgan Kaufmann.

[Kartha and Lifschitz, 1995] G. Neelakantan Kartha and Vladimir Lifschitz. A simple formalization of actions using circumscription. In C. S. Mellish, editor, *Proceedings of the International Joint Conference on Artificial Intelligence (IJCAI)*, pages 1970–1975, Montreal, Canada, August 1995. Morgan Kaufmann.

[Kartha, 1993] G. Neelakantan Kartha. Soundness and completeness theorems for three formalizations of actions. In R. Bajcsy, editor, *Proceedings of the International Joint Conference on Artificial Intelligence (IJCAI)*, pages 724–729, Chambéry, France, August 1993. Morgan Kaufmann.

[Kartha, 1994] G. Neelakantan Kartha. Two counterexamples related to Baker's approach to the frame problem. *Artificial Intelligence*, 69(1–2):379–391, 1994.

[Kautz and Selman, 1996] Henry Kautz and Bart Selman. Pushing the envelope: Planning, propositional logic, and stochastic search. In B. Clancey and D. Weld, editors, *Proceedings of the AAAI National Conference on Artificial Intelligence*, pages 1194–1201, Portland, OR, August 1996. MIT Press.

[Kautz, 1986] Henry Kautz. The logic of persistence. In *Proceedings of the AAAI National Conference on Artificial Intelligence*, pages 401–405, Philadelphia, PA, August 1986.

[Kortenkamp *et al.*, 1998] David Kortenkamp, Peter Bonasso, and Robin Murphy. *Artificial Intelligence and Mobile Robots: Case Studies of Successful Robot Systems.* MIT Press, 1998.

[Kowalski and Sadri, 1994] Robert Kowalski and Fariba Sadri. The situation calculus and event calculus compared. In M. Bruynooghe, editor, *Proceedings of the International Logic Programming Symposium (ILPS)*, pages 539–553, Ithaca, NY, 1994. MIT Press.

[Kowalski and Sergot, 1986] Robert Kowalski and Marek Sergot. A logic based calculus of events. *New Generation Computing*, 4:67–95, 1986.

[Kowalski, 1974] Robert Kowalski. Predicate logic as a programming language. In *Proceedings of the Congress of the International Federation for Information Processing (IFIP)*, pages 569–574. Elsevier, 1974.

[Kowalski, 1979] Robert Kowalski. *Logic for Problem Solving*, volume 7 of *Artificial Intelligence Series.* Elsevier, 1979.

[Kushmerick *et al.*, 1995] Nicholas Kushmerick, Steve Hanks, and Daniel Weld. An algorithm for probabilistic planning. *Artificial Intelligence*, 76(1–2):239–286, 1995.

[Kvarnström and Doherty, 2000] Jonas Kvarnström and Patrick Doherty. TALplanner: A temporal logic based forward chaining planner. *Annals of Mathematics and Artificial Intelligence*, 30:119–169, 2000.

[Kvarnström, 2002] Jonas Kvarnström. Applying domain analysis techniques for domain-dependent control in TALplanner. In *Proceedings of the International Conference on AI Planning Systems (AIPS)*, pages 369–378, Toulouse, France, April 2002. Morgan Kaufmann.

[Lakemeyer, 1999] Gerhard Lakemeyer. On sensing and off-line interpreting GOLOG. In H. Levesque and F. Pirri, editors, *Logical Foundations for Cognitive Agents*, pages 173–189. Springer, 1999.

[Levesque *et al.*, 1997] Hector Levesque, Raymond Reiter, Yves Lespérance, Fangzhen Lin, and Richard Scherl. GOLOG: A logic programming language for dynamic domains. *Journal of Logic Programming*, 31(1–3):59–83, 1997.

[Lifschitz, 1987] Vladimir Lifschitz. Formal theories of action (preliminary report). In J. McDermott, editor, *Proceedings of the International Joint Conference on Artificial Intelligence (IJCAI)*, pages 966–972, Milan, Italy, August 1987. Morgan Kaufmann.

[Lifschitz, 1990] Vladimir Lifschitz. Frames in the space of situations. *Artificial Intelligence*, 46:365–376, 1990.

[Lin and Reiter, 1994] Fangzhen Lin and Ray Reiter. State constraints revisited. *Journal of Logic and Computation*, 4(5):655–678, 1994.

[Lin and Shoham, 1991] Fangzhen Lin and Yoav Shoham. Provably correct theories of action. In *Proceedings of the AAAI National Conference on Artificial Intelligence*, pages 590–595, Anaheim, CA, July 1991.

[Lin, 1995] Fangzhen Lin. Embracing causality in specifying the indirect effects of actions. In C. S. Mellish, editor, *Proceedings of the International Joint Conference on Artificial Intelligence (IJCAI)*, pages 1985–1991, Montreal, Canada, August 1995. Morgan Kaufmann.

[Lin, 1996] Fangzhen Lin. Embracing causality in specifying the indeterminate effects of actions. In B. Clancey and D. Weld, editors, *Proceedings of the AAAI National Conference on Artificial Intelligence*, pages 670–676, Portland, OR, August 1996. MIT Press.

[Lloyd, 1987] John W. Lloyd. *Foundations of Logic Programming*. Series Symbolic Computation. Springer, second, extended edition, 1987.

[Lobo *et al.*, 1997] Jorge Lobo, Gisela Mendez, and Stuart R. Taylor. Adding knowledge to the action description language $\mathcal{A}$. In B. Kuipers and B. Webber, editors, *Proceedings of the AAAI National Conference on Artificial Intelligence*, pages 454–459, Providence, RI, July 1997. MIT Press.

[Lobo, 1998] Jorge Lobo. COPLAS: A conditional planner with sensing actions. In *Cognitive Robotics*, volume FS–98–02 of *AAAI Fall Symposia*, pages 109–116. AAAI Press, October 1998.

[Łukaszewicz and Madalińska-Bugaj, 1995] Witold Łukaszewicz and Ewa Madalińska-Bugaj. Reasoning about action and change using Dijkstra's semantics for programming languages: Preliminary report. In C. S. Mellish, editor, *Proceedings of the International Joint Conference on Artificial Intelligence (IJCAI)*, pages 1950–1955, Montreal, Canada, August 1995. Morgan Kaufmann.

[Martin and Thielscher, 2001] Yves Martin and Michael Thielscher. Addressing the qualification problem in FLUX. In F. Baader, G. Brewka, and T. Eiter, editors, *Proceedings of the German Annual Conference on Artificial Intelligence (KI)*, volume 2174 of *LNAI*, pages 290–304, Vienna, Austria, September 2001. Springer.

[Masseron *et al.*, 1993] Marcel Masseron, Christophe Tollu, and Jacqueline Vauzielles. Generating plans in linear logic I. Actions as proofs. *Journal of Theoretical Computer Science*, 113:349–370, 1993.

[McCain and Turner, 1995] Norman McCain and Hudson Turner. A causal theory of ramifications and qalifications. In C. S. Mellish, editor, *Proceedings of the International Joint Conference on Artificial Intelligence (IJCAI)*, pages 1978–1984, Montreal, Canada, August 1995. Morgan Kaufmann.

[McCain and Turner, 1998] Norman McCain and Hudson Turner. Satisfiability planning with causal theories. In A. G. Cohn, L. K. Schubert, and S. C. Shapiro, editors, *Proceedings of the International Conference on Principles of Knowledge Representation and Reasoning (KR)*, pages 212–223, Trento, Italy, June 1998. Morgan Kaufmann.

[McCarthy and Hayes, 1969] John McCarthy and Patrick J. Hayes. Some philosophical problems from the standpoint of artificial intelligence. *Machine Intelligence*, 4:463–502, 1969.

[McCarthy, 1958] John McCarthy. Programs with Common Sense. In *Proceedings of the Symposium on the Mechanization of Thought Processes*, volume 1, pages 77–84, London, November 1958. (Reprinted in: [McCarthy, 1990]).

[McCarthy, 1963] John McCarthy. *Situations and Actions and Causal Laws.* Stanford Artificial Intelligence Project, Memo 2, Stanford University, CA, 1963.

[McCarthy, 1977] John McCarthy. Epistemological problems of artificial intelligence. In R. Reddy, editor, *Proceedings of the International Joint Conference on Artificial Intelligence (IJCAI)*, pages 1038–1044, Cambridge, MA, 1977. MIT Press.

[McCarthy, 1980] John McCarthy. Circumscription—a form of non-monotonic reasoning. *Artificial Intelligence*, 13:27–39, 1980.

[McCarthy, 1986] John McCarthy. Applications of circumscription to formalizing common-sense knowledge. *Artificial Intelligence*, 28:89–116, 1986.

[McCarthy, 1990] John McCarthy. *Formalizing Common Sense.* Ablex, Norwood, New Jersey, 1990. (Edited by V. Lifschitz).

[McIlraith, 2000] Sheila McIlraith. An axiomatic solution to the ramification problem (sometimes). *Artificial Intelligence*, 116(1–2):87–121, 2000.

[Miller and Shanahan, 1994] Rob Miller and Murray Shanahan. Narratives in the situation calculus. *Journal of Logic and Computation*, 4(5):513–530, 1994.

[Moore, 1985] Robert Moore. A formal theory of knowledge and action. In J. R. Hobbs and R. C. Moore, editors, *Formal Theories of the Commonsense World*, pages 319–358. Ablex, 1985.

[Nilsson, 1969] Nils J. Nilsson. A mobile automaton: An application of AI techniques. In *Proceedings of the International Joint Conference on Artificial Intelligence (IJCAI)*, pages 509–520, Washington, DC, 1969. Morgan Kaufmann.

[Nilsson, 1984] Nils J. Nilsson. *Shakey the Robot.* SRI Technical Note 323, Stanford Research Institute, CA, 1984.

[Ortiz, 1999] Charles L. Ortiz, Jr. Explanatory update theory: applications of counterfactual reasoning to causation. *Artificial Intelligence*, 108:125–178, 1999.

[Pearl, 1993] Judea Pearl. Graphical models, causality, and intervention. *Statistical Science*, 8(3):266–273, 1993.

[Pearl, 1994] Judea Pearl. A probabilistic calculus of actions. In R. Lopez de Mantaras and D. Poole, editors, *Proceedings of the Conference on Uncertainty in Artificial Intelligence (UAI)*, pages 454–462, San Mateo, CA, 1994. Morgan Kaufmann.

[Peppas *et al.*, 1999] Pavlos Peppas, Maurice Pagnucco, Mikhail Prokopenko, Norman Y. Foo, and Abhaya Nayak. Preferential semantics for causal systems. In T. Dean, editor, *Proceedings of the International Joint Conference on Artificial Intelligence (IJCAI)*, pages 118–123. Morgan Kaufmann, Stockholm, Sweden, 1999.

[Petrick and Levesque, 2002] Ronald Petrick and Hector Levesque. Knowledge equivalence in combined action theories. In D. Fensel, D. McGuinness, and M.-A. Williams, editors, *Proceedings of the International Conference on Principles of Knowledge Representation and Reasoning (KR)*, pages 303–314, Toulouse, France, April 2002. Morgan Kaufmann.

[Pirri and Reiter, 1999] Fiora Pirri and Ray Reiter. Some contributions to the metatheory of the situation calculus. *Journal of the ACM*, 46(3):261–325, 1999.

[Prokopenko *et al.*, 1999] Mikhail Prokopenko, Maurice Pagnucco, Pavlos Peppas, and Abhaya Nayak. Causal propagation semantics—a study. In N. Foo, editor, *Proceedings of the Australian Joint Conference on Artificial Intelligence*, volume 1747 of *LNAI*, pages 378–392, Sydney, Australia, December 1999. Springer.

[Prokopenko *et al.*, 2000] Mikhail Prokopenko, Maurice Pagnucco, Pavlos Peppas, and Abhaya Nayak. A unifying semantics for causal propagation. In *Proceedings of the Pacific Rim International Conference on Artificial Intelligence*, pages 38–48, Melbourne, Australia, August 2000.

[Pylyshyn, 1987] Zenon W. Pylyshyn, editor. *The Robot's Dilemma: The Frame Problem in Artificial Intelligence.* Ablex, Norwood, New Jersey, 1987.

[Quine, 1982] Willard Quine. *Methods of Logic.* Harvard University Press, 1982.

[Reiter, 1991] Raymond Reiter. The frame problem in the situation calculus: A simple solution (sometimes) and a completeness result for goal regression. In V. Lifschitz, editor, *Artificial Intelligence and Mathematical Theory of Computation*, pages 359–380. Academic Press, 1991.

[Reiter, 2001a] Raymond Reiter. *Knowledge in Action.* MIT Press, 2001.

[Reiter, 2001b] Raymond Reiter. On knowledge-based programming with sensing in the situation calculus. *ACM Transactions on Computational Logic*, 2(4):433–457, 2001.

[Sandewall, 1972] Erik Sandewall. An approach to the frame problem and its implementation. In B. Meltzer and D. Michie, editors, *Machine Intelligence*, volume 7, chapter 11, pages 195–204. Edinburgh University Press, 1972.

[Sandewall, 1993a] Erik Sandewall. The range of applicability of nonmonotonic logics for the inertia problem. In R. Bajcsy, editor, *Proceedings of the International Joint Conference on Artificial Intelligence (IJCAI)*, pages 738–743, Chambéry, France, August 1993. Morgan Kaufmann.

[Sandewall, 1993b] Erik Sandewall. Systematic assessment of temporal reasoning methods for use in autonomous systems. In B. Fronhöfer, editor, *Workshop on Reasoning about Action & Change at IJCAI*, pages 21–36, Chambéry, August 1993.

[Sandewall, 1994] Erik Sandewall. *Features and Fluents. The Representation of Knowledge about Dynamical Systems.* Oxford University Press, 1994.

[Sandewall, 1995a] Erik Sandewall. Reasoning about actions and change with ramification. In van Leeuwen, editor, *Computer Science Today*, volume 1000 of *LNCS*, pages 486–504. Springer, 1995.

[Sandewall, 1995b] Erik Sandewall. Systematic comparison of approaches to ramification using restricted minimization of change. Technical Report LiTH-IDA-R-95-15, Department of Computer Science, Linköping University, Sweden, 1995.

[Sandewall, 1996] Erik Sandewall. Assessments of ramification methods that use static domain constraints. In L. C. Aiello, J. Doyle, and S. Shapiro, editors, *Proceedings of the International Conference on Principles of Knowledge Representation and Reasoning (KR)*, pages 99–110, Cambridge, MA, November 1996. Morgan Kaufmann.

[Scherl and Levesque, 1993] Richard Scherl and Hector Levesque. The frame problem and knowledge-producing actions. In *Proceedings of the AAAI National Conference on Artificial Intelligence*, pages 689–695, Washington, DC, July 1993.

[Scherl and Levesque, 2003] Richard Scherl and Hector Levesque. Knowledge, action, and the frame problem. *Artificial Intelligence*, 144(1):1–39, 2003.

[Schubert, 1990] Lenhart K. Schubert. Monotonic solution of the frame problem in the situation calculus: An efficient method for worlds with fully specified actions. In H. E. Kyberg, R. P. Loui, and G. N. Carlson, editors, *Knowledge Representation and Defeasible Reasoning*, pages 23–67. Kluwer Academic, 1990.

[Shanahan and Witkowski, 2000] Murray Shanahan and Mark Witkowski. High-level robot control through logic. In C. Castelfranchi and Y. Lespérance, editors, *Proceedings of the International Workshop on Agent Theories Architectures and Languages (ATAL)*, volume 1986 of *LNCS*, pages 104–121, Boston, MA, July 2000. Springer.

[Shanahan, 1989] Murray Shanahan. Prediction is deduction but explanation is abduction. In *Proceedings of the International Joint Conference on Artificial Intelligence (IJCAI)*, pages 1055–1060, Detroit, MI, 1989.

[Shanahan, 1995] Murray Shanahan. A circumscriptive calculus of events. *Artificial Intelligence*, 77:249–284, 1995.

[Shanahan, 1997] Murray Shanahan. *Solving the Frame Problem: A Mathematical Investigation of the Common Sense Law of Inertia.* MIT Press, 1997.

[Shanahan, 1999] Murray Shanahan. The ramification problem in the event calculus. In T. Dean, editor, *Proceedings of the International Joint Conference on Artificial Intelligence (IJCAI)*, pages 140–146, Stockholm, Sweden, 1999. Morgan Kaufmann.

[Shoham, 1987] Yoav Shoham. *Reasoning about Change.* MIT Press, 1987.

[Shoham, 1988] Yoav Shoham. Chronological ignorance: Experiments in nonmonotonic temporal reasoning. *Artificial Intelligence*, 36:279–331, 1988.

[Stein and Morgenstern, 1994] Lynn Andrea Stein and Leora Morgenstern. Motivated action theory: A formal theory of causal reasoning. *Artificial Intelligence*, 71:1–42, 1994.

[Störr and Thielscher, 2000] Hans-Peter Störr and Michael Thielscher. A new equational foundation for the fluent calculus. In J. Lloyd etal, editor, *Proceedings of the International Conference on Computational Logic (CL)*, volume 1861 of *LNAI*, pages 733–746, London (UK), July 2000. Springer.

[Thielscher, 1994] Michael Thielscher. Representing actions in equational logic programming. In P. Van Hentenryck, editor, *Proceedings of the International Conference on Logic Programming (ICLP)*, pages 207–224, Santa Margherita Ligure, Italy, June 1994. MIT Press.

[Thielscher, 1997] Michael Thielscher. Ramification and causality. *Artificial Intelligence*, 89(1–2):317–364, 1997.

[Thielscher, 1999] Michael Thielscher. From situation calculus to fluent calculus: State update axioms as a solution to the inferential frame problem. *Artificial Intelligence*, 111(1–2):277–299, 1999.

[Thielscher, 2000a] Michael Thielscher. Nondeterministic actions in the fluent calculus: Disjunctive state update axioms. In S. Hölldobler, editor, *Intellectics and Computational Logic*, pages 327–345. Kluwer Academic, 2000.

[Thielscher, 2000b] Michael Thielscher. Representing the knowledge of a robot. In A. Cohn, F. Giunchiglia, and B. Selman, editors, *Proceedings of the International Conference on Principles of Knowledge Representation and Reasoning (KR)*, pages 109–120, Breckenridge, CO, April 2000. Morgan Kaufmann.

[Thielscher, 2001a] Michael Thielscher. Inferring implicit state knowledge and plans with sensing actions. In F. Baader, G. Brewka, and T. Eiter, editors, *Proceedings of the German Annual Conference on Artificial Intelligence (KI)*, volume 2174 of *LNAI*, pages 366–380, Vienna, Austria, September 2001. Springer.

[Thielscher, 2001b] Michael Thielscher. Planning with noisy actions (preliminary report). In M. Brooks, D. Corbett, and M. Stumptner, editors, *Proceedings of the Australian Joint Conference on Artificial Intelligence*, volume 2256 of *LNAI*, pages 495–506, Adelaide, Australia, December 2001. Springer.

[Thielscher, 2001c] Michael Thielscher. The qualification problem: A solution to the problem of anomalous models. *Artificial Intelligence*, 131(1–2):1–37, 2001.

[Thielscher, 2002a] Michael Thielscher. Programming of reasoning and planning agents with FLUX. In D. Fensel, D. McGuinness, and M.-A. Williams, editors, *Proceedings of the International Conference on Principles of Knowledge Representation and Reasoning (KR)*, pages 435–446, Toulouse, France, April 2002. Morgan Kaufmann.

[Thielscher, 2002b] Michael Thielscher. Reasoning about actions with CHRs and finite domain constraints. In P. Stuckey, editor, *Proceedings of the International Conference on Logic Programming (ICLP)*, volume 2401 of *LNCS*, pages 70–84, Copenhagen, Danmark, 2002. Springer.

[Thielscher, 2004] Michael Thielscher. Logic-based agents and the frame problem: A case for progression. In V. Hendricks, editor, *First-Order Logic Revisited: Proceedings of the Conference 75 Years of First Order Logic (FOL75)*, pages 323–336, Berlin, Germany, 2004. Logos.

[Thrun, 2000] Sebastian Thrun. Probabilistic algorithms in robotics. *AI Magazine*, 21(4):93–109, 2000.

[Turner, 1997] Hudson Turner. Representing actions in logic programs and default theories: A situation calculus approach. *Journal of Logic Programming*, 31(1–3):245–298, 1997.

[Watson, 1998] Richard Watson. An application of action theory to the space shuttle. In G. Gupta, editor, *Proceedings of the Workshop on Practical Aspects of Declarative Languages*, volume 1551 of *LNCS*, pages 290–304. Springer, 1998.

[Weld *et al.*, 1998] Daniel S. Weld, Corin R. Anderson, and David E. Smith. Extending Graphplan to handle uncertainty & sensing actions. In J. Mostow and C. Rich, editors, *Proceedings of the AAAI National Conference on Artificial Intelligence*, pages 432–437, Madison, WI, July 1998.

[White *et al.*, 1998] Graham White, John Bell, and Wilfrid Hodges. Building models of prediction theories. In A. G. Cohn, L. K. Schubert, and S. C. Shapiro, editors, *Proceedings of the International Conference on Principles of Knowledge Representation and Reasoning (KR)*, pages 557–568, Trento, Italy, June 1998. Morgan Kaufmann.

Index